THE YOWIE

Previous Works by Tony Healy and Paul Cropper

Out of the Shadows: Mystery Animals of Australia

THE YOWIE

In Search of Australia's Bigfoot

Tony Healy and Paul Cropper

Foreword by Loren Coleman
Maps by Bryan Fox

Anomalist Books
*San Antonio * New York*

An Original Publication of ANOMALIST BOOKS

The Yowie: In Search of Australia's Bigfoot

ISBN: 1933665165

Book design by Ansen Seale

The authors wish to advise indigenous readers that this book contains the names and images of some deceased Aboriginal people.

For more information on the yowie and to view recent updates from the authors, go to www.yowiefile.com.

For information about Anomalist Books, go to anomalistbooks.com or write to:

Anomalist Books
5150 Broadway #108
San Antonio, TX 78209

Anomalist Books
PO Box 577
Jefferson Valley, NY 10535

Contents

This book is dedicated to the late Peter Chapple,
founder and president of the
Australian Rare Fauna Research Association (ARFRA).

Although many people helped us in our research (see Acknowledgements),
we are especially grateful to Graham Joyner, Neil Frost and Dean Harrison.
Without their unstinting contributions of data, advice, and encouragement, this book would
have been much slimmer and a great deal less interesting.

Foreword

Yowie? Yahoo? What could this book be about? A yodelling hillbilly? An internet company with a search engine? Perhaps these are unfamiliar words, in terms of cryptozoology (the study of hidden or unknown animals) to the American or European ear but they are intriguing, nevertheless. "Yowie" and "Yahoo", in fact, are terms used by Aussies to refer to an elusive Bigfoot or Yeti-like creature that allegedly lurks in the Australian backwoods. As you read this book, you may be amazed by how much evidence there is to support the notion of such creatures existing in eastern Australia.

Considering that Yowies have been reported by European settlers since at least 1848 and by Aborigines since time immemorial, an authoritative volume on the creatures has been long overdue. Luckily, two seasoned explorers have now stepped forward to help us decipher this mind-boggling mystery.

Having actively investigated strange animal reports in Australia and several other countries since 1975, Paul Cropper, a marketing manager at a major Australian telecommunications company, and Tony Healy, a retired government officer, are today two of Australia's leading cryptozoologists. Together they co-authored *Out of The Shadows*, a highly acclaimed, comprehensive summary of Australian zoological mysteries. In this new book they focus entirely on the Yowie phenomenon.

What will first surprise readers as yet innocent of Yowie lore is the incredibly lengthy history of this creature's interaction with humans in Australia. The Aborigines have many ancient traditions concerning the frightening "Hairy Man" of the mountains, which different tribes know by many different names, including *yahoo*. This book contains a great deal of that tribal lore, including tales of sporadic warfare between the Hairy Giants and Aborigines, not only in centuries past, but also in surprisingly recent times.

By the mid 1800s, Australia's white settlers had joined the Aborigines in reporting sightings of the huge, apelike creatures and their enormously long footprints. Thus, there have been hundreds, some say thousands, of sightings of Yowies and their tracks by surveyors, rangers, backpackers and others, largely in the south and central coastal regions of New South Wales and Queensland's Gold Coast hinterland. These sightings have continued unabated into the present.

Although most eyewitnesses describe hairy, ape-like creatures between five and eight feet in height, there are also many reports of *little* hairy hominids. Aboriginal people refer to the smaller type as *junjudees* (among many other names) and invariably insist they not juvenile Yowies, but an entirely different type of creature. This might seem rather confusing, but as Cropper and Healy unveil the massive amount of data in their Yowie and Junjudee files and examine various theories relating to the origin of both creatures, considerable light is shed on the matter. One particularly interesting theory that they explore is a possible relationship between Junjudees and the Hobbit-sized human remains of *Homo floresiensis* recently discovered on the island of Flores, just north of Australia.

Quite apart from the sheer volume of reported sightings, many of which have occurred at close range in broad daylight, the fact that many witnesses are trained observers such as surveyors, rangers, naturalists and members of Australia's elite Special Air Service Regiment, should give pause to the sceptical reader. Physical evidence, such as footprint casts, tree bites and "Yowie nests" – one of which was discovered by a team of leading scientists – is also presented and analysed.

Yowie, yahoo, dulagarl, thoolagarl, wawee, jurrawarra, noocoonah, tjangara – Australia's Great Hairy Man goes by many names – and is one of the world's greatest zoological or anthropological mysteries. Cropper and Healy are on an exciting quest, and with these two at the wheel, the reader can take part in an entertaining and very enlightening expedition into darkest Yowie Land.

Enjoy the trip.

Loren Coleman
Author of *Bigfoot! The True Story of Apes in America*

Explanatory Notes

Aboriginal terms

In documents dating from the 19th and early 20th centuries some Aboriginal terms for the yowie, like *doolagarl* and *jimbra*, are spelt in several different ways. Some writers capitalised or italicised terms like yahoo and hairy man and some didn't. In quoting earlier writers we have left all spelling exactly as we found it. We have, in fact, continued the grand tradition of inconsistency: we generally capitalise Hairy Man but don't capitalise yowie; sometimes, depending on the context, we italicise Aboriginal terms and sometimes we don't.

Cross-reference system

The main body of this book contains seven chapters that chronicle the yowie phenomenon from the early colonial era to the present day. The Catalogue of Cases contains summaries of every report we have collected in the past 29 years, complete with the source of each item.

We have cross-referenced key sections of the text in the main body of the book with items in the Catalogue of Cases. Hence, in addition to conventional endnotes, readers will frequently come across notations, such as [Case 207], which refer them to particular cases in the Catalogue. Similarly, notations following some cases in the Catalogue, e.g. [See Ch. 4], refer the reader to more detailed versions of the same cases in the main body of the book.

Abbreviations

ARFRA = Australian Rare Fauna Research Association

AYR = Australian Yowie Research.

BFRO = Bigfoot Field Research Organisation

GCBRO = Gulf Coast Bigfoot Research Organisation

And for the benefit of non-Australian readers:

NSW = New South Wales; VIC = Victoria; QLD = Queensland; TAS = Tasmania; SA = South Australia: WA = Western Australia; NT = Northern Territory; ACT = Australian Capital Territory; ABC = Australian Broadcasting Corporation; CSIRO = Commonwealth Scientific and Industrial Research Organisation

Weights and measures

Although Australia officially abandoned the imperial system of weights and measures some years ago, many older Australians still think in terms of pounds, ounces, yards, feet and inches. Young people tend to favour the metric system. Middle-aged people use both imperial and metric measurements in a rather haphazard way. Here and there, when it seemed useful to do so, we have converted metric to imperial and vice-versa. When quoting eyewitnesses, however, we have generally left their estimates of heights, weights and distances just as they expressed them.

We are confident that most readers are quite capable of converting imperial measurements to metric and vice-versa, but for the record:

1 centimetre = 0.39 inches
10 cm = 3.94 inches
20 cm = 7.7 inches
30 cm = 11.81 inches

1 metre = 3.28 feet
2 metres = 6.56 feet
3 metres = 9.84 feet
4 metres = 13.12 feet
5 metres = 16.40 feet
10 metres = 32.81 feet
50 metres = 164.04 feet
100 metres = 328.08 feet
1 kilometre = 0.62 miles

----

1 inch = 2.54 cm
6 inches = 15.24 cm
1 foot = 30.5 cm
2 feet = 61 cm
3 feet = 91 cm
4 feet = 1.22 m
5 feet = 1.52 m
6 feet = 1.83 m
7 feet = 2.13 m
8 feet = 2.44 m
9 feet = 2.74 m

1 yard = 3 feet
1 chain = 22 yards
1 mile = 1.61 km

----

1 pound = 0.45 kilogram
200 pounds = 90.72 kg
300 pounds = 136.08 kg
500 pounds = 226.80 kg
1 stone = 14 pounds

----

50 kilograms = 110.23 lbs
100 kilograms = 220.46 lbs
200 kilograms = 440.92 lbs
300 kilograms = 661.39 lbs

Maps

AUSTRALIA
Showing some of the places mentioned in the text.

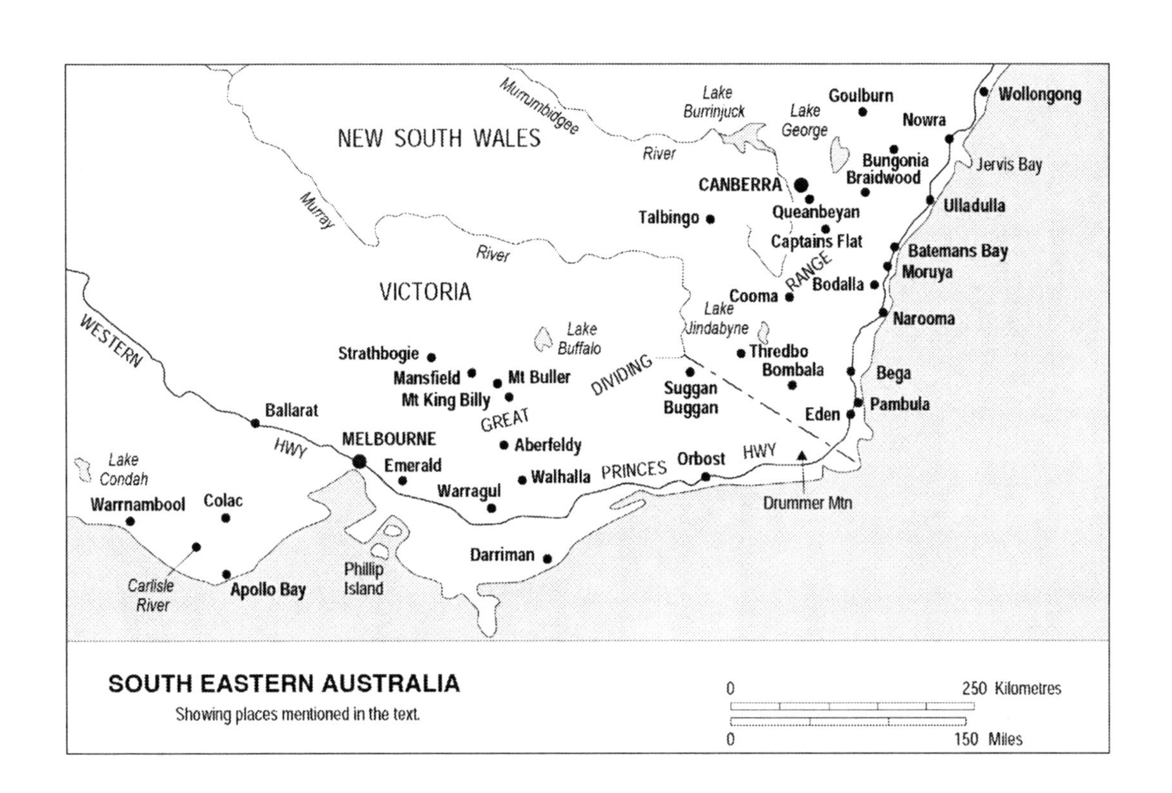

SOUTH EASTERN AUSTRALIA
Showing places mentioned in the text.

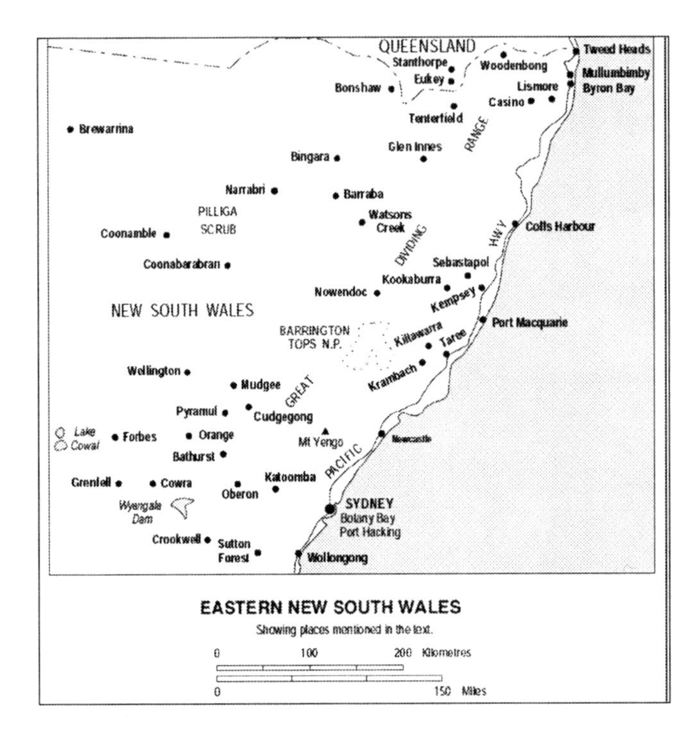

EASTERN NEW SOUTH WALES

Showing places mentioned in the text.

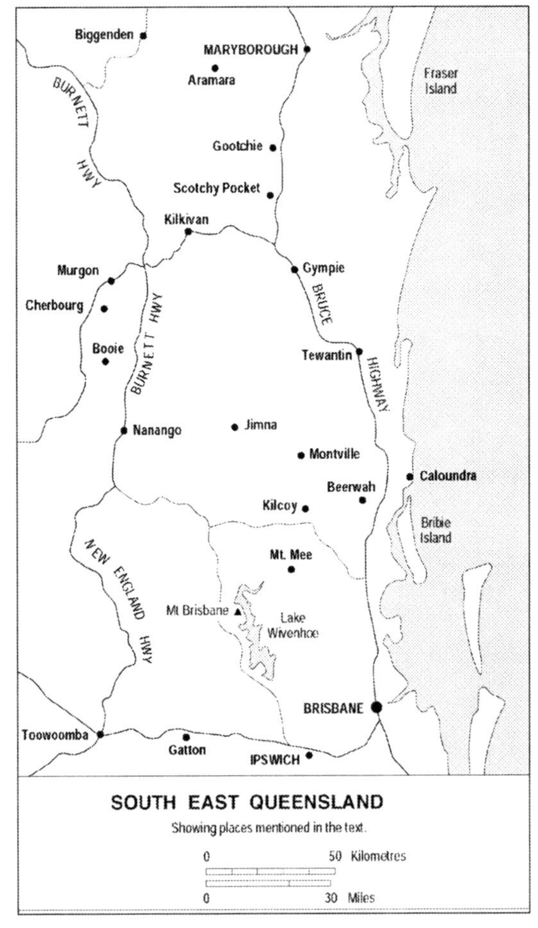

SOUTH EAST QUEENSLAND

Showing places mentioned in the text.

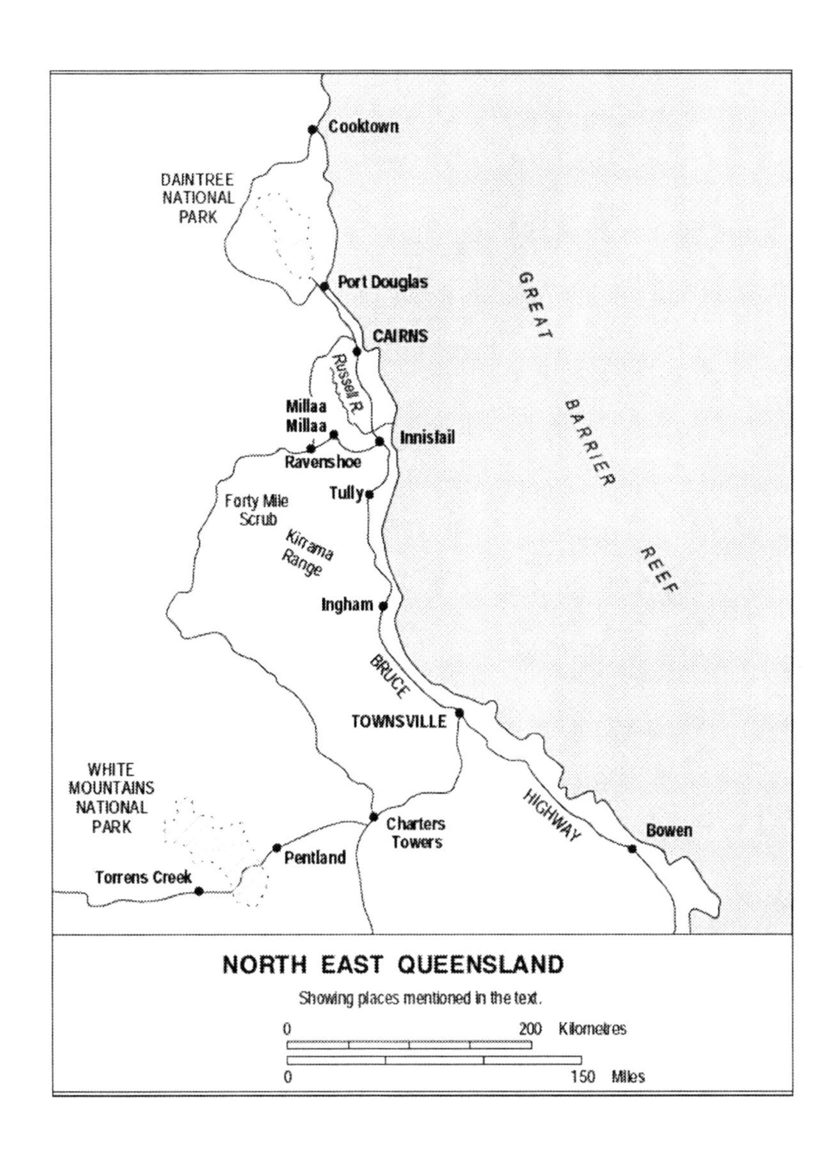

NORTH EAST QUEENSLAND
Showing places mentioned in the text.

Introduction

Gorillas In Our Midst

"... about twelve feet in front of me was this big, black, hairy man-thing. It looked more like a gorilla than anything."

Although Springbrook National Park is just 20 kilometres west of the glittering towers and crowded beaches of Queensland's Gold Coast, it might as well be on another planet. An immense maze of twisting gorges, razorback ridges and towering precipices, it is densely covered in tangled, leech-infested rainforest. Its cloud-wreathed peaks loom so high that, in winter, despite its sub-tropical location and proximity to the sea, light snow sometimes dusts its jungle canopy.

The park is more than 29 square kilometres in area, but it comprises only one small part of a great string of national parks and fauna reserves that stretch along almost the entire 300 kilometre length of the McPherson Range, which straddles the Queensland - New South Wales border from the Gold Coast in the east to the vicinity of Mt. Donaldson in the west.

Something of a "lost world" in the eyes of scientists, the rainforests of the McPherson Range provide sanctuary to several botanical curiosities, notably Antarctic Beech trees: gnarled survivors of the distant epoch in which Australia and its now-frozen neighbour were fused together as part of the super-continent Gondwanaland. The finest grove of such trees looms from the jungle almost on the edge of the mightiest cliff in Springbrook National Park, a breathtaking drop-off that marks the border between the states. Appropriately enough, it was there, almost in the shadow of those ancient, moss-covered survivors, that

Queensland National Parks and Wildlife Service ranger Percy Window came face to face with what was, to all intents and purposes, a living, breathing monster from the Dreamtime.

The encounter occurred in broad daylight on March 5, 1978, after Mr Window, who was clearing undergrowth on the trail to Best Of All lookout, heard grunting noises coming from the scrub. Although at first ordered by his superiors not to discuss it, he later recounted his experience to Frank Hampson of the *Gold Coast Bulletin* and to us:

"I thought a pig had gotten loose and was scrubbing amongst the trees. I went into the forest to see if I could find it. I heard the grunting again, but I couldn't find any tracks. Then something made me look up, and there, about twelve feet in front of me, was this big, black, hairy man-thing. It looked more like a gorilla than anything. It had huge hands and one of them was wrapped around a sapling."

He was close enough to see hair on the back of its fingers. The skin underneath was shiny and dark, and the fingers were "basically human-like". Mr Window, himself well over six feet in height, estimated the creature was over eight feet [almost 2.5 metres] tall. "It had a flat, black, shiny face, with two big yellow eyes and a hole for a mouth. It just stared at me and I stared back. I was so numb I couldn't even raise the axe."

It had a short, thick neck and was very muscular and solid. Its hands, feet and body were covered with short, black hair, as was part of its face. Although he noticed no genitalia, the ranger had the strong impression it was male.

As he stood there, paralysed with shock, he could feel sweat streaming down his back. "We seemed to stand there staring at each other for about ten minutes before it suddenly gave off a foul smell that made me vomit – then it just made off sideways and disappeared."

Some readers may find Mr. Window's report incredible, but his superiors in the National Parks and Wildlife Service and his neighbours in Springbrook village had no doubt he was telling the truth. His widow (he died in 1994) vehemently dismisses any suggestion that he would have invented such a story. The shock of the encounter, she recalls, affected the strapping, life-long outdoorsman so severely that he took to his bed for two days afterwards. She points out, also, that as a long-term resident of the area who valued the friendship of his neighbours, he had always tried to avoid publicity. When the story leaked out he was very worried that hunters might be drawn to the park. It was not in his nature to play practical jokes and it is difficult to believe he would deliberately frighten his own children with a fabricated story.

Mr. Window impressed us as a very genuine, down-to-earth person. Most open-minded readers, we believe, will concede that his words have a distinct ring of truth: "The point I want to make is that there is no fiction about my experience. I saw this beast in daylight, about two o'clock in the afternoon. Before, I might have agreed they were comic book stuff – but no more. They exist all right. The reason they aren't seen more often is that most people who go for bushwalks make a noise. I didn't make a noise because I was trying to stalk what I thought was a pig ... it's perfectly true: I met a yowie – and I've never been so scared in my life."

If Mr. Window were the only person ever to have reported seeing such a creature it would be reasonable to suggest he had experienced some form of hallucination. His report, however, is far from unique. As we will show in the following chapters, it is supported by a great deal of Aboriginal lore and by the testimony of hundreds of non-Aborigines who have reported similar hair-raising encounters from the early 1800s to the present day.

We are not scientists. We see ourselves primarily as chroniclers of this fascinating but long-neglected subject. Since the mid-1970s we have interviewed more than 120 eyewitnesses, walked miles of library corridors and crept, with cameras clutched tightly in shaky hands, through many wild and spooky places. The search has led us to many out-of-the-way corners of Australia and to many interesting people. It has been fascinating, tantalising, at times very frustrating, but most of all a great deal of fun.

In this book we aim to document the mystery from the earliest days of European settlement, through the colonial era and the early modern era to the present day. We will show that the reports of eyewitnesses have been, over the generations, remarkably consistent. Finally, we will assess the evidence and try to come to some conclusions about this truly mind-boggling mystery.

During the 1970s we found yowie reports extremely hard to come by but nowadays, because of the number of keen investigators involved, we can hardly keep up with the amount of data flooding in. Because we have now reached the point where we are struggling to stay abreast of developments, we have decided to put down in print everything we know about the phenomenon. Therefore, in addition to the seven chapters that form the main body of the book, we have included a Catalogue of Cases: a chronological listing of every sighting report we have collected in the past 29 years.

Some readers may conclude that yowies are real, flesh and blood animals; some may decide the phenomenon is largely sociological or psychological in nature; others may come to even stranger conclusions. Whatever their true nature, the hairy giants have been a part – but an oddly obscure part – of the Australian cultural landscape for a very, very long time.

These are the facts as we know them. Please make up your own minds.

Chapter 1

Aborigines and the Yowie - An Age-old Mystery

"A doolagarl is a gorilla-like man. He has long spindly legs. He has a big chest, long arms. His forehead goes back from his eyebrows. His head goes into his shoulders, no neck. They live now on Cockwhy and Pollwombra Mountains."
—Percy Mumbulla, quoted by Roland Robinson in "Three Aboriginal Tales", *The Bulletin*, October 13, 1954.

Although we have studied the yowie phenomenon for many years, we don't pretend to know all there is to know about Aboriginal yowie lore. We have found that while some Aborigines, like the late Percy Mumbulla, are happy to talk about the hairy giants, many others are a little reticent – and some refuse to discuss the subject at all. On some occasions, indeed, it is made perfectly clear to non-Aborigines that their interest in the subject is entirely unwelcome.

To us, any reluctance by indigenous people – who have had so many of their traditions misrepresented or ridiculed – to discuss the matter is quite understandable. Some Aborigines are annoyed that the subject is often dismissed as a joke by the media. Others are concerned that any further discussion of Aboriginal yowie lore might encourage irresponsible shooters.

Another factor is that there seems to be, among many Aboriginal groups, a spiritual aspect to the yowie phenomenon, which can result in it becoming something of a taboo subject, particularly in conversation with outsiders. Some Aborigines told Patricia Riggs of the *Macleay Argus,* who uncovered several fascinating stories in the Kempsey area, that to speak openly about the Hairy Man was to invite bad luck. Other items in our files indicate this belief is quite widespread.

Despite these sensitivities, quite a lot of traditional Aboriginal yowie lore was recorded in the 19th and early 20th centuries by explorers, timber-getters, graziers and anthropologists. This, combined with information kindly shared with us in recent decades, enables us to understand, in a general way at least, the Aboriginal view of the yowie.

It is clear that Aboriginal belief in the Hairy Man extends from Cape York Peninsula right down the east coast to Victoria. It also occurs in at least some parts of the Northern Territory, South Australia and Western Australia.

The ape-men were, and are, known by many names. In parts of north Queensland they are known as *quinkin* (or as a type of quinkin) and as *joogabinna*, in parts of north-east NSW as *jurrawarra* and *myngawin*, on the central NSW coast as *puttikan*, in south-east NSW and north-east Victoria as *gubba*, *doolagar*, *doolagarl*, *gulaga* and *thoolagal*. In South Australia they are known as *noocoonah* and possibly *mooluwonk*, in Western Australia as *jimbra*, *jingra*, *tjangara* and possibly *marbu*, and in the Northern Territory as *pangkarlangu*. Nowadays, when discussing the creatures with people outside their own language group, indigenous people generally use the term "Hairy Man" or, occasionally, recently coined terms such as "red-eye".

Aboriginal descriptions of yowies were, and are, reasonably consistent. In fact, when one considers that in the pre-colonial era the population was culturally diverse, divided into more than 220 language groups, and spread all over this vast continent, Aboriginal yowie lore could be said to be remarkably consistent.

Percy Mumbulla. (Lee Chittick)

The description given by Percy Mumbulla at the beginning of this chapter contains most of the features mentioned by many other Aborigines over the years. Other details are occasionally added. Sometimes, for instance, the creatures' hands are said to be equipped with long nails or claws.

Colonial-era Aborigines often mentioned the yowie's great size, its nocturnal lifestyle and its savage nature. During the 1840s, one South Australian pioneer wrote:

"The natives here have a tradition that a big black fellow, far higher than the ordinary size, walks about during the night, his object being to destroy good black fellows and their children – the latter articles being his favourite diet; and they will sometimes show a footprint, in size about three times as large as an ordinary foot, and in shape resembling the print of a man's step. He is said to walk

The giant *quinkin*

The Yalanji people of Cape York Peninsula believe in the existence of strange, usually malevolent creatures called quinkins. Quinkins come in a variety of shapes and most are said to be quite small. In earlier days, however, there was said to be one notable exception: the dreaded giant, Turramulli, who towered above the tallest trees. He resembled a huge hair-covered man but had only three clawed toes on each foot and three fingers, ending in fearsome talons, on each hand. His hideous head was set straight into massive shoulders.

Turramulli slept in a cave for months on end and ventured forth only at the beginning of the wet season, when he strode through the countryside, looking for kangaroos or people to devour.

During the 1970s, the bushman and writer Percy Trezise and his Aboriginal friend, Dick Roughsey, discovered two ancient cave paintings that appear to depict Turramulli. Some years later, they collaborated to produce a delightful children's book that tells the story of how, long ago, two Yalanji youngsters slew the fearsome giant.

Turramulli the Giant Quinkin, as depicted by Percy Trezise.

An Aboriginal rock painting of Turramulli, the giant quinkin. (Percy Trezise)

about principally during the night, for which reason they never stir out at that time. I asked one of the men why he did not kill this creature. Upon which he replied with much earnestness, 'Oh! me plenty run away – me too much frightened!' They give this being the name of Noocoonah." [1]

In the folk tales of the Yalanji people of Cape York, Turramulli, the giant *quinkin,* was so tall that it towered over the trees. Generally, however, descriptions of the "average" yowie given by both traditional and contemporary Aborigines tally quite well with those given throughout the last 150 years by non-Aboriginal eyewitnesses.

Aborigines say yowies are capable of climbing trees, and that they stomp the ground or whack tree trunks to announce their presence. They scream and growl and often produce a very foul odour. They are sometimes said to be man-eaters and are almost always said to live without benefit of fire or tools of any kind. The only people who say they carry clubs are the desert

Aborigines of central Western Australia and of the south-western Northern Territory. The Warlpiri and Pintubi of the Northern Territory are the only tribes who say the creatures use fire. [2]

In 1912, Mr. A.B. Walton, of Granville, NSW, recalled being told by Aborigines of the Braidwood District many years earlier, of the "big feller devil" they called *yahoo*. It was, they said, hair-covered, taller than a man and agile enough to climb trees. One yahoo, when disturbed by a party of Aborigines, strangled a woman and chased her companions away. [3]

Aboriginal lore sometimes hints that male yowies like to grab women for reasons other than homicide. Folklorist Aldo Massola tells one such story (which also contains the only reference we have seen to yowies having the magical power of flight) in his book *Bunjil's Cave*.

"*Dulugars* were very strong man-like hairy beings. They lived in the mountains behind Suggan Buggan, and when women ventured alone in the bush they came flying through the air and took them away. The women were released after a while. Big Charley and his wife were walking along a bush track [just before the beginning of the First World War] when they heard the three loud taps which always heralded a visit by the hairy men. Soon one of these creatures appeared, and Big Charley stood his ground and prepared to fight, while his wife ran into a cluster of gum trees for safety. The trees were close together, and the Dulugar was unable to reach the woman. He was also worried by the husband, and eventually retired. Big Charley, bleeding from a profusion of wounds, and his wife, with her dress almost torn from her where the Dulugar had got a hold, were able to reach Lake Tyers without any further molestation." [Case 43]

Big Charley is not the only man to fight a yowie and live to tell the tale. There are several accounts of Aborigines besting them in a fight and even killing them. A well-known and highly respected Ngunnawal elder, "Black Harry" Williams (c.1837 – 1921), told of seeing a large group of warriors kill one below the junction of the Yass and Murrumbidgee rivers, near the present site of Burrinjuck Dam, in about 1847. They dragged it down the hill by its ankles. He described it as "like a black man but covered all over with grey hair". [Case 2]

Surprisingly, given their skill as trackers, traditional Aborigines, as far as we know, rarely described the shape of yowies' feet in any detail. They generally said only that they were large and rather like those of humans. A notable exception was Turramulli, the hairy giant of far north Queensland, whose feet were said to be three-toed.

Harry Williams. (Courtesy of Merrilee Webb)

One other thing Aborigines occasionally said about the feet gives a bizarre twist, so to speak, to the whole picture: some stated that the creature's feet were *turned backwards* so that their tracks confused anyone

In 1985 Grahame Walsh discovered several ancient engravings of giant footprints in White Mountains National Park, Queensland. The location is now known as the Yahoo Site and the engravings, which are truly gigantic (the one shown here is 87 centimetres or 34 inches long), may well be related to Aboriginal yowie lore.

In recent years yowies have been seen in the same area on at least four occasions. [See Case 213]

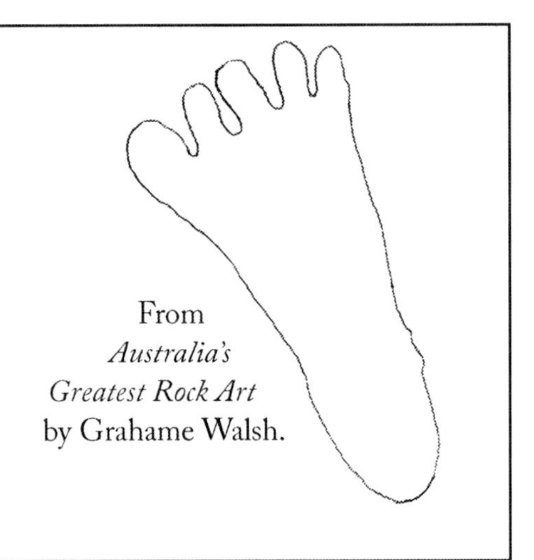

From *Australia's Greatest Rock Art* by Grahame Walsh.

attempting to follow.[4] Readers familiar with the yeti legend will remember that Himalayan Sherpas often say the same thing about the Abominable Snowman's feet.

Given those odd beliefs, it could be that some traditional Aborigines, like today's yowie researchers, didn't have a really clear idea of what the track of the rarely encountered creature should look like.

There is reason to suspect that yowies were less common in pre-colonial days than they are now. Eyewitness observations strongly suggest the creatures eat kangaroos [e.g. Cases 77, 131 and 213]. It has been estimated that before the British arrived in 1788 there were only about five million kangaroos in Australia. To get their share of that meat, yowies would have had to compete with hundreds of thousands of spear-wielding Aborigines. Today very few Aborigines live off the land, and because of the creation of thousands of dams, the country is teeming with up to 40 million 'roos. Pampered yowies of the modern era also have virtually all the other bush tucker – possums, wombats, goannas, snakes, and koalas – to themselves. Several Aborigines of the colonial and early modern eras, in fact, specifically mentioned that yowies were very rarely seen in the old days. Some people in the Kempsey area, for instance, said the creatures were seen only every 20 years or so.[5]

Yowie enthusiasts like the idea that yowies were very rare in the pre-colonial era because it helps to explain two inconvenient anomalies in Aboriginal lore. Firstly, as mentioned above, it could account for the apparent uncertainty about the shape of the creatures' tracks. Secondly, it might explain why some Aborigines also believed in other creatures that, though yowie-like, were supposedly not the same animal.

We refer to the dreaded *mumuga* and *yaroma*.

We would love to drop these two extremely weird creatures straight into the "Too Hard To Classify" basket. Unfortunately, however, both of them were frequently mentioned by Aborigines of the NSW south coast and are therefore inextricably bound up with the yowie mystery.

R.H. Mathews, a surveyor who worked in south-east NSW in the late 1800s, was one of the first to record the stories. He was told that yaromas, like yowies, resembled huge, hair-covered men. Their mouths, however, were so large that they could swallow a person whole. They had short legs, large feet and long teeth that they sharpened on rocks. They travelled in pairs, stood back to back and moved in a series of long jumps. At every jump their "genital appendages" struck the ground, making a loud, sudden noise.

The mumuga was just as odd, but had a slightly less eye-watering method of locomotion. It lived in mountain caves, was a man-eater, and was very strong. It had a hairy body, bald head, and very short arms and legs. Because it couldn't run very fast, it employed a novel method of stunning its prey: "when pursuing a blackfellow [it] flatulates all the time ... and the abominable smell of the ordure overcomes the individual, so he is easily captured. If the person who is attacked has a fire stick in his hand, the stink of Mumuga has no effect upon him."[6]

Freaky and fabulous though they are, the appendage-punishing yaroma and the foully flatulent mumuga are unquestionably related to the yowie legend. Sceptics might say that, with their wildly improbable appearance and habits, we should think of them as being purely mythological. That suggests, of course, that their cousin the yowie is also just a myth.

It is possible, on the other hand, that the yaroma and mumuga legends were triggered by real, but poorly observed encounters with yowies. As we will show later, data from our 300-odd reports indicate that yowies are largely nocturnal and that they shy away from firelight. After nightfall, traditional Aborigines had no form of light other than firebrands. It seems reasonable to suggest that over the centuries many thousands of Aborigines could have heard and smelled huge, hairy, smelly, vaguely man-like prowlers without getting a really good look at them. Since yowie visitations were exceedingly rare, and since Aborigines had no form of writing, some tribes could have, over many generations, come to believe that there were, in addition to the "normal" yowie, other yowie-like creatures: the weird mumuga and yaroma.

The sickening odour and loud thumping noises attributed to the mumuga and yaroma by some tribes have been attributed to the yowie by other Aborigines and by many non-Aboriginal eyewitnesses. As already mentioned, ranger Percy Window was

The yaroma, as imagined by R.H. Mathews.

overwhelmed by a devastating stench when he encountered the Springbrook yowie. Many other witnesses have reported similar sickening odours. Many also mention hearing loud thumping noises when yowies are near. The thumps, these days, are usually interpreted as heavy footfalls.

Despite their very weird aspects, we think of the mumuga and yaroma stories as being, essentially, a part (albeit a rather distorted part) of Aboriginal yowie lore.

Supernatural attributes

Yowie researchers are sometimes disconcerted to learn that some Aborigines attribute supernatural powers to the Hairy Man.

As mentioned earlier, some Victorian Aborigines apparently believed the creatures could fly through the air. Almost as odd is the belief, widespread in eastern Australia, that the Hairy Man can induce sleep. In an interview recorded in the 1970s, Percy Mumbulla, a fully initiated elder from south-east NSW, put it this way: "If you're walking up the mountain and you start to feel a bit sleepy, that's his presence. He'll try to put you to sleep but if you sit down and have a bit of a snooze, he'll get at you ... he has this mesmeric power that makes you drowsy and a bit confused about where you are and what you're doing, and then he can actually call you to him. Then you're in big trouble."[7]

As already mentioned, some Aborigines believe that to speak about the Hairy Man is to invite bad luck. Others say the creatures can induce illness. The 15-hectare Nunguu Miiral Aboriginal Area, also known as Pickett Hill, is one of the most prominent landmarks between Nambucca and Coffs Harbour, NSW. It is regarded as the most significant cultural and spiritual site of the Gumbaynggir people. The hilltop is a men-only area, and the Gumbaynggir have asked non-Aborigines not to go there. Russell Walker, a local elder, said a "Barga hairy man" protects the area. "A lot of people who go up there get sick. The older people tell the younger people not to go up there ... it's a sacred area."[8]

Researchers who are wedded to the idea of yowies being flesh and blood creatures find it easy to explain away the supernatural aspects of Aboriginal lore. They point out that Aborigines don't ascribe such qualities only to yowies; they see supernatural or spiritual qualities in virtually every other animal in Australia – and in all plants, streams, landforms and other aspects of creation. Furthermore, the researchers argue, it is only natural that tribal people would attribute some magical powers to an animal as big, fast, strong, terrifying and damnably elusive as the yowie.

Certainly, even wise old traditional Aborigines can be mistaken about some aspects of yowie behaviour. Percy Mumbulla once said that the *dulagarl* "keeps out of the road of white people, that's why they don't see him. He only appears to Aboriginal people".[9]

An ape by any other name

It is clear that, throughout the colonial era, many white pioneers believed *yahoo* was the most commonly used Aboriginal term for the hairy giants. That, however, may not really have been the case.

In 1726, in his satirical novel *Gulliver's Travels*, Jonathan Swift called a fictional race of monkey-men "yahoos". Although Aboriginal use of the word could conceivably have been coincidental (it consists, after all, of only two syllables) Graham Joyner, who has researched the matter in depth, suggests it is more likely Aborigines picked up the term from early settlers. He may well be right, but in any case, by the mid-1800s many eastern Australian Aborigines were using the term and many whites, believing the word to be of Aboriginal origin, were also using it to refer to the yowie. [10]

There is also some dispute about the term "yowie". Veteran researcher Rex Gilroy insists that it is an authentic Aboriginal term, once widely used in the Blue Mountains and beyond, meaning "great hairy man", but Graham Joyner disagrees. He points out that references to the word prior to about 1975, when Rex began to popularise it, are extremely difficult to find.

We are inclined to think Rex and Graham are both partly right. To us, the evidence suggests Rex is right in claiming *yowie* as an authentic Aboriginal term for the Hairy Men, but wrong in saying it was very widely used. The term was certainly applied to the ape-men at least 10 years before Rex began to popularise it. In 1962 or '63 P. J. Gresser wrote that the Aborigines of south-eastern Australia, particularly the mountain tribes, feared "the Yahoo or Yowie … an animal of large proportions whose body was covered with masses of long hair and whose feet were reversed, the toes being where the heel should be". [11]

The Yahoos of New Holland

In part four of Jonathan Swift's *Gulliver's Travels*, first published in 1726, Lemuel Gulliver finds himself marooned on a large island ruled by a race of wise and civilised horses, the Houyhnhnms (pronounced "Whinnums"). Interestingly, Swift placed Houyhnhnms Land just 25 kilometres off Australia's southern coast, in the Nuyts Archipelago. He may have chosen that location because it was, in his day, the absolute end of the known world. Oddly, to accommodate his imaginary island, he nudged the real Isles of St Francis to one side. Even odder, he wrote that Houyhnhnms Land contained, in addition to talking horses, a race of filthy, yowie-like creatures called yahoos.

They were man-sized, tailless and foul smelling. Their heads, chests and parts of their legs, feet and backs were covered in thick hair. Their skin was "a brown buff colour". They had "strong extended claws before and behind". Like yowies, they howled, roared and grimaced menacingly. Unlike yowies, they "had beards like goats". They had one foul habit that (mercifully) yowies do not share: while sitting in a tree above him, Gulliver wrote, they "began to discharge their excrements on my head".

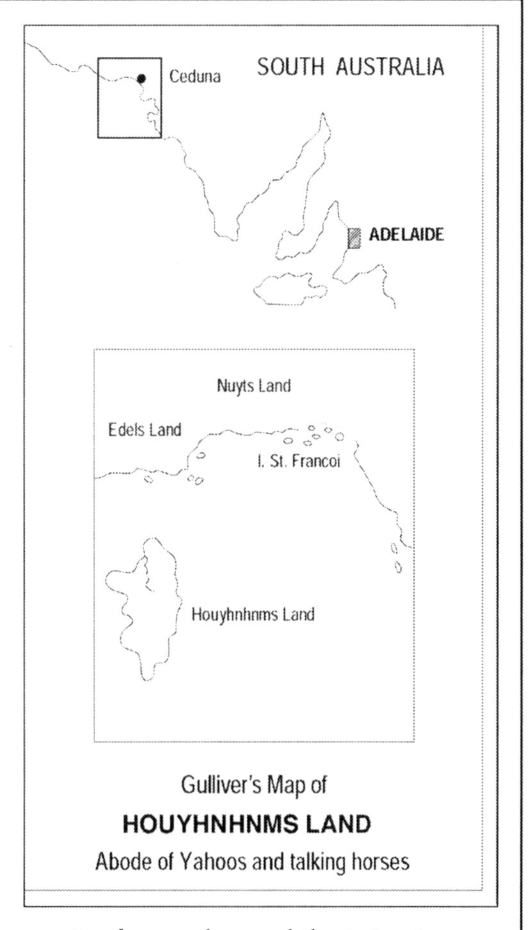

Gulliver's Map of
HOUYHNHNMS LAND
Abode of Yahoos and talking horses

Another thing that argues for the authenticity of "yowie" is that some very similar words clearly relate to the mysterious ape-men. *Macquarie Aboriginal Words*, for instance, contains the following: "In the Ngiyampaa language of central NSW the word *huwii* referred to the 'hairyman, a being that lives in caves'". Recently Don Bell, an Ngunnawal elder, informed us that his people, who occupied the Yass/Canberra region, use a very similar word to refer to the hairy wild men: they call them *wawee*.

At least as early as 1885 some white settlers used terms very similar to yowie and it seems reasonable to suppose they learned them from Aborigines. In 1976 Rod Knowles, who grew up a few kilometres west of Moruya in the 1930s and '40s, told us that old timers along the Deua River used three very similar-sounding terms when referring to the hairy giants. Some called the mysterious creatures *yahoos*, some called them *yowroos* and others called them *yowries*. Fred Howell, who encountered a hairy ape-man in the same general area the early 1930s, referred to it as a *yourie*. [Cases 62 and 49]

About 40 kilometres south of where Mr. Howell saw the ape-man is a locality – just a crossroads and a scattering of farms – called Yowrie (sometimes spelled Yourie). Nearby is a small watercourse called Yowrie River. Canberra archivist David Hearder has recently established that "the locality of Yourie came under that name as early as 1885 and at the same time the district was called Yowrie".

It is recorded that during the colonial era, Aborigines occasionally referred to the hairy giants as "wild blackfellows".[12] This serves to remind us of something that has been noted from time to time: whereas non-Aboriginal eyewitnesses are likely to say they have seen a man-like ape, Aborigines almost always refer to the creatures as *men* – enormous, hairy and primitive, but more human than animal. Kyle Slabb of Letitia Peninsula in northern NSW recently clarified this matter for us by explaining that his people, part of the Bundjalung tribe, did not think of the yowie as *entirely* human. The Bundjalung word for human is *beogil*, "but these things [the yowies] are called something different. The thing is, we never saw ourselves as apart from nature anyway ... a man is just like another animal of the bush. So you couldn't really say they [his ancestors] called the Hairy Man a person *or* an animal."

As mentioned earlier, in Western Australia the creatures were (and are) known as *jimbra*, *jingra*, or *tjangara* and it is interesting to note that 3,000 kilometres away, on the other side of the continent, the almost identical word *jingara*, while not used as an actual name for the creatures, is very closely associated with them. To the south-east of Cooma, near Twofold Bay, a steep, stark, rather eerie looking mountain called Egan Peaks or The Jingera, dominates a rugged, forested area that has produced several recent yowie reports. In the early 1970s, Graham Joyner discovered the following entry in "Aboriginal Dialects: Cooma Sub-District", *Science of Man*, August 23, 1904: "Jingara. A huge mountain, supposed to be haunted by a hairy man or Yahoo."

One recurring feature in Aboriginal yowie lore is that the Hairy Men spend most of their time in rugged mountain country and Egan Peaks/The Jingera is not the only mountain to be specifically mentioned as a yowie dwelling place. Others in south-east NSW are Pollwombra Mountain near Broulee, Cockwhy Mountain near Batemans Bay, Spring Mountain near Belowra and possibly Mt Dromedary (also known as Gulaga) near Narooma. It seems likely that a rocky hill near Bawley Point, which is now known as Monkey Mountain, was also traditionally associated with the yowie.

Egan Peaks, also known as The Jingera. Said to be an abode of the Hairy Man. (Tony Healy)

There are a great many yowie stories preserved by different Aboriginal groups throughout eastern Australia. One we like relates to what is now Jerawarrah Flora Reserve in heavily forested Washpool National Park, west of Grafton.

"Jerawarrah" is a corruption of the Bundjalung word *cherra-warra* or *jurrawarra*, which is said to translate as "large hairy man". The Bundjalung supposedly believed the *cherra-warra*, lurking in the trees, would sometimes swing down on a party walking single file and grab the first or last man. So in the colonial era, Aborigines, while guiding timber-getters to groves of prized red cedar, would always put white men at the head or tail of the group!

Rainy day yowies

One curious and perhaps significant part of the Turramulli legend of far north Queensland is that the hairy giant was said to appear most frequently in stormy weather. Interestingly, during the 1870s, white settlers in an area known as The Jingeras, to the south of Captains Flat, NSW, believed the same about the occasional appearances of the hairy "wild man". [Case 17]

Reg Birch, a Wyndham-based Aboriginal leader, told us that in north-west Western Australia the Hairy Man is traditionally associated with wild, stormy weather. Aborigines and some whites of the Batemans Bay, NSW, area say the same. It might also be worth noting that when the three young Aboriginal heroes of the book *Rabbit Proof Fence* encountered the giant, hairy, man-like *marbu* to the north of Moore River, WA, in 1931, a severe thunderstorm was brewing. [Case 55]

In her fascinating book, *Nature's Weather Watch*, Glenda Johns notes that one West Australian tribe believed a spirit called *yowie* controlled rainfall. She suggests that white pioneers picked up the idea and anglicised yowie to "Hughie" – which may explain the strange plea that drought-plagued Australian farmers still shout to the heavens: "Send her down, Hughie!"

Spirits of Heroes

Several Aboriginal stories describe skirmishes between yowies and Aborigines, but we know of two that refer specifically to large-scale conflict.

In 1965, while clearing scrub near Tyagarah, NSW, M. L. Mumberson discovered two curious spiral-shaped stone arrangements. One, consisting of 41 stones, ran clockwise, the other, consisting of 57, ran anticlockwise. They were each about three feet (one metre) in diameter and were 20 feet (six metres) apart. The round, flat, water-smoothed stones with which they were constructed ranged from approximately four inches (10 cm) to 14 inches (31cm) in diameter.

The stones were embedded edge-wise in the ground. The smallest were in the middle of the spirals with the stones becoming larger towards the end. Each ended with two or three much larger stones standing upright. Realising they were of significance and at risk of destruction where they stood, he moved both structures seven kilometres to Heritage Park at Mullumbimby and reconstructed them there in precisely their original form, cementing them in place.

Non-Aboriginal inhabitants of Mullumbimby found the spirals baffling until Koori elders approached the park managers with the following story, the text of which is now displayed near the stones:

"In the Dreamtime the people of this area [the Du-Rung-Bil or Durrumbul] were driven from their lands by the mountain people, the 'yowies'. They were dispersed far and wide over the north coast. Two of the local people reorganised them back together and under their leadership drove the yowies away and regained their lands and the source of the Sacred Red Ochre. The two people were male and female and the stones are their spirits ... they were buried at some

One of the Mullumbimby stone spirals.
(Tony Healy)

distance from the stones. The arrangement [to the north] is the male ... this story is passed down from the 'oldies' to the young and is still taking place. The current holder of the story told it personally to the park management and charged them with the care of the stones."

Another account of warfare between Aborigines and yowies is contained in William Telfer's *The Early History of the Northern Districts of New South Wales*, sometimes called the Wallabadah Manuscript, c. 1900. We will tell that story later, when we examine various theories about the origin of the yowies.

Littlefoot

Although most Aboriginal Hairy Man tradition lends strong support to the modern image of the yowie as a hairy giant, the subject is complicated slightly by the fact that many Aborigines also believe in very *small* hair-covered, man-like creatures. These entities, often called *junjudees*, but known by many other names, appear to fill more or less the same niche in Aboriginal Australia that leprechauns, fairies and elves do in Europe. Belief in these rather magical little hairy men may be stronger in some regions than in others. It certainly occurs, however, in many parts of eastern and central Queensland, NSW and Victoria.

The notion that one kind of unclassified hominid or pongid lurks in the Australian bush is difficult enough for most people to accept, but the suggestion that there could be *two* entirely different types running around out there seems completely outrageous. Small ape-like creatures, however, have been reported occasionally by whites as well as by Aborigines.

Most researchers assume these small creatures are juvenile yowies rather than representatives of a completely different species, and that certainly seems logical. On the other hand, it would be insanely arrogant to ignore what Aboriginal people, with their thousands of years of tradition, have to say on the matter, so we try to keep an open mind.

Most yowie researchers find Aboriginal belief in the magical little junjudee rather inconvenient. It tends to cloud the issue somewhat and provides welcome ammunition for sceptics. We, however, having encountered "little people" beliefs in various other countries, welcome the junjudee phenomenon as another intriguing mystery to be investigated. Without, for the time being, dismissing the little critters as myth, declaring them to be juvenile yowies or straining to fit them into some other zoological or anthropological category, we simply collect everything we read or hear about them and see where it leads us.

Because our "littlefoot" file is now quite sizeable, because testimony of whites as well as Aborigines has to be considered, and because the recent discovery of junjudee-sized skeletons (*Homo floresiensis*), only 700 kilometres north of Australia, has added an exciting new dimension to the mystery, we will put the matter aside for the moment and deal with it in a later chapter.

Despite the distracting junjudee, mumuga and yaroma legends, and despite some inconsistencies in their descriptions of the yowie, we think it is fair to say that a great many eastern Australian Aborigines believed – and still believe – in hairy giants very similar to those reported by non-Aboriginal people since the early 1800s.

Ron Heron is a Bundjalung elder. In 1991, when he was Head of the College of Indigenous Australian Peoples at Southern Cross University, he summarised what contemporary Aborigines in north-east NSW believe about the Hairy Man:

"The *jurrawarra* are estimated to be about seven feet tall and covered in hair. They have a strong smell similar to that of a wet cattle dog. They are supposed to eat mostly vegetation and some small animals. It is said they will not harm humans. They can run at great speed over the most rugged landscapes. Some people say they can make themselves invisible or change their appearance to fit into the local forest or environment. They call out 'Yaw, Yaw', to coax people further and further into the bush. The older people would tell us they also make a grunting sound but they cannot talk." [13]

As we conclude this chapter it is important to remind ourselves that, prior to 1788, because of their thousands of years of isolation on the island continent, Australia's Aborigines could have had no knowledge whatsoever of monkeys, chimpanzees, orang-utans or gorillas.

Chapter 2

The Colonial Era 1788-1901
Hairy Times on the Wild Frontier

"But the Hairy Man was permanent, and his country spread from the eastern slopes of the Great Dividing Range right out to the ends of the western spurs. He had been heard of and seen and described so often and by so many reliable liars that most people agreed there must be something. So now and then...search parties were organised and went out with guns to find the Hairy Man, and to settle him and the question one way or another. But they never found him." – Henry Lawson, "The Hairy Man", in *Triangles of Life and Other Stories* (Melbourne 1913)

In a letter to the *Australian Town and Country Journal*, December 9, 1882, H. J. McCooey told of meeting, just a few days earlier, in the bush between Batemans Bay and Ulladulla, NSW, an "Australian Ape":

"My attention was attracted by the cries of a number of birds which were pursuing and darting at it ... it was partly upright, looking up at the birds, blinking its eyes and distorting its visage and making a low chattering ... Being above the animal and distant less than a chain [22 yards] I had ample opportunity of noting its general appearance."

The creature was nearly five feet tall, tail-less and "covered with very long black hair which was dirty red or snuff colour about the throat and breast. Its eyes, which were small and restless, were partly hidden by matted hair. The length of the arms seemed to be strikingly out of proportion ... it would probably weigh about 8 stone ... it was a most uncouth and repulsive looking creature, evidently possessed of prodigious strength". The bemused bushman eventually threw a stone at the animal, which rushed off into a ravine, pursued by its squawking persecutors.

McCooey went on to say that he knew "half a dozen men at Batemans Bay who have seen the same, or at any rate, an animal of similar description" and claimed that "the skeleton of an ape, 4ft in length, may be seen at any time in a cave 14 miles from Batemans Bay, in the direction of Ulladulla".

It seems old H. J. experienced the kind of sighting we have always hoped for: of a smallish yowie in broad daylight, at a nice, safe distance. Despite his eyebrow-raising assertion that a readily accessible cave contained an ape skeleton (was it, perhaps, the skeleton of an Aboriginal or European child?) there are a couple of things about his report that we like. Firstly, the location fits well with local Aboriginal lore: about halfway between Batemans Bay and Ulladulla is Cockwhy Mountain. As noted in the previous chapter, Aboriginal elder Percy Mumbulla declared that yowies lived there until at least the 1950s. Cockwhy Mountain is, in fact, about 14 miles (22.5 km) north of Batemans Bay – so it could also be the site of McCooey's mysterious cave. Secondly, only about eight kilometres north of Cockwhy Mountain is a rather unusual looking hill, covered with trees and capped with a chaotic jumble of huge boulders. In the early 1990s, local Aborigines attempted to prevent any development on or near the hill. For reasons we were never able to pin down, they wanted the area classified as a "site of significance". We find it interesting that the hill is known locally as "Monkey Mountain".

Monkey Mountain. (Tony Healy)

Another thing that makes us inclined to believe McCooey's story is that the site of his alleged encounter lies within about 20 kilometres of four particularly interesting modern reports [Cases 77, 85, 132 and 208]. Those reports – one involving a rare sighting of male and female yowies together, one involving two witnesses and another involving a yowie hunting wallabies – will be covered in chapters three and four.

McCooey's report (which was unearthed in the 1980s by the tireless researcher Graham Joyner) is a good example of the "Australian Ape" or "Australian Gorilla" reports that appeared occasionally in newspapers and journals during the 19th century. His story, however, was far from the first European reference to the yowie. For that we have to go back to the early colonial era.

Although it is difficult to pinpoint the earliest actual sighting report by a non-Aborigine, it is certain that Europeans became aware of Aboriginal yowie lore in the first couple of decades of the 19th century, if not earlier. Captain Peter Cunningham, a Royal Navy surgeon who farmed in the Hunter Valley near Mt. Sugarloaf in the 1820s, may have been the first European to write about the phenomenon. In his memoir, *Two Years in New South Wales*, published in 1827, he mentioned that the local Awabakal people believed in a fearsome man-eating creature called *puttikan*, which resembled a tall man with a hairy body and long mane. Its feet were reversed to confuse trackers and its skin was so tough that spears could not pierce it. It roamed by night and devoured children, but was afraid of fire. Cunningham believed fear of *puttikan* was the reason his Aboriginal neighbours never travelled at night and always slept close to a fire. On still summer evenings, he heard strange cries that were attributed to the hairy giant, echoing through the mountains, so despite the bizarre reversed feet detail, it seems he was inclined to give the legend some credence. [14]

It is clear, however, that the next person to write about yowies was quite sceptical. When Aborigines warned him about such creatures in the forests to the north of Newcastle in the 1830s, Alexander Harris assumed he was being hoaxed. In his memoir, *An Emigrant Mechanic, Settlers and Convicts or Recollections of Sixteen Years Labour in the Australian Backwoods*, he wrote:

"The river [probably Williams River] on the banks of which we now were, rises and for a long distance winds to and fro among the mountains of the country of Durham ... It is now well settled, but at that time we were there spoiling it of its cedar, only here and there amidst the lonely wilderness was there to be found a settler's farm or stockman's hut. The blacks were occasionally, but not often troublesome. The stories they used to tell us about the brush thereabouts being haunted by a great tall animal like a man with his feet turned backwards, of much greater, however, than the human stature, and covered in hair, and perpetually making a frightful noise as he wandered alone, made me sometimes doubt whether they were themselves really terrified, or were merely endeavouring to scare us away; but I very strongly incline to the latter opinion." [15]

Alexander Harris.

Although Harris was sceptical, there is evidence that by the 1830s Australian naturalists were seriously debating the reality or otherwise of "Australian apes". In an 1842 edition of the *Australian and New Zealand Monthly Magazine*, Graham Joyner spotted the following:

"The natives of Australia ... believe in the imaginary existence of a class which, in the singular number, they called Yahoo or, when they wish to be anglified, *Devil-Devil.* This being they describe as resembling a man, of nearly the same height, but more slender, with long white straight hair hanging down from the head over the features, so as almost entirely to conceal them; the arms as extraordinarily long, furnished at the extremities with great talons, and the feet turned backwards, so that, on flying from man, the imprint of the foot appears as if the being had travelled in the opposite direction. Altogether, they describe it as a hideous monster, of an unearthly character and ape-like in appearance.

"On the other hand, a contested point has long existed among Australian naturalists whether or not such an animal as the Yahoo existed, one party contending that it does, and that from its scarceness, slyness and solitary habits, man has not succeeded in obtaining a specimen, and that it is most likely one of the monkey tribe." [16]

Ludwig Leichhardt – yowie tucker?

These days very few recent reports reach us from South Australia or Western Australia, so it is strange that two very early references relate to those states (or colonies, as they then were).

In *The History of Australian Exploration From 1788-1888*, Ernest Favenc tells how, in 1851, two squatters named Oakden and Hulkes, while searching for good grazing land to the west of Lake Torrens, South Australia, were told by Aborigines that ape-like creatures were sometimes encountered in the area. He also tells how Aborigines warned Messrs. Dempster, Clarkson and Harper about similar animals in south-west Western Australia in 1861:

"[The Aborigines] gave an account of the *jimbra* or *jingra*, a strange animal, male and female, which they described as resembling a monkey, very fierce, and would attack men when it caught one singly. Thinking there might be a confusion of names, the explorers asked if the *jimbra* or *jingra* was the same as the *ginka* – the native name for devil. This however, was not so, as the natives asserted that the devil, or *ginka*, was never seen, but the *jimbra* was both seen and felt."

The natives also told of their belief that, some years previously, three white men had died far away to the east at some large inland lake – possibly killed by the *jimbras*. Favenc adds that some Western Australian Aborigines believed the three white men were the last survivors of Ludwig Leichhardt's doomed 1848 expedition, and that *jimbras* not only killed them – but also ate them. He notes that "Whatever may have been the origin of the native tradition about the deaths of the three white men, which Forrest [Sir John Forrest, explorer and later premier of Western Australia] afterwards investigated, it must seem strange that the natives should in the *jimbra* have described an animal (the ape) they could not possibly have ever seen." [17]

Ludwig Leichhardt.
Some Aborigines said the great explorer was devoured by the ape-like jimbra.

For many years veteran yowie hunter Rex Gilroy has made passing references to a yowie sighting near Sydney Cove in 1795. Sometimes he says the witnesses were kangaroo hunting at the time, but to date, as far as we know, he has not revealed the source of the story. The oldest reference to hair-covered wild men in our own files actually pre-dates Rex's story by about five years, but is, unfortunately, an obvious hoax. It is worth looking at, however, not only as a curiosity in itself, but also because it might – possibly – have been inspired by warnings about the yowie by Aborigines at Botany Bay.

The story is contained in an eighteenth century English handbill entitled:

A defcription of a wonderful large WILD MAN or monftrouf GIANT, BROUGHT FROM BOTANY BAY.

(In the original, as was the custom of the time, "f" was used where modern English uses the letter "s". In the following extract, for the sake of legibility, we have used the modern form.)

"THERE have been various reports concerning this most surprising wild man, or huge savage GIANT, that was brought from Botany Bay to England, numbers of People arguing and disputing his enormous Size, but to prevent further contending, the following is sufficient to satisfy the Reader as many Thousands have seen him in Plymouth, where he was landed alive and in good health.

"This surprising monstrous giant was taken by a crew of English sailors when they went on shore to furnish themselves with fresh water at Botany Bay. To their surprise they beheld at a distance three of the most surprising tallest and biggest looking naked men that have been seen in the memory of this age, turning towards them, which much affrighted the sailors, caused them to make expedition on board the ship for the safety of their lives, leaving the casks of water and a quantity of good old rum which they had in a cag to refresh themselves and make merry.

"When the three savages got to the seaside they stared at the ship for a long time with wonder and admiration, one of them having got the cag of rum, he tasted, spit it out and shook his head, another did the same, but the third drank plentifully, and began to jump about in a frightful wild manner, shouting and making a hideous noise. The other giants went off and left this one ... who drank to such excess that he dropped on the ground and lay as if dead the sailors went on shore ... bound him fast

The 'Monstrous Giant' of Botany Bay – a woodcut from the handbill.

with ropes and with much fatigue got him on board the ship, where they secured him with iron chains, where he slept upwards of 24 hours ... He showed not the least token of illness at sea. He came in the ship Rover, Capt. Lee, to England from Botany Bay, and landed at Plymouth, November 29, 1789.

"He is much tamer, and not so savage in temper as might be expected. He is 9 feet 7 inches high, 4 feet 10 inches broad, a remarkable large head, broad face, frightful eyes, a broad nose and thick lips like a black, very broad teeth, heavy eye-brows, hair stronger than a horse's mane, a long beard strong as black wire, body and limbs covered with strong black hair, the nails of his fingers and toes may be properly called talons, crookt like a hawk's bill, and as hard as horn. In short, he is viewed with admiration and astonishment on account of his huge size."

Although the story is almost certainly a complete fabrication, it is possible that it was inspired by genuine reports of hairy ape-men in the area. We recently became aware of another very early – and somewhat more plausible – yowie report from a location only five kilometres to the south of Botany Bay. In his autobiography *Life and Adventures (of an Essexman)* Captain William Collin said that he and a friend named Massie received dire warnings about a giant, hairy yahoo while camped on the southern shore of Port Hacking in 1856. The story is of considerable interest for a number of reasons. Not only is it one of the earliest yowie cases where the supposed eyewitness is identified, but it is also our only colonial era case in which a protagonist mentions the fictional yahoos of *Gulliver's Travels*.

In 1856 there was only one white family, the Gogerlys, living at Port Hacking [at a spot still known as Gogerlys Point]. The patriarch of the clan, Charles Gogerly, was, as Captain Collin put it, "a curious old gentleman".

William Collin.

Collin and his mate were at the bay to gather seashells for lime making, an occupation that could be quite lucrative: "We did very well, selling the shell to schooners and ketches which carried it to Sydney. For some reason or other Gogerly did not like us in his vicinity, and ingeniously worked up a scheme which he vainly thought would frighten us away. One afternoon Gogerly sent down two of his boys in an old log canoe, to tell us that their father had seen a Yahoo, or wild man of the woods; it was about 12 feet high, they said, and carrying a staff 20 feet long. He warned us that we were not safe from the creature, as it was seen close to our tents." Later, old Mr Gogerly told them how, "on hearing a noise ... he used his spyglass ... and scanned the shore, till his eyes rested on the monster, which he declared was looking at my mate and myself, as we gathered shells on the beach".

Massie took the warning seriously: "My mate ... who was a great reader, and who had no doubt dived at one time or another into 'Gulliver's Travels', said he knew such things as yahoos existed; and as there were a number of deep gullies about this was no doubt a likely place for them." Although Collin insists, in his memoir, that he always thought Gogerly's story was an invention, it is clear from his actions that he was not entirely convinced of that at the time:

"'If we could only trap this Yahoo', I told my mate, 'we should not need to trouble any further about gathering shell'. We loaded our guns, and took them ... to find the Yahoo. We certainly did find some remarkable tracks, which had not been made by a human being. What they were I had no idea. Massie ... would not sleep on the shore, feeling safer in the boat. I eventually lay down with my gun alongside me. The place was infested with dingoes ... which made noise enough for half a dozen Yahoos. In fact, I made up my mind one night that the creature was upon us, and fired, only with the result of setting a dingo yelling most pitifully. We shifted camp ... and were not further disturbed by either wild dogs ... or the mythical Yahoos." [Case 5]

Charles Gogerly's house. (Tony Healy)

One of the most intriguing aspects of Collin's story is that on the northern shore of Port Hacking, only 1.5 kilometres across the water from the spot where Gogerly supposedly saw the yahoo, is an inlet called Yowie Bay. At first glance, this "lexi-link" seems very significant. At one time we would have taken it as evidence that the local Aborigines used the word *yowie* to refer to the hairy giants and that the creatures had long been associated with the area.

Archivist David Hearder, however, has recently shown that this tantalising conjunction of "yahoo" and "yowie" is probably just a coincidence – albeit a rather remarkable one. In earlier times, in parts of Britain, female sheep – ewes – were sometimes called "yoes" or "yows" (pronounced like "rose"). The term found its way to Australia and can still be heard in a lyric ("the bare-bellied yoe") from the song "Click Go The Shears". Lambs, as the offspring of ewes or yows, were often referred to as "eweys" or "yowies" (pronounced like "joeys"). A NSW Government Geographic Names Bureau data base that David accessed states that "sheep were bred [at Yowie Bay] by Thomas Holt (1811-88) in the 19th century and he employed some shepherds from Yorkshire. Since 'yowie' is a Yorkshire word for lamb, this is the most likely explanation for the name".

Case closed ... or is it? David also found a digital warehouse of all historical (non-current) parish maps of NSW. Notations on one of those maps indicate, as David puts it, that "Yowie Bay was ... named by assistant Surveyor Robert Dixon – apparently using local Aboriginal names – in 1827..."

That seemed to put a different slant on the matter. Thomas Holt was only 16 years old in 1827. It seems unlikely he would have been, at that age, supervising herds of sheep and mobs of bucolic, dialect-sprouting Yorkshire men in the wilds of Sutherland shire. The suggestion that Surveyor Robert Dixon used local Aboriginal words in his nomenclature is also very encouraging. Frustratingly and confusingly, however, other notations on the same map indicate that "Yowie Bay was originally called Ewey Bay" – which of course more or less confirms that the bay was named after little woolly critters (Tom Holt's or someone else's) rather than big hairy ones.

So, reluctant as we are to let go the notion that part of Sydney (Yowie Bay – the suburb which now nestles around the inlet) is named after a race of giant, drooling apemen, we have to concede that the evidence does suggest it is named after nothing more exciting than sheep. The coincidence of the two key words, yahoo and yowie, both appearing at Port Hacking, however, really is quite remarkable. Consider this: since Gogerly's day, many other people have reported encounters with hairy apemen around Port Hacking, and one recent sighting occurred very close to the old homestead at Gogerlys Point.

During the 1840s the area now occupied by the suburb of Heathcote East (eight kilometres south-west of Gogerlys Point) was the site of the short-lived village of Bottle Forest. In later years, as the abandoned hamlet succumbed to nature, numerous sightings of hairy giants occurred in and around its ruins.

In the *St George Call* of June 8, 1907, an anonymous correspondent recalled how, one rainy night in the 1860s, he and some friends took shelter in the village's largest intact structure, Bottle Forest House. One fellow, identified only as Pat, heard a noise, went outside to investigate, and returned to say he'd encountered a yahoo that stood more than 12 feet tall, and which had "starfish-like" feet.

In *From Bottle Forest to Heathcote*, historian Patrick Kennedy records another incident from the same era: "A Spaniard, who had a camp near Bottle Forest House ... had a frightening experience one dark night, when one of his usually savage dogs ran to him in terror at the sight of a 'hideous yahoo'. The Spaniard seized his gun and was frightened out of his wits when he encountered the 'yahoo' moving among the trees."

Mr. Kennedy notes that yahoo sightings were not confined to the immediate vicinity of Bottle Forest. From the late 1860s onwards, residents of the area between Sutherland (seven kilometres to the north) and Waterfall (five kilometres south) began "hearing strange and fearsome noises at night from 'The Thing'. 'The Thing' was also known as the 'wild man of the bush'. 'The Thing' apparently resembled a tall hairy creature that was neither man nor beast". At times its cry sounded "like someone screaming in pain". According to *The Sutherland Shire Historical Society Bulletin* of July 1975, "The Thing" continued to make occasional appearances until about 1910.

Today, the Princes Highway, which links all the settlements between Waterfall and Sutherland, is effectively the western boundary of the 160 square kilometre Royal National Park. So it is interesting to note that contemporary residents of those settlements live right on the edge of the same wilderness in which the colonial-era "Thing" supposedly lurked. Even more interesting is the fact that some locals claim to have seen similar creatures quite recently.

A 23-year-old Helensburgh man told us how he and a mate came within six metres of a huge bipedal ape-man in broad daylight in Royal National Park in the year 2000. They returned to the site on several occasions and stalked the creature, discovering its lair, many footprints and droppings. The full story of that particularly interesting episode is told in the Catalogue of Cases [Case 233]. Suffice to say, at this point, that what they saw very much resembled the yahoos of old.

Our colleague Steve Rushton, who grew up in the area and first alerted us to its Hairy Man tradition, remembers hearing stories of the creatures being seen on the southern shore of Port Hacking in the early 1960s.

One interesting modern sighting – particularly in terms of location – occurred in 1985 at the Anglican Youth Department Conference Centre on the southern shore.

Adam Bennett, then aged 10, was sharing a cabin with four other boys and a teacher when they were all woken, at 1 am, by a series of deep, throaty growls and high-pitched screams. When they stepped outside the noises intensified, giving the impression that several wild creatures were calling to each other. Despite being quite frightened, a couple of boys including Adam ventured further into the scrub, heard a noise, swung their torch-beams around and found they were within two or three metres of a hideous, hair-covered "dwarf". It stood erect, apparently stunned, while the boys stood staring, equally stunned.

Although only about one metre tall, it was bulky and powerful looking. Covered in long, light brown hair, it had a large head, flat face, a nose "like an Aboriginal", small, round ears and huge, terrifying eyes shining red in the torchlight. As if things were not already frightening enough, the horrible little goblin then emitted another growl followed by a nerve-rending scream that sent the boys running back to their cabin, wetting their pants all the way. [Case 159]

It is tempting to assume the boys encountered a juvenile yowie. Be that as it may, even today, nearly 20 years after the event, Adam cannot tell the story without experiencing cold chills and butterflies in the stomach.

As mentioned earlier, the most interesting thing about his sighting is its location. The church land on which it occurred is bounded on one side by Royal National Park and on the other by Port Hacking – and encompasses Gogerlys Point and the old Gogerly homestead. The encounter with the little ape-man, therefore, may have occurred quite literally within a stone's throw of where old Gogerly saw the huge yahoo 129 years earlier.

In several reports made by Europeans during the 1870s and 1880s it is stated or implied that yowies were seen by other colonists 20 or 30 years earlier. One such account appears in the *Lismore Northern Star* of May 17, 1878:

"About thirty years ago a shepherd in W. Sutton's employ averred that he had seen a hairy man in a scrub north of Cunningham's Creek, but his story was treated as childish. However, he persisted to the day he died that it walked upright and was covered in hair, and the dogs that hunted everything else ran back from this frightened with their tails between their legs."

A more detailed early report appeared in *The Empire* (Sydney), on April 17, 1871:

A SUPPOSED GORILLA AT ILLAWARRA –

"The following particulars have been supplied to us by Mr. George Osborne, of the Illawarra Hotel, Dapto, concerning a strange looking animal, which he saw last Monday, and which he believes was a gorilla. It is to be hoped successful means may be adopted to capture the animal, (alive if possible), as it is quite evident it is one of the greatest natural curiosities yet found in the colony. Together with the interest attached to the peculiarity of this strange 'monster in human form', there is something very remarkable and suggestive in the fact that he should have presented himself to Mr. Osborne, while that gentleman was going his rounds, collecting the census. The following are Mr. Osborne's remarks concerning the animal:

'On my way from Mr. Matthew Reen's, coming down a range about a half mile behind Mr. John Graham's residence, at Avondale, after sunset, my horse was startled at seeing an animal coming down a tree, which, I thought at the moment to be an Aboriginal, and when it got to within about eight feet of the ground it lost its grip and fell. My feelings at the moment were anything but happy, but although my horse was restless I endeavoured to get a good glimpse of the animal by following it as it retreated until it disappeared into a gully. It somewhat resembled the shape of a man, according to the following description:

'Height, about five feet, slender proportioned, arms long, legs like a human being, only the feet being about eighteen inches long, and shaped like an iguana, with long toes, the muscles of the arms and chest being well developed, the back of the head straight, with the neck and body, but the front or face projected forward, with monkey features, every particle of the body except the feet and face was covered with black hair, with a tan-coloured streak from the neck to the abdomen. While looking at me its eyes and mouth were in motion, after the fashion of a monkey. It walked quadruped fashion, but at every few paces it would turn around and look at me following it, supporting the body with the two legs and one arm, while the other was placed across the hip. I also noticed that it had no tail.'

"It appears that two children named Summers saw the same animal or one similar in the same locality about two years ago, but they say it was then only the size of a boy about thirteen or fourteen years of age. Perhaps this is the same animal that Mr. B. Rixon saw at the Cordeaux River about five or six years ago. The query is, 'Where did it come from?'"

The sightings by the Summers children and by Mr. Rixon would therefore have occurred in about 1869 and 1865 respectively.

In his memoir, *Old Convict Days*, William Derrincourt recorded his daughter's account of a yowie she met while lost in the bush

New England pioneer William Telfer (1841-1923) was a self-educated man who wasn't overly concerned with the niceties of punctuation. This is an extract from his memoirs:

"i had an Experience myself of this gorilla or hairy man in the year 1883 i was making a short cut across the bush from Keera to Cobedah via top bingera ... had to camp for the night made a camp on a high bank of the creek lit a fire and made myself comfortable my dog laying down at the fire alongside me i sat smoking my pipe the moon rose about an hour after when you could discern objects two hundred yards away from the camp i heard a curious noise coming up the creek opposite the camp over the creek i went to see what it was about one hundred yards away he seemed the same as a man only larger the animal was something like the Gorilla in the Sydney museum of a darkish colour and made a roaring noise going away towards top Bingara the noise getting fainter as he went along in the distance i started at daylight Getting to Bells mountain at about 9 oclock Mr Bridger lived there stopped and had breakfast i was telling them about the

night before they said several people had seen the gorilla about there he was often seen in the mountains towards the Gwyder and about mount Lyndsay ... some people think they are only a myth but how is it they were seen by so many people in the old times fifty years ago." [Case 27]

ANOTHER GORILLA.—It is siad by persons frequenting the neighbourhood of Belgrave, that a gorilla has made its appearance in that vicinity. A short time ago a camp of blacks were so scared by the appearance of the alleged monster, that they left their camp, and hastened with all possible speed to Warneton, and refused to return. When asked for a description of the animal they saw, they said, "That fellow run on four legs, and stand up and run on two legs; him got plenty fellow hair all over." Two young men are also said to have been riding along through the bush between Belgrave and Warneton, when the supposed gorilla rushed through the bush near them, and so frightened the horse (a very quiet one) which one of the young men were riding that it was with great difficulty that he could keep his seat, and prevent the horse from bolting. A short time after, so the story goes, a person residing in the same neighbourhood, hearing his bull-dog barking and making desperate efforts to break his chain, evidently wishing to get at something he saw in the bush, let the dog loose. The dog, a very savage brute, immediately tore away in a furious manner towards the bush, but in a short time he was seen beating a speedy retreat, with his courage evidently cooled. He took refuse in the house, and could not be persuaded to leave it. A party of young men, it is said, formed a sort of expedi-

Queanbeyan Age July 27, 1871

AN EXTRAORDINARY ANIMAL — Mr. Prosser, manager at Messrs. Amos and Co.'s sawmills at Amos's Siding, near Sutton Forest, has just informed us (*Scrutineer*) that a most peculiar animal has been seen by two men, Patrick Jones and Patrick Doyle, residents of Sutton Forest, in the bush between Cable's Siding and Jordan's Crossing. Mr. Prosser himself has seen the footprints; they are 3 feet apart, and the impression made by the feet is similar to that of an elephant. The animal is described as being 7 feet high, with a face like a man, and long shaggy hair, and makes a tremendous noise. Fourteen of the men from the mill, fully armed, intend starting on Saturday next to endeavour to capture this "wild man of the woods." Mr. Prosser assures us there is no exaggeration about this affair, and every one at the mill believes in the existence of this strange creature.

Sydney Morning Herald, Oct. 12, 1877

near Braidwood, NSW. Although no specific date is given, clues in the text strongly suggest the incident occurred during the 1860s. If so, it would pre-date Osborne's sighting by at least a couple of years.

On the third day of her ordeal Miss Derrincourt encountered, as she put it, "something in the shape of a very tall man, seemingly covered with a coat of hair and looking as frightened of me as I was of him. While he stood gazing at me, without attempting to get nearer, I heard at a distance a peculiar cry, between a laugh and a bark, which my companion of the scrub answered in the same manner and, after seeming to consider for a few moments he leisurely walked or shuffled off, greatly to my relief. I was afterwards told it was what the people here call a 'Yahoo' or some such name".

Because the notion of hairy apemen roaming the Australian bush seems so mind-bogglingly improbable, we always try to interview eyewitnesses face to face. In the case of colonial era reports, we try to get as close as possible to the original source. The mere fact that something was reported in a 19th century newspaper does not, of course, necessarily mean it really happened. Nevertheless, we have always found the old clippings give at least a *sense* of authenticity – and several, we believe, have a distinct ring of truth.

It occurs to us that many readers, sceptical or otherwise, might find it interesting to examine a few early reports in exactly the same form as they originally appeared. In viewing them, the reader can, at

MORE ABOUT THE PYRAMUL HAIRY WONDER.

A CORRESPONDENT, on whose good faith we rely, the same who sent us some particulars of the reported appearance in a western district of a strange creature resembling a hairy man—gives the following additional information on the subject:—

It is now about 18 months since I first heard of Pat King's adventure with the hairy man. I thought as little about it as my neighbours, until I got the recital from his own lips. I fancy I am pretty sharp in detecting a falsehood, in a certain link between the voice and the eye, but I could see no reason to doubt the story. Moreover, the character of the whole family is above reproach. I have since seen the young man's sister, who tells me that when her brother Tim ran home and told about the sight he had seen he was as white as a sheet, and gave a better description in a few more particulars than his brother. She likewise reports another meeting with this strange thing. A settler's daughter having gone for the cows, an older sister, thinking she was long away, went out to assist her. On turning the corner of a bush fence, about a quarter of a mile from the hut, she suddenly stood face to face with the stranger. No doubt both were frightened, as they stood watching each other, until the sister called out that she had all the cows, when the hairy creature turned about and walked leisurely away. This last adventure, like the three black crows, took all shapes ere it reached our neighbourhood two years ago. We always doubted the existence of this strange animal, but after conversing with some of the actors, and hearing the recital from neighbours who live beside them, we see no reason to discredit it any longer.

The Freeman's Journal, Apr. 13, 1878

least, be reassured the yowie is not merely an elaborate 20th or 21st century hoax.

We believe the majority of colonial era yowie reports tally quite well with modern eyewitness testimony. We would like to be able to claim that *all* of them do, but have to admit that, just as some Aboriginal lore contains yowie descriptions that deviate from the norm, a small number of 19th century reports contain exceedingly odd details that are jarringly out of whack with our treasured notion of what yowies should look like. Although we would prefer to consign these inconvenient reports to the "Too Weird" basket, it would be less than honest to do so. So, for better or for worse …

Don't mess with Pig Man

This strange story, which appeared in the *Lismore Northern Star* on May 17, 1878, under the heading "An Australian Man of the Woods", describes an incident that supposedly occurred near Cudgegong, NSW, a couple of years earlier.

"Pat Wring … heard his kangaroo dogs bark from 10 am to 4 pm down the inaccessible cliffs … Pat's surprise may easily

The Wild Man of Snowball.

On the 3rd of October last young Johnnie McWilliams was riding from his home at Snowball to the Jinden P.O., Braidwood. When about half-way the boy was startled by the extraordinary sight of a wild man or gorilla. The boy states that a wild man suddenly appeared from behind a tree, about thirty yards from the road, stood looking at him for a few seconds, and then turned and ran for the wooded hills a mile or so from the road. The animal ran for two hundred yards across open country before disappearing over a low hill, so that the boy had ample time to observe the beast. The boy states that he appeared to be over six feet in height and heavily built. He describes it "as a big man covered with long hair." It did not run very fast and tore up the dust with its nails, and in jumping a log it struck its foot against a limb, when it bellowed like a bullock. When running it kept looking back at the boy, till it disappeared. It was three o'clock in the afternoon, and the boy describes everything he saw minutely. The boy is a truthful and manly young fellow, well acquainted with all the known animals in the New South Wales bush and persists that he could not have been mistaken.

For many years there have been tales of trappers coming across enormous tracks of some unknown animal in the mountain wilds around Snowball. Of course, these tales were received with doubt, and put down as clever romancing on the part of the 'possum hunter, but the story of Johnnie McWilliams is believed by all who know the boy as a true tale. The proof of the existence of such an animal in New South Wales should be of some interest to the naturalist.—Braidwood "Dispatch"

The Queanbeyan Observer, Nov. 30, 1894

be imagined when his eyes looked down on a hairy monster standing upright, a body apparently as round as a horse, arms as round as a man's thigh, three claws on each foot. It stood, to the best of his belief, about 4 feet high. The head resembled a pig's, but turned upwards, and he threw into the air the only dog that ventured within its reach. Pat could see the milk-white hair under his armpits.

"Fearing the dog would be killed, as it fell on the rock about sixty yards away each time it was thrown up, he threw about 14 lbs weight of a stone, which struck the animal without doing any damage. The animal was at the foot of the rocks on which Pat stood, and in two springs or strides it sprang or strode in an upright position and then commenced to climb monkey-fashion.

"Pat saw no more as he thought it was time to run for his life; he never looked back. His heart beat so audibly that he fancied it was the quick stamping of the strange thing behind him. The dog died shortly after, but not a hair of the strange creature could be found, though the dog's hair and blood was plentiful on the rocks."

The hair-raising encounter supposedly occurred in the vicinity of Cunningham Creek, about five miles (8 km) from the place where the unnamed shepherd saw a Hairy Man 30 years earlier. [Case 3]

Although young Pat's description of the pugnacious porker is bizarre in the extreme, it is just possible to imagine that he viewed a "normal" yowie from such an odd angle – directly overhead – in such frightening circumstances that he couldn't give an accurate description.

Be that as it may, the next story, which appeared in *The Argus* on October 25, 1849, isn't so easy to rationalise:

"We are informed by Mr Edwards, the managing clerk, at the office of Messrs. Moor and Chambers, that during his late trip ... to capture the runaway Hovenden, that while on ... Phillip Island, he and his party were astonished at observing an animal sitting upon a bank in a lake. The animal is described as being from six to seven feet long, and in general appearance half man, half baboon. Five shots were fired ... upon the first shot whistling past him, he appeared somewhat surprised, and shook his head apparently in disapprobation of the proceeding; at the second, he grinned fiercely and showed an uninviting set of teeth; at the third he backed towards the water; the fourth was answered by a half growl, half shout, which made the 'welkin ring', and the fifth ... was replied to, by a spring into the air, and a contemptuous fling out of the hind legs and a final disappearance in the placid waters of the lake. A somewhat long neck, feathered like the emu, was the peculiar characteristic of the animal."

Although this strange and irritating story seems at first glance to scream "Hoax!" we have always been quite tantalised by it. Despite the "long, feathered neck" and the journalist's jokey tone, it remains one of the very earliest references to man-sized ape-like creatures in Australia. It is possible the story was based on a real incident. In a long postscript, full of ponderous colonial-era humour, the editor of *The Argus* admits that he invented the detail of the creature's leaping into the air and flicking its legs. As the story came to *The Argus* via the *Melbourne Morning Herald* it is possible the other inconvenient detail of the long, feathery neck was also added somewhere along the way. [Case 4]

Thankfully, references to long feathery necks and pig faces are the exception rather than the norm in colonial era yowie reports. The brave old pioneers almost always said just what most modern-day witnesses say: that the creatures looked like giant, hairy, long-armed ape-men with flat faces and short necks. When their descriptions varied, it usually involved (as it did with tribal Aborigines and as it does with modern witnesses) the shape of the creatures' feet.

We will relate several more 19th century reports in later chapters and in the Catalogue of Cases. For the moment, however, we will end this chapter with a bang: the supposed shooting of a yowie in the Brindabella Mountains, just west of Canberra, in about 1885.

In *An Alpine Excursion*, published in 1903, veteran journalist and clergyman John Gale referred to the many Hairy Man or yahoo reports that came from the area surrounding the present site of the national capital during the 1800s. One story in particular convinced him of the creatures' reality. It was told to him by his friends William and Joseph Webb, both "strongminded, experienced, and educated men":

"They were out in the ranges preparing to camp for the night. Down the side of a range to the eastward, and with only a narrow gully separating them from the object which attracted their attention, they first heard a deep guttural bellowing and then a crashing of the scrub. Next moment a thing appeared walking erect, though they saw only its head and shoulders.

"It was hirsute, so much of the creature as was visible, and its head was set so deep between its shoulders that it was scarcely perceptible. It was approaching towards their camp. Now it was in full view, and was of the stature of a man, moving with long strides and a heavy tramp. It was challenged: 'Who are you? Speak, or we'll fire'. Not an intelligible word came in response; only the guttural bellowing. Aim was taken, the crack of a rifle rang out along the gully; but the thing, if hit, was not disabled; for at the sound of the shot it turned and fled.

"The two gentlemen, filled with amazement and curiosity, but not alarm, went to where they had seen and shot at this formidable-looking creature ... There were its footprints, long, like a man's, but with longer, spreading toes; there were its strides, also much longer than those of a man; and

Below: William (left) and Joseph Webb. (Merrilee Webb)
Right: The Flea Creek yowie, as sketched by 'an eyewitness'.

there were the broken twigs and disordered scrub through which it had come and gone. They saw no blood or other evidence of their shot having taken effect."

The descendants of William and Joseph Webb, who still own land in the area, think the incident probably occurred near Flea Creek, at the foot of the Webb Range in the northern part of the Brindabellas.

At least two other people, including young Alexander McDonald and his father John, were present during the encounter. Some time later, a member of the party sketched the creature "from memory", complete with spurs, horns and cheerfully grinning visage. Even though a fair bit of artistic licence was obviously employed in its creation, Alexander kept the sketch all his life and insisted the story was true. [Case 28]

Canberra, a mere crossroads in 1885, now has a population of 330,000, but the Brindabella Range – most of which is now part of the vast Namadgi/Kosciuszko National Park system – is still very wild country. Every couple of years someone vanishes without a trace in the Brindabellas and, every now and then, as in the old days, giant, hairy ape-men are seen.

Chapter 3

The Early Modern Era
1901-1975
The Yowie Vanishes

"I've lived with this for sixty years - and now I can finally tell someone about it."
– Clyde Shepherdson, Ipswich, Queensland, May 2001

Our records suggest that during the colonial era a large minority, possibly even the majority, of non-urban white Australians were at least vaguely aware of the Hairy Man / yahoo / yowie phenomenon. It seems strange, therefore, that by the middle decades of the 20th century awareness of the yowie among non-Aborigines was virtually zero. There are several possible reasons for this strange collective amnesia.

By the time the colonies united to form the Commonwealth in 1901, the great gold rushes were over and most of the hundreds of thousands of prospectors had returned from the backblocks to the cities, where they were no longer likely to hear outrageous stories of hairy giants. In the first few decades of the new century, as farming became more mechanised and as city-based secondary industry grew, the nation went through other great changes. Paradoxically, by the early 1900s, Australia, the most sparsely populated continent on earth, was also the most urbanised: more than a third of the population lived in the cities and many outback settlements had become ghost towns.

By the 1920s the continent had been fairly well explored, the last of the bushrangers had been dead for 40 years, armed resistance by Aborigines had ceased except in parts of the far north and north-west, and the Australian bush no longer seemed an entirely wild, alien place. Axemen had cleared great swathes of it and naturalists had, it seemed, collected, dissected and catalogued just about everything that slithered, scurried or bounded through it. The possibility that thousands of huge gorilla-like creatures could be lurking out there somewhere had begun to seem increasingly remote.

We suspect that after about 1900, as many country newspapers disappeared and city-based papers grew, many eyewitness reports that might have been printed by small town proprietors were dismissed out of hand by big-city editors who simply *could not* believe them. Until the mid-1950s few Australians owned motorcars – let alone four-wheel-drive vehicles – and trail bikes were unknown. Bush roads were appallingly bad and few urban people ventured far off the beaten track. Most inland passengers and freight went by rail and a great deal of intercity travel was done by sea. Regular passenger steamer services linked all the major cities until the late 1950s.

We now know that awareness of the yowie phenomenon persisted among country folk in many isolated locations, but during the mid-1900s the vast majority of white Australians hadn't heard even the slightest suggestion that such creatures might exist.

Another factor in the yowie's "disappearance" was probably the dire situation of Aborigines during those years. Decimated, dispossessed and largely confined to remote missions and reserves, they were, as far as most white Australians were concerned, out of sight and out of mind. Communication between the races was very limited. The complexity and antiquity of Aboriginal culture was not appreciated. Their legends and spiritual beliefs were usually dismissed as the rankest superstition.

Despite the difficult circumstances, some tribes managed to maintain their cohesion and their culture, but many virtually disappeared, with the disorientated survivors losing their families, their language and even their names. Under the circumstances, it is amazing that they preserved as much of their culture as they did. We now know that one tradition preserved within many groups was the age-old belief in the Hairy Man. But during those grim decades, when they were so used to having their beliefs derided, many Aborigines, quite understandably, seem to have decided to keep the yowie tradition – and, possibly, quite a few other matters – to themselves.

So for all those reasons – and, no doubt, for others we are unaware of – the yowie/yahoo virtually ceased to be mentioned in Australian newspapers from about the turn of the century. It was not until 1975 that the Hairy Man lurched back into the news.

Having said that, we now hasten to mention two notable exceptions: a couple of extremely interesting reports that appeared in major Sydney newspapers in late 1912 (and which were re-discovered by Graham Joyner in 1976). It is probable the papers featured these stories so prominently because the people involved were well-known, reputable citizens.

On October 23, 1912, the noted poet and bushman Sydney Wheeler Jephcott wrote to the *Sydney Morning Herald* about of a series of incidents near Creewah, NSW:

"After nearly 50 years in the 'bush' with every sense alert to catch the secrets of the wilds, up until a few days ago not the faintest scintilla of first-hand evidence had reached me that any animal of importance remained unknown in our country. But about 10 days ago, when riding through the jungle which lies on the eastern slope of Bull Hill (about 12 miles south-east of Nimitybelle railway station), I noticed on a white gum trunk a series of scratches such as could be made with the point of a dessert spoon. These scratches were in a series of three on one side meeting a single scratch coming from the opposite direction, being exactly such as would be made by three fingers and the thumb of a great hand with abnormally strong and large nails. Beginning at a height

Sydney Wheeler Jephcott.
(John Oxley Library)

of about three feet six inches, the series of scratches rose to a height of about seven feet. All ... were made by a right hand, suggesting that the creature which made them shared a peculiarity of mankind.

"From these indications I judged that some animal unknown to science was at large in this country, but took no further action. However, on Sunday (October 12), I heard that George Summerell, a neighbour of mine, while riding up the track which forms a short cut from Bombala to Bemboka, had that day, about noon, when approaching a small creek about a mile below 'Packers Swamp', ridden close up to a strange animal, which, on all fours, was drinking from the creek. As it was covered in grey hair, the first thought that rose to Summerell's mind was: 'What an immense kangaroo'. But, hearing the horse's feet on the track, it rose to its full height, of about 7ft, and looked quietly at the horseman.

"Then stooping down again, it finished its drink, and then, picking up a stick that lay by, it walked steadily away up a slope ... and disappeared among the rocks and timber 150 yards away. Summerell described the face as being like that of an ape or man, minus forehead and chin, with a great trunk all one size from shoulders to hips, and with arms that nearly reached to its ankles.

"I rode up to the scene on Monday morning. On arriving about a score of footprints attested to the truth of Summerell's account, the handprints where the animal had stooped at the edge of the water being especially plain. These handprints differed from a large human hand chiefly in having the little fingers set much like the thumbs (a formation explaining the series of scratches on the white gum tree).

"A striking peculiarity was revealed, however, in the footprints: these, resembling an enormously long and ugly human foot in the heel, instep, and ball, had only four toes – long (nearly 5 inches), cylindrical, and showing evidence of extreme flexibility. Even in the prints which had sunk deepest into the mud there was no trace of the 'thumb' of the characteristic ape's 'foot'.

"Beside, perhaps, a score of new prints, there were old ones discernable, showing that the animal had crossed the creek at least a fortnight previously. After a vexatious delay, I was able, on the Wednesday afternoon, to take three plaster of Paris casts – one of a footprint in very stiff mud, another in very wet mud, and a third of the hand with its palm superimposed on the front part of the corresponding foot. These I have now forwarded to Professor Davis, at the university, where, no doubt, they can be seen by those interested." [Case 42]

The Jephcott/Summerell report interests us for many reasons. Plaster casting of yowie footprints is rare enough, but casting of handprints is very rare indeed (the casts, sadly, seem to have disappeared). Claw marks such as those described by Jephcott have been reported since, but rarely.

The location of Summerell's sighting is significant because two interesting incidents have occurred nearby in recent years. During one of those incidents, in late 1977, just 35 kilometres south-east of the Summerell site, Kos Guines of Frankston, Victoria, shot at what he assumed was an escaped gorilla. In 1997, part-time ranger Chris McKechnie almost ran over another yowie less than 10 kilometres to the north-west. [Cases 116 and 216]

Few people knew the coastal mountains and gorges of south-east NSW as well as the surveyor Charles Harper (1840-1930) who worked throughout the region for about 50 years, based first at Bombala and then at Moruya. On hearing of the Summerell incident, he wrote to the Sydney *Sun* with the following account of his own yowie sighting, which the *Sun* featured prominently on November 10, 1912:

"In various parts of the southern district of this state on the coastal slopes, and at various times, extending over a long period, I have met men (and reliable men at that) who unhesitatingly assert that they had seen this hairy man-shaped animal at short distances. They were so terrified at the apparition and the hideous noise it made that they left their work as timber-getters, and at once cleared out from the locality, leaving their tools and work behind them. The description of this animal, seen at different times by different people in several localities, but always in the jungle, invariably coincided. At the risk of being considered by your readers the reincarnation of Ananias or the late Thomas Pepper, I will describe this animal as once seen as briefly as possible.

"I had to proceed some distance into the heart of these jungles for a special purpose, accompanied by two others, and two large kangaroo dogs with a strain of the British bulldog in each. On the night of the second day, about 9 pm ... we heard a most unusual sound, similar to the beating of a badly-tuned drum, accompanied by a low, rumbling growl. The dogs were supposed to be able to tackle anything. But in this case they seemed utterly demoralised; they would not bark, but whined, and made to come into the tents.

"The horrible sounds gradually drew nearer and our thoughts flew to escaped tigers ... We had no firearms, only a scrubhook and an axe ... after much coaxing I induced one of my companions who had a large bundle of leaves and dry kindling to ... place them on the smouldering camp fire ... they flickered up into a big blaze, illuminating the scrub ... when a most blood-curdling sight met our gaze.

"A huge man-like animal stood erect not twenty yards from the fire, growling, grimacing, and thumping his breast with his huge hand-like paws. I looked round and saw one of my companions had fainted. He remained unconscious for some hours. The creature stood in one position for some time,

Charles Harper. (Courtesy of Graham Joyner)

sufficiently long to enable me to photograph him on my brain.

"I should say its height when standing erect would be 5ft. 8in. to 5ft. 10in. Its body, legs, and arms were covered with long, brownish-red hair, which shook with every quivering movement of its body. The hair on its shoulder and back parts appeared in the subdued light of the fire to be jet black, and long; but what struck me as most extraordinary was the apparently human shape, but still so very different.

"I will commence its detailed description with the feet, which only occasionally I could get a glimpse of. I saw that the metarsal bones were very short, much shorter than in the genus homo, but the phalanges were extremely long, indicating great grasping power by the feet. The fibula bone of the leg was much shorter than in man. The femur bone of the thigh was very long, out of all proportion to the rest of the leg.

"The body frame was enormous, indicating immense strength and power of endurance. The arms and forepaws were extremely long and large, and very muscular, being covered with shorter hair. The head and face were very small, but very human. The eyes were large, dark and piercing, deeply set. A most horrible mouth was ornamented with two large and long canine teeth. When the jaws were closed they protruded over the lower lip. The stomach seemed like a sack hanging halfway down the thighs, whether natural or prolapsus I could not tell. All this observation occupied a few minutes while the creature stood erect, as if the firelight had paralysed him.

"After a few more growls, and thumping his breast, he made off, the first few yards erect, then at a faster gait on all fours through the low scrub. Nothing would induce my companions to continue the trip, at which I was rather pleased than otherwise, and returned as quickly as possible out of the reach of Australian gorillas, rare as they are."

The 'Bombala Anthropoid', as depicted by newspaper artist Will Donald in 1912.

Although the sketch that accompanied Mr Harper's story looks rather comical, it is evident the artist, Will Donald, was genuinely attempting to draw the creature as described by the surveyor. All the animal's characteristics are depicted more or less as described, but the end result looks rather odd.

Among the wealth of information in Harper's report some details, such as his description of the creature's grotesque, sagging stomach, are almost unique [but not quite - see Case 243]. Although several recent witnesses have reported sounds suggestive of chest thumping, Harper is the only witness we know of who actually observed it. His assertion that the creatures fibula [shin] was much shorter than a human's and that the femur [thigh] was "very long, out of all proportion" is also, we believe, unique.

Fascinating as these details are, his story is equally interesting because of the number of details it contains that have been echoed in the testimony of more recent eyewitnesses – most of whom have never heard of Charles Harper.

These details include:

- The extreme fear reaction of his fierce dogs, which "were supposed to be able to tackle anything".

- His assertion that acquaintances of his had, on encountering yowies, "left their work as timber getters, and at once cleared out ... leaving their tools and work behind".

- That the creature moved bipedally but also on all fours.

- Even the strangest detail of Harper's story, his assertion that one of his companions fainted and remained unconscious for some hours, was reported again (albeit a long, long way away) in 1973. [See Ch. 7]

After the Jephcott/Summerell and Harper stories, the yowie mystery received barely a mention in any major newspaper for more than 60 years. As a result, our file for the years 1913 to 1975 was, for a long time, very thin. Recently, however, several long-forgotten reports have been discovered in obscure journals, memoirs and in the files of local historical societies. In recent years, also, many now elderly people, who had previously been reluctant to do so, have talked publicly about yowies encountered during "the forgotten years".

Although Bob Mitchell experienced two good sightings of yowies while riding with two mates through the Queensland/ NSW Border Ranges in 1928, it was not until 1980, when he was 76 years of age, that he told his story in the Brisbane *Sunday Mail.* The first sighting occurred near Palen Creek: "It was about 10 am – the yowie was standing in a clearing not far from us and in that light there was no mistaking it ... It was about seven feet [2.1 metres] tall, with a black human face and a gorilla-like body covered in thick brownish hair. It showed no aggression; it just looked at us for a moment, then turned and disappeared into the bush. It had really big feet and could move fast." A few weeks later, the men were camped near Widgee Mountain, about 32 kilometres from their first sighting: "We saw another yowie – it too just looked at us for a moment, then disappeared." [Case 48]

Cecil Thomson was born into a large family that owned farms near Stanthorpe, Queensland. After serving in World War Two he had a varied career as a farmer, teacher and salesman. Over the years he has interviewed several friends and relations who have encountered yowies in the Stanthorpe/ Eukey area. In 1934, when he was 12 years old, he saw one of the creatures himself. When he told us about it 63 years later, the incident still seemed fresh in his mind.

On the day in question he and his brother Ernie had worked for hours carrying bags of peas up to the homestead from the river flat. They'd left one bag there, and after tea their father told them to go back down and fetch it.

"It was in the dusk. We saw a big form bending over the bag, and it straightened up. I thought it was George Wells or someone dressed up in a suit having a joke, so I yelled out, 'Whoever you are, you don't scare me!' It took no notice at first, so I threw a clod of earth and it took that sort of forward and sideways movement, with that sort of guttural 'Woonk, woonk, woonk.' And it had taken off by this time across a drain … and I took off too, and passed Ern, and we got back to Grandfather's place out of breath.

"It was anything between five foot six and six feet, covered with dangling hair, but not on the face. It looked like a person dressed up in a suit, but of course it wasn't. It had dangling arms, and the forward and sideways movement was a bit like Charlie Chaplin used to walk – sort of a waddle, but not exactly, because there was a sort of stamp or stomp at the same time." This occurred off Sugarloaf Road, just out of Stanthorpe, "near the two-mile peg." [Case 59]

In the previous chapter we presented a dramatic 1871 report from the Kempsey district. During the mid-1970s many more Kempsey-area stories, most of which fall into the early modern era category, were brought to light by Patricia Riggs, associate editor of *The Macleay Argus*. Her interest in the yowie was sparked by a dramatic story she heard from a former timber worker, George Gray, in September 1976.

In 1968, when he was 55 years old, Mr. Gray was working at Kookaburra, an isolated saw-milling settlement on the Carrai Plateau, about 50 kilometres west of Kempsey, where he camped in a hut surrounded by dense scrub. He told Ms. Riggs that one dark night he woke to find he was being attacked by a hair covered ape-man. It is tempting to assume that the creature, which was only about four feet tall, was a juvenile yowie. Whether or not that was the case, it was extremely broad, powerful, and apparently intent on dragging the terrified timber worker out of his house. Mr. Gray defended himself for 10 desperate minutes until his assailant abruptly released him and ran out the door. [Case 83]

Although the old bushman seemed very sincere, Ms. Riggs found his amazing story almost impossible to accept in isolation. So she appealed through the *Argus* for other reports. Soon a wealth of material poured in from other local people, both Aboriginal and non-Aboriginal. Meanwhile, George Gray appeared, along with yowie researcher Rex Gilroy, on the Mike Walsh television show in Sydney. As a result, even more eyewitnesses decided to go public.

It soon became clear that yowie encounters had been occurring regularly for many years all over the Kempsey district, and, according to one informant, George Gray was not the only person to narrowly escape abduction.

Mrs. Mamie Mason of South West Rocks was born into the Aboriginal community at Burnt Bridge, about four kilometres south-west of Kempsey (and only 6 km east of the site of the 1871 report) where she was raised on stories of the Hairy Man. Her grandparents often described yowie vocalisations and told how they used to leave out honey and other food to keep the creatures away from their camps.

Sometime in the early 1900s, two-year-old Chris Davis of Burnt Bridge was carried off by a yowie. Like all her contemporaries, Mrs Mason knew the story well:

"On this day the little boy's mother stopped to have a drink of water at the creek. It was just on dark and next thing … he was gone. They listened. They could hear him screaming. They crossed over the gullies and … saw the Hairy Man had Chris … it was going to the cave … but the father got to the child. The Father pulled the child out of its arms. That was Old Man Davis, the father. The child was almost insane, they said, when they got it. You don't make up those things. You never forget them …"

When she was a girl of 10, Mrs Mason (then Mamie Moseley) saw a Hairy Man herself, quite close to where Chris Davis had been abducted.

"My two young cousins, Tom Campbell and Zelma Moseley, were with me ... we were coming home ... just on dark. I had a little dog with me. We were going past the lantana and I could hear this growling ... and the sticks cracking. That little dog was yelping and going on ... its hair was standing on end. I pushed [the little ones] under the fence just as it came through the lantana. They rolled down the embankment ... and I stood there and this thing ... it had hair all over it and the smell was something terrible ... something like a pig, but not quite ... a sour, stale stench, like something rotten. It came towards me ambling, moving its arms. It had long arms. About down to its knees. Long hands ... and it stood looking at me. I could see it had a face ... not like an animal and its hair was coming down over its eyes and the eyes were glaring. It sort of had me hypnotised. My little dog butted me on the legs ... I jumped the fence, rolled down ... and ran. The other two kiddies ... started screaming. My grandfather, John Moseley ... he ran down and shot the gun off. I couldn't talk ... the little dog crawled under the house ... wouldn't come out until the next morning."

It wasn't only Aborigines who saw yowies around Kempsey in that era. Old timer Sam Chapman told the *Argus* about something that occurred in 1923 when he and a mate, Jack Brewer, were felling timber at the head of Nulla Nulla creek, about 50 kilometres north-west of town. One day, Jack walked into their camp, kicked out the fire and announced he would never go into that forest again. He had seen a Hairy Man walking along a log. "He never did go back, I had to get some contractors to come in and finish the job." [Case 44]

A close encounter with a yowie must indeed be a profoundly unsettling experience. Time and time again, witnesses state that their sighting was one of the most memorable events of their lives.

Mrs. Melba Cullen, who saw a yowie in 1930, was interviewed by the *Argus* in September 1976 and by the authors in 2001. On both occasions, she stressed that the experience was still vividly engraved on her mind. So clear is her memory of the incident that she is confident the sketch she did for us, 71 years after the event, is a reasonable likeness of the creature. She was 12 years old at the time, walking near her home on the Maria River, about 10 kilometres south of town.

"I loved collecting flowers and ... I walked out of the bush into a clear patch ... I heard heavy footsteps a few paces behind, but I thought it was my brother looking for me ... I looked back ... there was a big stump just through the fence near me ... suddenly I saw this huge Hairy Man looking around that big stump. He was about seven feet tall and very broad shouldered. He had long, tan-coloured hair all over him. The hair on his face was about as thick as the hair on a dog ... I took one look at him and ran away screaming. I swear to this day the thing ... was a real Hairy Man. It wasn't a kangaroo and wasn't an ape or a monkey. It stood up straight like a man." [Case 53]

Although Mrs. Mason, Mrs. Cullen and most other witnesses described yowies that were human sized or bigger, at least one other person saw a smaller creature similar to the one that attacked George Gray.

In 1995, we talked to Kevin Davis who, as a high school student in the late 1940s, got a very close look at a 1.2 metre [3ft 8in] ape-man in a gully near the present site of Kempsey airport. Although it was so short, the creature was extremely broad and heavy looking.

As he stood staring from a range of only six or eight paces, Kevin noted the creature's flat face, snub nose and yellow canines. Like the yowie encountered by ranger Percy Window, this creature had yellow eyes. Kevin also insisted it had "droopy ears", a detail we never heard before or since. Its body was covered in 3-inch [7.5 cm] hair, but the hair on its head was shorter. It was standing beside a lilly pilly tree and may have been eating its fruit. One huge clawed hand, twice the size of a man's was clutching a branch. As the stupefied scholar stood staring, the golden-eyed goblin broke and ran. It left several four-toed tracks.

Melba Cullen's sketch.

The site of Kevin's encounter lies about halfway between Warneton and Belgrave Falls – just where the "gorilla" was twice seen in 1871. [Case 16] With regard to the "yellow eyes" it may be worth pointing out that many gorillas have light brown, honey-coloured eyes and that some Madagascan lemurs – nocturnal animals – have large, glaring eyes with striking yellow irises.

It seems the Kempsey area was an absolute hotbed of yowie activity in the early 20th century. In fact it is still quite a "hot spot", and has produced several interesting modern reports. In 1995 one of the authors (Paul) investigated a sighting and track find that occurred within 10 kilometres of several much earlier reports. [Case 203]

Some dramatic yowie encounters occurred in south-east NSW towards the end of the early modern era.

In the winter of 1971, as 22-year-old Jim Banks and his mate Stan Hunt were spotlighting for rabbits on Wild Cattle Flat Road, to the south of Captains Flat, they noticed an unusual pair of eyes. Reflecting red, and seemingly very high off the ground, they were between the road and the boundary fence, some distance ahead. Something, it seemed, was peering around a tree at them. "We said, 'What's that?' It wasn't a kangaroo."

Driving forward, they saw a large creature about 50 metres away on the other side of the fence, running across a cleared paddock that rose sharply towards the edge of the bush. It seemed to be over seven feet [2.13 m] tall and was covered with "dark brown to grey hair ... maybe four to six inches [10-15 cm] long". The neck was either very short or non-existent. It didn't have a tail. Its head was down and its shoulders were hunched as it "plodded away, like a big heavy front row ... the gait didn't correspond with anything I'd ever seen before ... it was clumping, like it was very heavy".

Some readers might be dismayed by what happened next. Because he was sure the creature wasn't human, Jim decided to fire "to see how it reacted". It was an easy target: "I couldn't miss." Raising his .22 magnum semi-automatic, he pumped two bullets into its broad back. Although he had more shells in the magazine he immediately stopped shooting because of the creature's startling reaction. Throwing up its arms, it let out a scream like nothing the men had ever heard. It was wild and totally unnerving: "an unearthly sort of squeal – real high-pitched". The creature kept running. The men called it a night, went home, and never hunted that stretch of road again. [Case 90]

At the time neither Jim nor Stan had heard the slightest thing about the yowie legend. They had no idea that the site of their encounter was in the heart of an area – The Jingeras – that was, in colonial days, notorious as a haunt of the Hairy Man. [Cases 17, 24, 29 and 32]

Another incident occurred a few months later, 70 kilometres to the east.

Early one morning in 1972, George Birch of Bowral was driving a truck from Nowra to Bega on the Princes Highway. Just on daylight, he stopped near Cullendulla, six kilometres north of Batemans Bay, to check the load. As he was doing so he heard a strange "screeching" noise from the forest on the western side of the road.

Walking to the embankment and looking down into a gully, he was astonished to see two strange animals staring back at him from a small clearing about 20 metres away. Although they were covered in brown-to-black hair, the hair was rather short, "not wild-looking or untidy". The overall impression was of two hairy people rather than two apes, but they had flat noses and arms that were longer than those of a human. Their posture was rather stooped and one creature appeared to have breasts.

Although they were shorter than George (who is six feet three inches tall), he cheerfully confesses they gave him a terrible fright: he is certain his hair actually stood on end. He quickly moved towards the truck and, glancing back, saw the animals walking away through the bush. In his haste he almost ran into the front of the truck and, once inside, sat for ten minutes before regaining his composure. [Case 85]

Mr Birch had no way of knowing it, but about 10 years earlier another remarkable yowie encounter had occurred only two kilometres away.

"It had a wallaby over its shoulder"

Sawmill owner Laurie Allard is a lifelong resident and respected citizen of Batemans Bay. When they were in their early twenties, he and a mate, Bill Taylor, often went wallaby shooting. Somewhere between about 1958 and 1962, in the bush not far from Cullendulla, they ran into a yowie.

They usually took several dogs with them and, to avoid accidents, would stand on tree stumps about 200 yards apart and send the dogs ahead to flush out the game.

"There was a gully about 150 or 200 yards in front of us", Laurie recalled in 1999, "the ground was pretty clear, and I looked down and there was this big grey thing running up the gully, and I thought it was my mate. I knew that didn't add up, but at the time we were the only people in the area, so you just take it for granted that it must be him, running to a new position. It took really large strides and I thought, 'Boy, he's really running today!' Later I realised it was running *too* fast; a man could not run up that gully at that speed.

"Anyway, it ran around the head of the gully, took a turn and came back to where I was standing on my stump, stopped about 60 or 70 feet away, and just stood there. And it had a wallaby over its shoulder. The wallaby was over its back and it was holding the black tail in front of its chest with both hands. It was no taller than about five foot six [168 cm] and was grey all over, from head to foot. I couldn't say if it was long hair or short hair, but it was grey, about the same colour as the clothes Bill used to wear – about the same colour as a grey kangaroo. It stood looking straight at me."

Laurie acknowledges that the shock of suddenly seeing such an odd creature face-to-face sent him, temporarily, into a state of confusion. Although it obviously wasn't another human or a kangaroo his mind simply would not accept what his eyes were seeing. "Funny thing: it gave me a really hairy, scary, feeling. It was stockier than a man and shorter in the neck, but I didn't really absorb it. And I thought it was my mate – I still had it half in mind it was him – and I spoke to it: 'What have you got – have you got a wallaby, Bill?'

"After watching me for maybe a minute he turned around and ran away into the scrub. Then Bill came over to me and said, 'what were you doing running around in the bush?' and – I was still confused – I said, 'what were *you* doing running around?' He hadn't seen this thing but he'd heard it charging around – and he had the same scared feeling. All the dogs came in, and they were that terrified that they were in amongst our legs – they *would not* leave us. We called it a day and left."

Although he still works in the bush and has been to many remote corners of the south coast hinterland, Laurie has never so much as glimpsed another yowie. "I often think of them, though," he says, "and I feel really privileged to be one of the few people to have seen one up close." [Case 77]

Another early modern era report came from the outskirts of Lismore, in north-east NSW. That story took 42 years to emerge.

In July 1977, a 52-year-old man told Lismore *Northern Star* reporter Gary Buchanan that he had seen a yowie on his grandfather's farm in South Lismore in 1935: "I was standing on the verandah ... when I saw a man walking across the paddock from the direction of the hills ... as it walked towards the house my grandfather's horse started to kick up a hell of a fuss. When [his grandfather] saw what it was he pushed me inside, blew out the lamp ... then grabbed his rifle ... we all watched through a small window as the creature walked past ... [about 25 yards away and clearly visible in the moonlight] ... its head didn't seem to have a neck, but was sitting straight on its shoulders. It also looked as if it had a hunched back, but it was standing up straight. It was much thicker around the shoulders and chest than a man ... my grandfather told me it was the same creature he had seen up in a gully on the property only a few years earlier. He had ridden into the gully to pick some guavas when he saw the creature come down one side of the gully, cross a small creek, then climb up the other side of the hill." [Case 60]

"You'd better get out of here!"

Clyde Shepherdson, who we quoted at the beginning of this chapter, experienced his yowie encounter between Nanango and Maidenwell, Queensland, in 1938 or '39, when he was 13 or 14 years old:

"There was a lot of heavy vine scrub up there in them days – it's all gone now. Me and a mate, Clarrie Parsons, had a couple of single-shot .22 pea rifles and early in the winter we used to go shooting little red-backed scrub wallabies. You had to be pretty quiet or you wouldn't see them, so we were [moving quietly] and I say to this day that's the only reason we came across him.

"It was deep into the scrub, a couple of miles from the nearest road, and we were creeping through, bending down [under the vines] and we got to a little open space and stood up and there, about 20 feet [6 metres] away, this great big yellow thing was standing there with its back up against a fair lump of a tree – and I think he got a bigger fright than we got!

"He put his hands up on either side of his head ... with his palms out towards us and snarled, more or less to say 'You'd better get out of here!' His fingers were curled, like a claw – like it was going to grab you or rip you open. I suppose we could have shot him but we never even had it in our mind ... we just thought, like, 'the quicker we get out of here the better!' We just wandered off slowly, walked side-on, keeping our eye on it, didn't turn our backs. We didn't want to run because he might have took after us. It kept its back against the tree, followed us with its eyes until we got out of sight, into the scrub. Once we got out of the thick scrub we just bolted!

"It was a rusty yellow ... camel-colour would be the closest. I've still got him imprinted on my mind to this day. It would be a good six foot; it was shaped more or less like a person only very broad across the shoulders; fairly nuggetty; solid. And it had a fair amount of fur on it – more like fur than hair. It was pretty rough looking but not too dishevelled, not too dirty. Fairly long

Clyde Shepherdson. (Tony Healy)

arms ... its legs were furry all the way down; they were pretty wide.

"It seemed to have a fairly broad forehead ... it looked real ape-like – definitely an animal impression. The face was fairly flat and broad. You see those drawings of stone-age men: it looked a bit like one of them. It had a short neck. We saw its teeth when it growled ... they were more fangs than teeth ... fairly long – savage. It had fairly big ears, covered with hair but sticking out – you couldn't miss them. A flattish nose. The skin looked dark on the face but it had a lot of hair on it – scruffy-looking hair on its head.

"It was a pretty cloudy and damp day, but it was after lunch, in full daylight. He was out in the open – there's no two ways about that. I'll never forget it – never!"

Between themselves, for years afterwards, Clyde and Clarrie referred to the creature as "the yella hairy thing" and although Clyde heard second-hand stories about the yowie, he never met anyone else who had actually seen one. He has described his experience to a lot of people over the years, "but you know what they tell you – that you've had a few drinks ..." Consequently, he has been very heartened by the revival of interest in the subject and by the spate of modern reports of creatures similar to "the yella thing".

"I've lived with this for 60 years", he concluded, with unmistakable sincerity, "and now I can finally tell someone about it." [Case 63]

Chapter 4

The Modern Era

In 1975, after more than 60 years in the wilderness (so to speak), the yowie lurched bashfully back into the limelight.

In that year, articles about the hairy giants began to appear occasionally in the big city tabloids, in some regional newspapers and in popular magazines such as *Australasian Post* and *People*. The man behind the articles was Rex Gilroy, a self-taught naturalist, then about 30 years of age.

Over the years we have had an uneasy relationship with Rex and, if we could have written this book without mentioning him, we would have done so. To write about the yowie mystery without a discussion of his contribution, however, would be like writing a history of Australian exploration without mentioning, say, Ludwig Leichhardt. In the following account of his activities we have tried very hard to be fair, to give him all due credit.

Growing up in the rugged Blue Mountains west of Sydney, Rex heard bits and pieces of yowie lore and by his teenage years had become quite intrigued with the subject. This casual interest, he says, was turned into near-obsession by a dramatic experience he had on August 7, 1970. He claims that, while resting below the Ruined Castle Rock formation near Katoomba on that day, he was lucky enough to observe a hairy, primitive-looking hominid cross a small clearing just a few metres in front of him.

Fired with a determination to prove the creature's existence, he began spending a great deal of time out in the bush searching for evidence. In 1975, he established the Australian Yowie Research Centre and

Rex Gilroy. (Tony Healy)

began sending articles and press releases to the newspapers. Although editors of the quality broadsheets seemed determined to ignore him, stories by or about him began appearing regularly in the popular press. He was occasionally interviewed on radio and television.

Throughout the late 1970s, he issued constant appeals for information and was, as far as the press and public were concerned, Australia's one and only yowie hunter. If you encountered a yowie in those days it wasn't a matter of "who ya gonna call?" – you called Rex. As a result, he soon accumulated an enormous amount of yowie lore. As early as 1976 he claimed to have a staggering 3,000 reports on file.

He continued to live right in the heart of the Blue Mountains, an area that is, as we will show later, the very hottest of all yowie hot spots. A skilled bushman and accomplished naturalist, he operated a small natural history museum there for more than 25 years. With such a superbly located base, high public profile and wealth of data at his fingertips, he should have been able to, by the late 1970s, convince a large slice of the population that the phenomenon was worthy of serious attention. Unfortunately, he never really managed to do so. Although many people were intrigued by his press releases, very few journalists or mainstream scientists took the matter seriously.

Part of the problem was that, despite his claims of having huge numbers of reports and of knowing a great deal about the creatures' habits, his press releases and magazine articles tended to be repetitive and frustratingly lacking in documentation. He seemed unwilling to accept the notion that extraordinary claims require extraordinary proof.

His outlandish assertion that giant ape-men roamed the bush wasn't the only thing that bothered the boffins. At his Mt. York museum, for example, he had on display what he claimed was a fossilised *Gigantopithecus* footprint, which he'd found near Kempsey, NSW. The remains of *Gigantopithecus*, a huge hominid that has been extinct for about 500,000 years, have been found only in southern China, Vietnam and northern India. Since those remains consist only of jawbones and teeth, it is very difficult to see how he could identify the object as the footprint of one of the creatures.

Rex's unsubstantiated assertions about *Gigantopithecus* and about even stranger matters (such as his discovery of several "Egyptian pyramids" in Queensland, his detection of a UFO base under the Blue Mountains and his claim to be one of a select band of "extraordinary geniuses" who are chosen and guided by space aliens)[18] totally spooked the scientific establishment. Those wilder claims also caused many ordinary mortals (such as ourselves), who were quite tantalised by the notion of an Australian yeti to remain, for a time, cautiously sceptical of anything he said.

In 1995 he published *Mysterious Australia,* which contained a single chapter about the yowie, and in 2001, his opus, *Giants From The Dreamtime. Giants* runs to 380 pages and contains about 300 yowie reports, many of which are very interesting. Frustratingly, however, it doesn't contain an index, footnotes, maps, picture credits or even a single eyewitness sketch, apart from his own.

We now know that despite his foibles, Rex was basically right – about the yowie, at least – and the boffins were wrong. The yowie mystery *is*, as he always maintained, a genuine phenomenon worthy of serious attention. Anyone who had the moral courage to stand up and publicly declare his belief in the yowie in the sceptical 1970s and who has had the physical courage to venture repeatedly, often alone and always unarmed, into as many isolated, spooky and dangerous areas as he has, really deserves a lot of credit.

No fair-minded person would deny that Rex really is, as he often claims to be, "the father of Australian yowie research". Nobody can deny, either, that during the mid-1970s he single-handedly dragged the yowie out of almost total obscurity. However, because of his uneasy relationships with the scientific establishment, many of his fellow researchers and the serious media, it might be said that he dragged the hairy giant only as far as the tabloids.

Fortunately, in 1977, an entirely different kind of researcher arrived on the scene.

Graham Joyner is a quiet spoken, erudite man whose main interest, before he became intrigued by the yowie mystery, was the history of science. In the early 1970s, while employed as an archivist in Canberra, he unearthed several references to yowies (or yahoos, as he prefers to call them) in old documents and 19th century newspapers. Realising he had stumbled on something important, which had apparently been forgotten by the general public and overlooked by science, he researched the subject further and, in August 1977, published *The Hairy Man of South Eastern Australia*.

That scholarly document, which is regarded by many in the field as a model of cryptozoological research, contained 29 early references to the yowie, dating from 1842 to 1935. Twelve of these items contained fascinating, and quite consistent, descriptions by eyewitnesses, most of whom were identified by name. Other entries showed quite conclusively that many Aboriginal groups in south-eastern Australia had had a strong belief in the hairy giants for at least 150 years and, quite possibly, since time immemorial. Aborigines knew the Hairy Man by various names, several of which Graham listed in the booklet.

Because every item in the collection was thoroughly documented, because Graham's own brief comments were modest and to the point, and because he would always cheerfully share the fruits of his research with others, he succeeded in making the Hairy Man an acceptable subject for discussion in some sections of the scientific community. Graham had quietly proved what Rex had simply asserted: that Australia's Hairy Man tradition dates back a very, very long way and is worthy of serious investigation. All yowie researchers owe him a great debt.

In the years since *The Hairy Man* was published Graham has continued his research, unearthing several more 19th century reports that he has again put on the public record via scholarly articles in the *Canberra Historical Society Journal* and *Cryptozoology*, the journal of the International Society of Cryptozoology.

During the late 1970s, thanks to the efforts of Rex and Graham, many eyewitnesses who had previously remained quiet, sometimes for decades, went public with their stories, small-town journalists and local historians began to uncover more and more interesting early cases and, for better or worse, the likes of ourselves were drawn to the mystery.

The very slow but steady trickle of new reports continued throughout the late 1970s and '80s. It increased considerably after our first book *Out of the Shadows* was published in 1994, and again when Rex's *Mysterious Australia* (1995) and Malcolm Smith's excellent *Bunyips and Bigfoots* (1996) appeared. Then suddenly, in the late 1990s, the trickle became, in relative terms, a raging torrent. The reason for the sudden flood was that some researchers had begun to utilise the Internet. One man in particular, Dean Harrison, used it to great advantage. In fact, the website that he set up attracted so many reports, at times, that he could scarcely manage them.

Dean's life-changing introduction to the yowie phenomenon occurred in south-east Queensland in July 1997, when he was 27 years of age. The event was unexpected, unwelcome and utterly unforgettable: a horrendous encounter with a huge, roaring ape-like creature that chased him through the scrub seemingly intent on ripping him to pieces. We will tell the full story of that remarkable incident later. Suffice to say, for the moment, that it was a particularly frightening instance of yowie aggression – the kind of story that makes you wonder a little about the fate of some of the people who have disappeared without trace in the Australian bush.

Dean Harrison. (AYR)

Dean's experience would have stopped the hearts – or at least comprehensively cleared the bowels – of lesser men, such as ourselves, but its only effect on Dean was to instil in him a burning desire to know more about the massive animal that had taken such an unwelcome interest in him. Knowing it could only have been a yowie, he began looking for information about the creatures and soon became a leading figure in yowie research.

In early 1999, while based in Sydney, he placed advertisements in newspapers requesting eyewitness reports or other information and within a short time had pinpointed two or three areas of apparently quite intense activity in the Blue Mountains. He began spending all his spare time, up to three nights a week, in these "hot spots", and soon experienced several more yowie encounters.

It was while in Sydney, also, that Dean set up his first yowie-related website. Intending simply to document his own findings and to begin a database of witnesses and sightings, he expected only a moderate level of public interest. The response, however, was quite overwhelming: he received not only thousands of "hits" to the site and many eyewitness reports, but also many offers of assistance from aspiring yowie hunters.

Some months later, he returned to Queensland and founded Australian Yowie Research, an organisation that now has members in several states. Since then, he has led groups of his colleagues, outfitted with night-vision devices and other high-tech gear, on many lengthy, well organised and productive expeditions all over eastern Australia.

Although Dean's well-publicised activities have been responsible for bringing forth most of the yowie reports from the late 1990s onwards, several other researchers have brought considerable new information to light.

Cryptozoologists Steve Rushton, Malcolm Smith, Dave Glen, Rob Millar, Jeanie and Nigel Francis and Brett Green have documented many yowie reports in south-east Queensland. Paul Compton and naturalist Gary Opit have done a great deal of valuable research in northern NSW and Geoff Nelson has documented many excellent reports around Taree. Blue Mountains residents Michael Williams, Rebecca Lang, Jerry and Sue O'Connor and Neil and Sandy Frost have collected a great deal of valuable new data. Neil, in particular, who spoke of his experiences on ABC TV's *Quantum* program, has contributed greatly to the pool of information.

Another notable yowie researcher burst onto the scene in May 1995. His debut was memorable. Clad in a baggy gorilla suit, he strode across our TV screens as a voice-over informed us that, having seen a similar hairy creature in the Brindabella Mountains a year earlier, he now intended to prove the existence of the yowie. He wanted to be known only as Tim the Yowie Man.

For some time after his dramatic entrance from stage left – or, should we say, from left field – we didn't know quite what to make of Tim. Although he seemed interested in the yowie phenomenon, he seemed just as interested in becoming a media star in the style of The Bush Tucker Man and The Crocodile Hunter. Hence, as Tim the Yowie Man, he wore a distinctive Akubra complete with "Tim the Yowie Man" hatband and khaki bush clothes emblazoned with shoulder patches. He issued constant press releases and engaged in many amusing publicity stunts.

Tim was young, bright, photogenic and media-savvy. He had a clear goal in mind and was soon well on his way to attaining it. Within just a few weeks he had blitzed the national media, attracted sponsorship, stitched up deals with various magazines, newspapers and radio stations and was zooming around Australia reporting on a wide range of mysteries.

"Nonplussed" is probably the correct term for describing our reaction to Tim in the early stages of his media blitz. There we were, veterans of more than 20 years of yowie research who fondly imagined we had the right to occasionally pontificate on the subject, being ignored by the media while this brash Johnny-come-lately with his flashy clothes, big hat and witty sound bites grabbed all the limelight. We had to admit, however, that, as we had rarely sought media attention, we could hardly accuse Tim of hijacking it. In any case, whatever mild resentment we may have harboured quickly dissolved when we got to know him. Tim turned out to be an engaging character, a good bloke to have a beer with, and someone who, despite his jokey public persona, was genuinely interested in the yowie enigma.

His life-changing introduction to the phenomenon occurred as he was walking down from the summit of Mt. Franklin, in the Brindabella Mountains, on May 1, 1994. The sun had only just set, and visibility was very good. As he approached the deserted ski chalet (now destroyed) he suddenly became aware of a huge form 80 to 100 metres away.

"There in full view ... was this massive black, hairy ape-like creature. It had long arms, no neck and was moving from the south to the north at a slow pace. I couldn't make out its legs because of the thick bracken, but I had a clear, side-on view of the rest of its body. My immediate reaction was to run ... but for some reason I couldn't. So I started walking backwards. Slowly. Very slowly. After about a dozen steps, with the creature now dipping below the curve of the spur, I turned and bolted back to my car as fast as I could." [Case 197]

Although he then had a very good conventional job, Tim felt that an active, adventurous, young bloke like himself, if he displayed enough chutzpah and took a few risks, could achieve a lot more – or at least have a lot more fun. So he decided to have a crack at solving some of Australia's and the world's major mysteries and if he managed to become rich and famous at the same time – well, he wasn't averse to that idea either. So he bought his gorilla suit, Akubra and khakis, prepared a bunch of press releases and went for broke. The rest is history ...

Although it was a yowie that catapulted him into his present career, Tim now gets involved in such a variety of other mysteries (such as Blessed Virgin Mary apparitions, Tasmanian tiger sightings, haunted houses and ships) as well as other unrelated media capers that he now devotes only part of his time to yowie hunting. He has, nevertheless, managed to do a lot of fieldwork and has turned up several interesting reports, which he kindly shared with us.

Tim the Yowie Man and Steve Rushton. (Tony Healy)

So Tim is now something of a national institution, liable to turn up just about anywhere at any time, giving the unique Tim the Yowie Man slant on a wide variety of offbeat stories. His most recent triumph, for which he deserves the cryptozoological equivalent of the Victoria Cross, was his victory in a David-and-Goliath court battle with the massive Cadburys chocolate empire. Incredibly, that foreign-owned company, which markets a line of sweets called Yowies, attempted to prevent Tim from copyrighting his Tim the Yowie Man persona, claiming that their juvenile customers would confuse the cryptozoologist with its chocolate.

Hot spots

Although the hundreds of new reports that have come to light in the last 25 years or so fall short of proving the yowies' existence, they do enable us to discern certain patterns in the data.

The most obvious patterns are geographic. Although there is a sprinkling of reports in other parts of Australia, including the arid inland, the far west and even Tasmania, it seems Henry Lawson's assertion that the Hairy Man's range extended "from the eastern slopes of the Great Dividing Range right out to the ends of the western spurs" is broadly correct. The vast majority of yowie reports do emanate from the forested mountains of eastern Australia, from Cape York right down to Victoria. The only thing we would add is that in some areas the yowie's range extends well to the east of the dividing range. Quite often, in fact, it extends right through the coastal ranges to the sea.

Another thing that emerged from the rapidly accumulating data was that within the yowies' general range there seemed to be certain hot spots that produced an unusually high number of reports. Although we rather like the idea of these apparent hot spots, we should say that some might be, in reality, no more yowie-infested than many other parts of eastern Australia. The large number of reports they produce might simply reflect the presence of keen local investigators.

The Kempsey area, for instance, seems to be absolutely red-hot – until we remember that Kempsey is blessed with a newspaper that is willing to take the yowie mystery seriously. Most of the many local reports came to light as a direct result of journalist Patricia Riggs' stories in the *Macleay Argus*. Rex Gilroy spent time there and was instrumental in publicising the area's yowie tradition; Kempsey resident Dave Reneke added to the tally of reports, as did yowie hunter Geoff Nelson.

Similarly, the high level of apparent yowie activity in Queensland's Sunshine Coast hinterland may simply be a result of the investigative skills of local cryptozoologists Steve Rushton, Malcolm Smith, Gary Opit, Nigel and Jeannie Francis; Rob Millar, Dave Glen and Brett Green. Dean Harrison and his energetic team have also concentrated heavily on that region.

Another hot spot – the far south coast of NSW – may also owe part of its yowie-infested reputation to the number of keen investigators who have tramped its rugged terrain. As mentioned in chapter three, the pioneering yowie investigator Sydney Wheeler Jephcott collected many early reports in that area. More recently Graham Joyner has carried out research there, as has Canberra academic Helmut Loofs-Wissowa. One of the authors (Tony) has spent a great deal of time there, as have two resourceful young Canberrans, Tim Power and Martin Stallard.

In some areas, however, the incidence of reports is so high that it must be due to something other than the mere presence of active investigators. Be that as it may, for our immediate purposes the hot spot phenomenon is actually quite handy. Because it would be impossible to discuss, in this chapter, every modern-era report, we have decided instead to focus on material relating to just three of the hottest hot spots.

In so doing, we hope to show that there really is some coherence to the yowie phenomenon. Readers will notice that the accounts of modern-day witnesses contain many very similar details, despite the fact that they come from widely separated regions. They will also notice that details from obscure 19th century reports are often echoed in recent stories, that Aboriginal yowie lore often lends support to non-Aboriginal testimony, and that recollections of old country people often corroborate the testimony of gob-smacked "city folk".

The hot spots we have chosen are the far south coast of NSW, the Gold Coast hinterland and the Blue Mountains.

Hot: The Far South Coast of New South Wales

"...the blacks from Nowra down to Orbost in Victoria all know of the existence of this doolagarl."

– Roland Robinson, *Black-Feller White-Feller,* 1958

Although the NSW south coast encompasses the entire eastern seaboard from Sydney down to the Victorian border, the area we intend to focus on comprises only the southern part of that region. That area, known as the far south coast, extends 150 kilometres from Batemans Bay down to Eden and reaches inland roughly 50 kilometres to the escarpment, where the land suddenly jumps up about 1,000 metres [3,000 feet] to the high plains of the Monaro.

Compared with much of eastern NSW, the far south coast is quite sparsely settled. The combined population of its four largest towns – Batemans Bay, Moruya, Narooma, Bega and Eden – amounts to fewer than 50,000 souls. Those towns, furthermore, are all strung out along the coast-hugging Princes Highway. Because almost all of the region's other few hundred people – residents of scattered villages and farms – also live within a few kilometres of the sea, immense swathes of the hinterland are completely uninhabited ... at least by humans.

One huge area, bounded by the Kings Highway in the north, the Snowy Mountains Highway in the south, the Princes Highway in the east and the escarpment in the west – an area of approximately 4,500 square kilometres [1,737 sq miles] – is almost completely devoid of settlement and almost completely covered in forest. The Deua and Wadbilliga National Parks sprawl over nearly 2,000 square kilometres of it and most of the remainder is made up of state forests and nature reserves. It is crossed by only one dirt

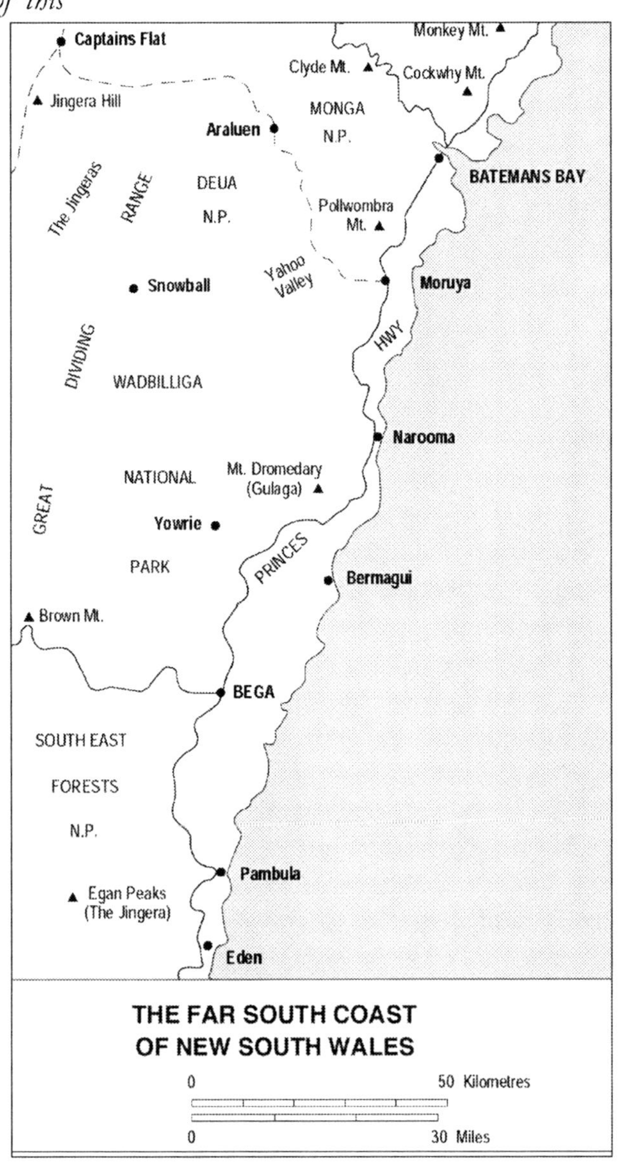

road (Araluen to Moruya), a couple of four-wheel-drive tracks and a few fire trails.

The hinterland south of the Snowy Mountains Highway through to the Victorian border has a few more roads and settlements, but it is still, compared to most places on planet Earth, very sparsely populated indeed. National parks and state forests cover two thirds of it and much of the terrain is extremely rugged.

All in all, the hinterland has changed little since surveyor and yowie eyewitness Charles Harper [Case 35] described it in 1912: "from the head of the Clyde River extending southerly to the Victorian border, the eastern slope consists of excessively broken, lateral ridges, deep gorges, and dense jungles, extremely difficult of access for man or beast; therefore its primeval solitude is very rarely disturbed."

As mentioned in chapter one, the Aborigines of the NSW south coast have long known of the Hairy Man. The creature, said Percy Mumbulla, was "a gorilla-like man. He has long spindly legs ... a big chest and long swinging arms. His forehead goes back from his eyebrows. His head goes into his shoulders, no neck". Mumbulla's friend, Roland Robinson, observed "the blacks from Nowra down to Orbost in Victoria all know of the existence of this *doolagarl*". [19]

As we saw in chapters two and three, white settlers on the south coast and its hinterland also experienced encounters with "Australian apes" and "gorillas" throughout the colonial and early modern eras. Until Graham Joyner published his booklet *The Hairy Man of South Eastern Australia* in 1977, however, very few outsiders knew of those reports. It is interesting to note, therefore, that the yowie's appearance and behaviour, as described by contemporary witnesses, tallies very well with the testimony of both traditional Aborigines and white pioneers.

"We sort of couldn't believe it. But it was there – and it was very real."

The hairy giant that Peter and Belinda Garfoot of Newcastle, NSW, encountered on Clyde Mountain in 1996 certainly matched the descriptions given by the old timers, both black and white. It happened as they and their children were driving to the coast along the Kings Highway via Braidwood.

"It was pretty reasonable weather", Peter recalls, "a warm time of year ... we were travelling with a car and trailer." In the early afternoon they reached the edge of the high country at Clyde Mountain, where the road suddenly drops for hundreds of metres, twisting and turning through rain forest all the way. Peter knew that the nausea-inducing descent could also put quite a strain on vehicles, so, about half way down, as soon as he reached the first reasonably straight, level section of road, he pulled over. The spot was a kilometre or so below the notorious hairpin known as Government Bend (or as a very persistent graffiti artist would have it, "Government Bend Over").

"I thought I'd better give the brakes a rest and check the trailer. So I got out and had a look and a smell around and thought 'Yeah, they're ok. We'll just give it a minute'. And I sat in the car for a second and looked in the mirror and saw this ... I don't know what you'd call it ... a creature ... or man-like creature." It was large, hairy, upright, and it was walking across the road.

"I said 'Heck, what on earth is that?' or something, and my wife looked back through the rear window [and] I turned around too." The children, crammed in the back seat, were too small to see what was going on. "It was maybe 150 metres back, up the hill slightly, but plain as day. We saw him from [when he was] in the middle of the road through into the bush." There was no other traffic and Belinda had the distinct impression "that it had waited till we'd gone past before crossing. It was like it wasn't expecting us to stop."

"It crossed diagonally," Peter continued, "walking slightly away from us to the other side. At that spot the trees were well back from the road because they had been doing roadwork probably only months before. He was walking across to a wider area, like a turning spot for trucks. He actually sort of looked over his shoulder as he kept going. I didn't [notice if] the whole body turned but it was a natural movement. We had the impression it didn't want to be seen. Yet ... he did look over his shoulder, and it would have been obvious we were there. He was not moving leisurely. He was going at a reasonable pace to get across and back into the bush. We watched it for 5 or 10 seconds – it wasn't just the last fleeting glimpse. You get the feeling it's a man, too, you know. I don't know why, but you do." It was covered from head to toe in light brown or fawn coloured hair that seemed rather matted, like that of some goats.

Peter, then about 40 years old and with good eyesight, estimates the creature weighed about 126 kilograms [278 lb] – one and a half times his own weight. "I'm not too bad on heights and sizes. I wouldn't call him monstrous, but something like seven foot, possibly a bit taller." Although it was somewhat less than King Kong-sized it still appeared much too large to be a man. Belinda was quite emphatic: "It was too tall. It was *big* – that caught my eye. We were close enough to see it was something odd."

The arms seemed slightly longer than a human's, but Peter thought the face was rather man-like. "The impression was that it was lacking hair there. It wasn't a full-on hairy face. Not like, say, a gorilla." Not surprisingly, at a range of 150 metres, ears were not apparent, and Peter "didn't see much of a neck, if at all, what with the hair ... from the back you didn't see any neck."

It didn't slump and its general bearing added to the impression that it was very man-like: "because of the uprightness ... very definite uprightness and a very strong walk". But while its gait was similar to a human's Peter gained the firm impression that it "was not like a person's – not quite". The main difference was that it didn't seem to swing its arms, but it was also different in some other, indefinable way.

Peter and Belinda hadn't felt at all afraid while the creature was in view. Nevertheless, as soon as it disappeared into the bush their first reaction was to look at each other and say, almost in unison, "Let's get out of here!"

As readers will have noticed, throughout their testimony the Garfoots, reflecting the dilemma experienced by so many eyewitnesses in other places and earlier times, referred to the creature as both "it" and "him". On balance, though, they felt the huge creature was slightly more animal-like than human. "To me", Belinda concluded, "it was like an animal but the way it moved was human."

Despite their obvious sincerity, Peter and Belinda's statement is like every other individual yowie report: if it stood alone it would count for nothing. When correlated with the many very similar reports from the same area, however, its value can be appreciated. Although they were totally unaware of it at the time, the site of their encounter lies within just a few kilometres of several other reports from the colonial, early modern and modern eras.

As mentioned in chapter one, south coast Aborigines have long believed that hairy giants haunt the area. Cockwhy Mountain, just north of Batemans Bay, was said to be a favourite haunt of the *doolagarl.* H.J. McCooey's remarkable "Australian ape" encounter of 1882 occurred in the same area, as did George Birch's 1972 sighting of two yowies. [Cases 25 and 85]

Laurie Allard's face-to-face encounter with the wallaby-killing yowie, as described in the preceding chapter, also took place just north of Batemans Bay. Since then, Laurie has heard of several other sightings in the area. Just on sundown one evening in the early 1990s, "A", a young employee of his, was sitting in his vehicle on Big Bit Road, a dirt track about eight kilometres north of town when he noticed, in his rear view mirror, a huge, dark, upright figure lurching towards him. Although the encounter was brief ("A" took off immediately in a cloud of dust) he was adamant the creature could only have been a yowie. A couple of years later, Ms. "B" told Rod she'd driven past a big hairy ape-man as it stood, late one night, beside the Kings Highway at Sheep Station Creek, between Batemans Bay and Nelligen.

Roaming in the gloaming

A couple of other events in the vicinity of Batemans Bay indicate that yowies aren't afraid of getting their big feet wet. One afternoon in 2000, as ferry captain Andy Crole and his wife Meryl were relaxing at a lonely campsite on tree-covered Big Island, in the Clyde River, they heard something large crashing through the scrub. "There were big heavy branches snapping and cracking", Andy recalls, "whatever was doing it was pretty big and strong". Later, around sunset, they experienced the unpleasant sensation that something was watching them from the shadows. Their little dog felt it too: uncharacteristically silent, it huddled around Andy's feet.

Shortly after they retired things got even scarier: as they lay, terrified, something overturned pots and pans right outside their tent – "making a hell of a racket" – then ran back into the scrub. "This was bigger and heavier than anything I've heard in my life", says Andy, " it ran right past the tent two-legged, and the whole ground shook beside our heads – that's how close the footsteps were. We did have a camera with a flash, but there was no way known I was going to get out of bed at 10:20 pm to go running through the bush trying to get a photo of whatever was in front of me. I'd rather be a poor ferryman than a dead one, thanks very much!

Ferry Captain Andy Crole approaching Big Island. (Tony Healy)

"My wife reckons it was a dulagarl. I never used to be a big believer in them ... when we'd go four-wheel-driving she'd tell stories about them, because she's part Aboriginal. I'd have a bit of a giggle, but she'd say to me, 'One day, Croley, you're going to cop it!' And I guarantee, that night on Big Island – that thing *was* a dulagarl."

It seems that 23 years earlier another Clyde River island felt the yowie's heavy tread.

On September 7, 1977, *The Moruya Examiner* reported that oyster farmer Kevin Connell had found three giant footprints on Budd Island, just upstream from the bridge at Batemans Bay. There were two prints of a left foot and one of a right, and they were more widely spaced than those of a tall man. About a pace further on was a wooden board that had been used to retain a step consisting of crushed oyster shell. The board had evidently been trodden upon and the step broken down. The tracks were five-toed, about 18 inches long, almost six inches wide and pressed very deeply into the sand. Another oysterman, Alan Small, who, despite his name, weighed 210 pounds [95 kg] couldn't make an impression anywhere near as deep.

Mr. Connell, Mr. Small and their colleagues Bill Johnson and Leo Latta were at a loss to explain the phenomenon. Mr. Small pointed out that if the tracks were a hoax the prankster must have been quite ingenious, as the surrounding sand was completely unmarked. [Case 112]

We heard of the incident shortly after it happened, but because Budd Island is only 500 metres from the main street of Batemans Bay and because we then knew of no yowie activity so close to town, we dismissed the story as a probable hoax. Although largely scrub-covered, the island couldn't possibly sustain or conceal a hungry, hulking hominid for very long. Only 400 metres long and 200 wide, it contains several boatsheds and jetties and from dawn to dusk sees the comings and goings of dozens of busy oyster farmers.

The island is, however, within about 60 metres of the Clyde's southern bank, at the mouth of swampy McLeods Creek, which emerges from Mogo State Forest. Yowies have been reported in Mogo State Forest and it is conceivable that an unusually adventurous ape-man might have followed the creek down to the swampy estuary and then, in the dead of the night, waded or swum out to the island. While we may never know whether that scenario is correct, we do know that at least one yowie has been seen at the McLeod Creek estuary, right on the western edge of Batemans Bay township.

In April 2003 Tanya Bowen told us she'd seen a huge, hair-covered upright creature near the creek in about 1979 – approximately two years after the Budd Island track find. She was about 11 years old at the time and was playing on the edge of the wetlands, "where the bypass road now is". At about 3:30pm she glanced towards a nearby rise and saw something moving quickly uphill. Her first thought was that it must have been someone in an ape suit, "but then I saw it was much too fast and too big". Although it was some distance away, she "could see it as plain as day". It was brown to black in colour, took huge strides, and was "very big, maybe seven to eight feet tall. It was running uphill faster than a man could and was chasing something; I couldn't see what – maybe a dog or a 'roo".

Although the sighting lasted only a few seconds Tanya feels she will remember it for the rest of her life. Immediately after the creature disappeared into the scrub she ran to her grandmother's house and poured out the story. At the time the old lady told her not to worry about it and quickly changed the subject, but Tanya now knows her grandma was quite familiar with the Hairy Man legend.

Tanya Bowen. (Tony Healy)

Tanya is a sixth generation citizen of Batemans Bay and, like most long-term residents, has heard many references to possible yowie activity. Some years ago a cousin was pig hunting in the McLeods Creek wetlands when his dogs suddenly "went crazy". He then became aware of a terrible odour and left hurriedly. Her father, too, thought he might have seen evidence of yowies. Once, near Buckenbowra, he came across the carcass of a kangaroo whose head had seemingly been "broken off" by some immensely powerful animal.

Yahoos, yowroos, yowries

To the west of Batemans Bay, the Buckenbowra and Mogo State Forests meet the north-eastern boundary of Deua National Park in a very steep and wild region drained by the serpentine Deua River. In the early 1900s the Knowles family owned land beside the lower reaches of the Deua, and Rod Knowles, who grew up there in the 1930s and '40s, often heard old timers speak of something they called the yahoo, yowroo or yowrie. In September 1976, he told *The Queanbeyan Age* that, twice in 20 years, he found footprints in an area about 15 miles inland from Moruya that "could not be identified as being those of any known animal." Like the Budd Island tracks, they were five-toed and "of considerable depth in river sand ... the distance between them was up to seven feet [213 cm]".

When we interviewed him Mr. Knowles added a significant detail: he had found both sets of footprints in a large gully that comes down to join the Deua from the west. For as long as he could remember it had been known as Yahoo Valley.

Another old-timer, Fred Howell, drove bullock teams from the south coast up to the tablelands during the Great Depression. In later years, he often told his grandson, Billy Southwell, about encountering a Hairy Man at a place called The Three Ways, apparently in the vicinity of the Deua. The creature climbed onto a wagon and threw grain bags around. Old Fred sometimes referred to the creature as a "yourie". [Case 49]

Interestingly, about 35 kilometres south of the upper Deua is another, much smaller stream – the Yowrie [sometimes spelt Yourie] River. On September 17, 1976, journalist John Leach reported in the *Bega District News* that "giant man-tracks" had recently been discovered at Verona, just south of Cobargo. Although Mr Leach found (as we did) the story difficult to verify, it is worth noting that Verona lies only 18 kilometres south-east of the locality of Yowrie, on the Yowrie River.

It seems almost every town, village and district of the far south coast has some yowie connection. The surveyor Charles Harper, for instance, whose hair-raising "Australian gorilla" encounter was described in chapter three, was based for much of his 50-year career in Moruya. Pollwombra Mountain, long believed by local Aborigines to be a haunt of the *dulagarl*, is just north of Moruya.

The Narooma area has also produced at least one interesting report, albeit from an anonymous informant. In late July 1978, a man phoned the *Moruya Examiner* to say that he'd sighted strange "bear-like" creatures on three different occasions "north west to south west of Narooma". (He asked the *Examiner* not to divulge the precise spot as he did not want the creatures hunted.)

He claimed to have seen single animals on the first two occasions and finally, on July 25, a group of three. His 16-year-old son was with him during two of the sightings. The creatures were bipedal, stooped, had heads sitting right on their shoulders without discernable necks and were covered in rusty-brown hair. The tallest stood two metres [6ft 7ins] high. So far so good: the informant was describing to a "T" the average yowie-in-the-street. One additional detail, however, seemed to come straight out of left field: the creatures' eyes, he insisted, were big, beautiful, and "doe-like". They were also very penetrating. [Case 124]

It seems hard to reconcile those alluring, doe-eyed creatures with the huge, roaring, rock apes that have scared the tripe out of so many other people but, since the witness reportedly described them in that way, we are obliged to let his words stand.

Because the yowie phenomenon is a matter that severely tests the credulity of many people, we have tried in this book to avoid focusing on incidents where witnesses choose to remain anonymous. That having been said, one of our best south coast stories involves witnesses who absolutely insist on anonymity. It is simply too good a case for us to skim over.

Yowieville

"Sue" and "Jim" live a couple of hundred metres apart in a small township located right on the edge of thousands of square kilometres of rugged bushland. Because both run demanding businesses, they are adamant that even the name of their village be kept confidential.

During an unusually extended spell of wet weather in 1998, Sue noticed that something had been disturbing her three Shetland ponies. Night after night their blankets were dragged halfway off their backs and on one occasion a blanket was completely removed and thrown over a bush. As a result, she made it her habit to check on them whenever she could.

The animals were tethered across the road from her house, in a small scrubby paddock beside a creek that flows out of the vast forest to the west. The area was dimly lit by a nearby streetlight. One rainy night, as she began to walk across the field, she noticed a large, grey shape about 10 metres to her left. For a moment she thought it was the trunk of a large gum tree, but quickly realised there were no trees of that size in the immediate area. She took a few more steps. To her dismay, the "tree trunk" moved along with her. She then saw, to her further dismay, that the pale form resembled the lower half of a huge, hairy, manlike creature. She could make out what looked like two forearms, held close to the torso, and a pair of massive legs. Overhanging foliage obscured the upper body. Despite the rain, an unpleasant smell "like a burnt mattress" hung in the air.

As she realised what she was looking at, Sue was struck by a feeling of overwhelming terror. The fear was so intense that she felt almost as if something sinister had physically clamped onto her spine with an ice-cold grip. Although she was, she admits, almost wetting herself with fear, she sensed that to run away screaming might have been a bad mistake. Instead, she forced herself to continue walking steadily towards the horses before turning away in a wide loop and heading back across the road.

She discovered later that some of her Aboriginal neighbours were unsurprised by the occurrence. Long periods of wet and stormy weather, they said, seemed to drive the Hairy Men out of their usual haunts in the deep forest. During such times they had often been seen right on the edge of – and even within – the village.

Sue feels sure that, a few weeks after her first experience, she came close to the same, or a similar creature. It happened as she was riding along a forest track about three kilometres west of the village. An expert horsewoman, she had a very strong rapport with the animal she was riding. A highly trained, very valuable stallion she had worked with for 18 years, it was a truly remarkable beast, tall, beautiful and exceptionally intelligent. There was, as Banjo Paterson would say, "courage in his quick, impatient tread".

About mid-afternoon, on an uphill section of the track, the horse, which had thus far enjoyed the excursion, suddenly acted very strangely. Slowing almost to a halt, it lowered itself into an odd semi-crouch and snapped its head sharply to the right, eyes bulging. Sue could feel waves of fear running through its body, and, seconds later, she too was struck by unreasoning terror. It was the same overwhelming, literally spine-chilling sensation she'd experienced on seeing the yowie near her house. She turned the horse and gave it its head. It galloped, full speed and non-stop all the way home.

Sue was so traumatised by the second incident that she never ventured deep into the forest again. Yowies, however, still occasionally visit her village. She and her son have noticed the strange "burnt mattress" smell on several occasions. Sometimes it seems to waft in from across the creek on the edge of the forest; at other times it seems unnervingly close, seemingly right outside their house. Her son has seen large red eyes reflecting torchlight, and strange, non-human tracks have been discovered both in the nearby bush and within the village. Occasionally, when the "burnt mattress" smell is strongest, Sue's dogs clamour to be let into the house. On those occasions she and her son aren't just terrified – they feel physically ill, with headaches and stomach pains.

On the yowie track

While Sue has experienced only one actual sighting, Jim has seen yowies on many occasions. When we first interviewed him, in fact, he confessed that he was almost embarrassed to say how many times he'd seen them: "Over the course of about seven years," he said, "it must have been *dozens* of times."

One reason for his unique track record may be that he is a man of regular habits. Every weekday between 1991 and 1998, after spending long hours in his workshop, he would unwind by going for a horse ride – almost always taking the same route through the forest. All of his encounters occurred on one particular section of the trail. To begin with, he heard only heavy, thumping noises. Although they seemed unusually regular and continuous, he naturally assumed kangaroos were responsible. One evening, however, he glimpsed a huge, hairy, upright creature loping through the scrub. It was running parallel with the track, as though it was trying to stay abreast of him.

Thereafter he encountered what seemed to be the same creature and one or two others on many occasions. The episodes were intermittent – sometimes months would go by without incident – and the creatures generally stayed well back in the trees. Although he packed a 30.30 rifle as insurance, he gradually became almost used to the experience, as did his horse. On one occasion, when he attempted to ride a different horse down the trail, the animal reared and threw him to the ground.

Jim, in fact, believes that the reason he saw the creatures so often was that they were curious about his horses. The theory is well worth considering: as we know, Sue also encountered yowies while riding or attending to horses.

Throughout his lengthy experience with the creatures their pattern of behavior was fairly consistent. They would run parallel with him through the bush but, whenever he stopped, would "freeze", partly or entirely concealed by tree trunks. As he and his horse became used to them, the hairy giants also, apparently, became more confident. Gradually, they began to move closer and closer to the trail, sometimes approaching to within 30 metres. Their ability to merge with the forest, however, was quite remarkable: although Jim experienced dozens of partial sightings, he never once, in seven years, obtained an entirely unimpeded view. On several occasions he clearly saw shoulders, arms and legs, the side of a head, but never a face.

His best sighting contained almost comical elements. One creature attempted to hide behind a tree that was much too small. As a result both shoulders, both arms, both legs and much of the chest were left in plain view. Essentially, all that remained hidden was the face. It was reminiscent, Jim says, of a naïve toddler playing peek-a-boo. That creature, like the others, was about seven feet tall, its shoulders were massive and its arms looked every bit as broad as Jim's legs. It appeared to be covered from head to foot in hair.

Because of his repeated encounters over the course of several years, Jim noticed something that other witnesses have not observed: the creatures' hair seemed to change colour from season to season. In the warmer months it was brown but in winter it looked quite gray. While it is possible different creatures frequented the area during different seasons, he is fairly confident, because of their consistent behaviour, that he only ever encountered the same two or three individuals.

His unique interaction with the hairy giants ended in 1998 after one of them moved to within 25 metres of the trail. At that, his horse, which had until then kept its nerve, became distinctly unsettled, as did Jim himself. Deciding he'd pushed his luck quite far enough, he headed for home and never rode the "yowie track" again.

Although all of his actual sightings occurred on one particular stretch of trail, Jim saw signs of the yowies elsewhere. Once, near the site of Sue's second experience, he found a three-inch-thick sapling snapped over and twisted about six feet above the ground. At its foot lay an enormous, stinking turd. On another occasion, he and a friend discovered a few huge human-like, but distinctly non-human, footprints in a creek bed. During the course of his daylight encounters he never smelt the foul odour sometimes reported by other witnesses. On some nights, however, he has noticed the "burnt mattress" smell near the village.

Although he could have shot at the creatures on many occasions, Jim never did so. At least one other south coast yowie wasn't so lucky. The incident is of particular interest because at the time it happened – late 1977 – the shooter had never even heard of the yowie phenomenon.

The monster of Monk Farm

During the 1970s Kos Guines of Frankston, Victoria, and his family spent several Christmas holidays on the NSW south coast. One sunny afternoon in December 1977, Mr. Guines took his sons to Monk Farm, a long-abandoned, overgrown property 16 kilometres inland from Pambula. They planned to do a little rabbit shooting. Not everything, however, went quite according to plan. Just after sunset, with plenty of light remaining, Mr. Guines was walking very quietly down a gully when he was startled by a sudden crash in the bracken:

"I swung round, startled – anticipating perhaps a kangaroo – and saw the back of a huge black creature like a gorilla making off from only 10 metres … I brought up my shotgun and had a shot at it. No way I'd miss from that range. But it made no noise – just loped off into a cavity in the scrub."

Although upright, the creature wasn't particularly tall – only about the height of a small man – but much, much broader. The detail that stayed most vividly in Mr. Guines' mind was the way its dome-shaped head seemed to sit directly on its huge shoulders, as if it had no neck at all.

After 20 years in Australia, Mr. Guines, originally from Greece, was familiar with all of the larger native animals, so he assumed the creature was an escaped gorilla. Although it seemed to have departed, he hurried to stand guard over his sons as they cleaned rabbits in a nearby creek. He thought later the creature might have blundered into him while seeking to avoid the boys.

Over the course of many years, several other people have reported sightings of gorilla-like animals in the same area. The 1912 incident involving George Summerell occurred just 35 kilometres to the north-west, and two good sightings were reported near Eden, only 20 kilometres to the south, in the 1980s and 1990s. [Cases 42, 168 and 200]

One interesting aspect of Mr. Guines' experience is that just six kilometres south of where he shot the "gorilla" a steep, stark, rather eerie looking mountain called Egan Peaks, or The Jingera, looms high above the surrounding bush. As mentioned in chapter one, colonial-era

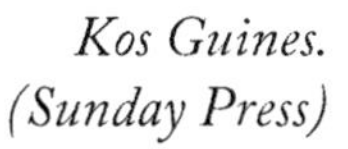

Kos Guines. (Sunday Press)

Aborigines believed The Jingera to be an abode of the yahoo.[20]

Because we began this hot spot section with a story from Clyde Mountain, where the Kings Highway drops from the tablelands to the south coast hinterland, it seems appropriate to end it with a similar story from Brown Mountain, where the Snowy Mountains Highway climbs back up to the high country from the coast.

"An Australian native gorilla"

Late one night in 1997, Chris McKechnie of Pebbly Beach almost ran over a yowie on Brown Mountain. The encounter was brief, but Chris, a part-time ranger with good eyesight, a calm disposition and a good knowledge of local wildlife, managed to absorb quite a lot of detail in the few seconds it lasted. For the occasion, he was the ideal eyewitness.

It happened as he was driving from the tablelands down to the coast. One of his companions, Lisa Stack, was dozing in the front passenger seat and the other was asleep in the back. At around 11 pm they entered Glenbog State Forest, where the road loops around the shoulder of the mountain before zigzagging down towards Bemboka.

"Because it was fairly foggy", Chris told us, "I was driving carefully, going about 60 kph [37 mph] around a curve, and there, in the middle of the road, was this 'whatever'. He was walking, just cruising, across the road with his head down, hands down. He was right in front of me ... about 40 or 50 feet ahead. And when the lights hit him he suddenly woke up, saw the car, turned, dropped on all fours and tore straight up the bank – a cutting which you'd find pretty difficult to climb, because it went straight up. It was all over in about four seconds.

"It was about seven feet tall and pretty well uniformly covered with hair. I honestly can't remember the exact colour ... fairly light, maybe a gingery-brown or light tan. It was curled together in thick lengths about eight inches long, not quite like a dreadlock, but matted in the early dreadlock stage all over, although I can't really say if its face was hairy. I didn't get a good look at its face and didn't see any eye shine – no red-eye effect at all. There was no defined neck, it sort of formed right into its shoulders, but I don't know whether that was just because the hair filled in that area and obscured it. It had big, rounded shoulders, not a defined figure. Very solid. Big legs. Big long arms; it looked ape-ish in that regard. I don't know whether that was because it was hunched over, but they appeared to hang down towards the knee. I didn't get a look at the feet.

"He'd been sort of strolling. When I first saw him I thought he looked stoned, like he was in his own world. I was surprised: you'd expect a creature like that would be fairly alert – they're not seen very often. You wouldn't expect to see it like that. It occurred to me later that maybe there's some kind of eucalyptus which at that time of the year produces a drug – and that he'd been chewing it.

Neither of his companions saw the creature: "When I saw it, I just yelled, 'Check *that!*' and by the time they lifted their heads up we'd driven past where he'd been, and they were saying, 'What? What?' And I just said straight out: 'I've seen an Australian native gorilla!' I considered stopping to look for tracks but the girls freaked out a bit – they didn't like the idea of me getting out of the car.

"I'd heard of yowies before but had never heard exactly what they were supposed to look like. Recently though, I brought the subject up with an Aboriginal friend, Mick Darcy, when he was showing us some sacred sites on Mt. Dromedary. He described what I'd seen to a 'T' and said that his people had known of them for years and had seen many of them, particularly a little to the south, in north-eastern Victoria. The term he used for them was *doolagar*."

Like several other recent incidents mentioned in this hot spot section, Chris's sighting occurred just a stone's throw (if a yowie throws the stone, that is) from the site of a much earlier event. Just ten kilometres to the south-east is Packers Swamp, where George Summerell encountered the seven-foot-tall ape-man in 1912.

The far south coast of NSW, it seems, really is the Land of the Giants.

Hotter: The Gold Coast Hinterland

"Perhaps next time you're driving down a dark and lonely country road, you'll turn a corner and there, right in front of you, something – some looming, hairy nightmare – will step ... out of the shadows."

Those are the final lines of our previous book, *Out of the Shadows – Mystery Animals of Australia,* written in 1994. We employ them again here because it would be difficult to find a better way of describing what actually happened to Aaron Carmichael seven years later, on Gilston Road in the Gold Coast hinterland.

The night in question, March 2, 2001, was very dark and rainy – the kind of conditions that would keep most people indoors. Seventeen-year-old Aaron, however, had just fitted new headlights to his Holden Commodore station wagon. "They were the brightest I could get. That's why I was out there – to try them out." From his home at Ashmore, near Surfers Paradise, he drove inland to Hinze Dam, up in the hills between Numinbah and Nerang State Forests, close to the Canungra Land Warfare Centre. At about 11 pm, while driving home along sparsely settled Gilston Road, he received the fright of his life.

It happened on a rather serpentine section of the road that has a high embankment on the right hand side. On the

Aaron Carmichael indicates the yowie's height. (Paul Cropper)

left is a steep, scrubby slope that drops down to the Nerang River.

"I came around a right hand bend, going about 70 kph [43 mph], and this thing like a big hairy man came across the road from the left, taking really big steps ... like it tried to beat me across the road. And I swerved to the right [and lost sight of it] and I was fully sideways ... and hit something, and it threw me down towards the steering wheel. [The collision made] an extremely loud bang and it seemed like I went over something, because the actual car got airborne – a half metre off the ground – and then I skidded off into the gravel and hit the embankment, then spun around ... did about a 180. It all happened so quick."

His headlights broken, Aaron sat there in pitch darkness, stunned and, although he had only glimpsed the hairy roadrunner, very scared. "No person could have a build like that, it was, like, *big*. Just too big – about eight feet tall. Extremely broad ... really long sort of hair; blacky-browny colour. Scraggy. Wild. I'd never seen anything like it ... it was not anything normal." And it was still there.

"I had my window down. I wound it straight up because I knew this thing was still around. And then, as I was reversing back, there was a really loud bang – like somebody kicking the side of the car. So I just got out of there. I got down the road and one light came back on."

When Aaron arrived home and poured out his story, his mother and sister, fearing a pedestrian might be lying there injured, insisted on driving him straight back to the accident site. On arrival they quickly discovered what looked like large footprints to the left of the road, but didn't look much further. As Aaron's mother told us later, there was something very creepy about the place.

"We all felt uncomfortable out there. I felt there was something watching us from the bush. I felt, 'I have to get out of here!' We only stayed about three or four minutes."

Understandably, Aaron couldn't sleep that night. Instead he began searching the Internet for information about yowies and was soon viewing Dean Harrison's website. Realising that the bodywork of Aaron's car might yield vital tissue samples, hair or even handprints, Dean dropped everything and sped to Ashmore. He got there pretty quickly, but he wasn't quite quick enough. The Commodore was Aaron's pride and joy. On the morning after the collision he had washed it.

"I feel so stupid ... I was out there, washing all the mud and everything off it. I didn't see any blood or anything. I did hit some hair on the tailgate ... I blew it off with the pressure cleaner and then realised what I'd done."

Aaron and Dean searched every inch of the car, top and bottom, without finding suspicious organic material. As Aaron pointed out, however, on his journey home after the collision he'd driven through an absolute deluge. The only visible traces of the yowie encounter – apart from broken headlights, grill and bumper, were a half a dozen shallow scratches, perhaps claw marks, on the right hand rear window and door column.

Just 10 kilometres south of the site of Aaron's adventure is Springbrook National Park. As described in the introduction, ranger Percy Window's hair-raising, stomach-turning, face-to-face yowie encounter of March 1978, took place in that park, which is at the eastern end of the mighty McPherson Range. A tangled mass of ridges, ravines and jungle-clad plateaus that begins west of Stanthorpe, the range extends eastwards, twisting and turning, for more than 300 kilometres before stopping just short of the western suburbs of the Gold Coast, only a few kilometres from the sea.

Dean Harrison, Aaron Carmichael and the damaged vehicle. (Dean Harrison)

Yowie encounters have been reported for many years all through the range, sometimes in dense scrub, sometimes on mountain roads and sometimes on remote rural properties. In 1976 and '77 two sightings actually occurred within the residential section of Woodenbong, a small town situated about midway along the range. We will detail those particularly interesting incidents later. [Cases 105 and 111] The Stanthorpe/Eukey area near the western end of the range has also produced a number of good reports, particularly in the early modern era. [Cases 34, 36, 59 and 61]

Yowies, in fact, have been reported all along the McPherson Range and in many other parts of south-east Queensland and north-east NSW. The area that has produced the most reports in recent years, however, is a lot smaller and, for the purposes of this exercise, more manageable. The area we think of as the Gold Coast hinterland hot spot extends 45 kilometres from the NSW border up to Ormeau and about 25 kilometres inland to Springbrook, Canungra and Tamborine.

Although old timers say yowies have been seen in the Springbrook area for several generations, there has never been such an intense outbreak of sightings as that which occurred just before and after Percy Window's encounter. Between October 1977 and March 1978, at least seven other sightings were reported there.

When word of his adventure got around, another Springbrook resident told Mr. Window that she, too, had seen a yowie. Hers, however, was a female, complete with long, pendulous breasts. Soon after that,

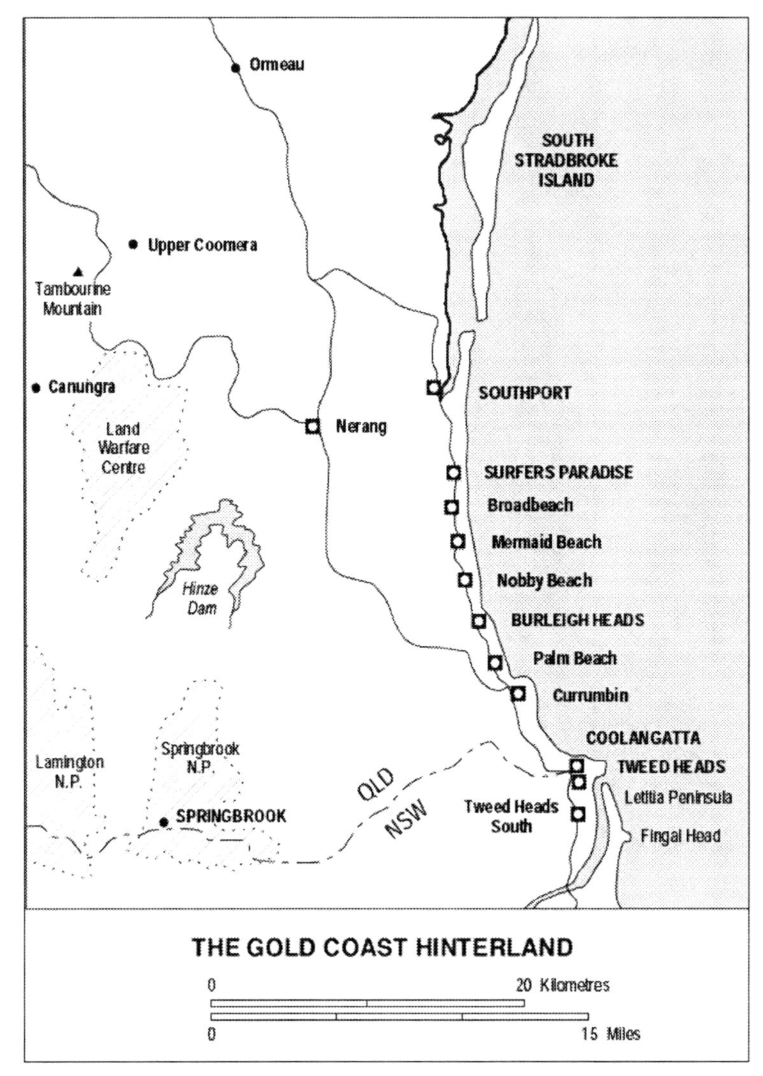

THE GOLD COAST HINTERLAND

Mr. Window experienced a second, much briefer sighting of an apparent male yowie, close to the entrance of the park. Some time later, a Sydney woman revealed that she and her boyfriend had been frightened by a 10-foot-tall, smelly ape-man at the site of Mr. Window's first experience, the Antarctic Beech grove, in January 1978 – two months before his own close encounter. [Case 122]

Another incident, which occurred less than two kilometres from that ancient grove, is of particular interest because it involved up to 20 eyewitnesses and because the principal witness, Bill O'Chee, later became a high-profile federal politician.

Pardon me boys, is that the Koonjewarre yahoo?

According to the *Gold Coast Bulletin* of November 17, 1977, students of the Southport School, after returning from a two-day camp near Springbrook, claimed they had repeatedly sighted a huge, bipedal, hair-covered creature. One of the many students who saw the animal was young Bill O'Chee, who told the *Bulletin*:

"About 20 of us saw it. It was about three metres tall, covered in hair, had a flat face and walked to one side in a crab-like style. It smashed small saplings and trees like matchsticks as it careered through the bush. We spotted it several times and once watched it through binoculars ... We first saw the yowie at 12:30 pm on October 22 and last saw it just before we returned to Southport on the afternoon of October 23."

In 1993, when we dug this clipping out of our files and telephoned Senator O'Chee for confirmation, we didn't know what his reaction would be. Many politicians, we suspected, would lie themselves blue in the face rather than be connected to such a story. We wondered if Bill O'Chee, Australia's youngest senator and international standard athlete, would have the guts to stand by what he'd said 16 years earlier.

He didn't let us down. Although he obviously rued the day he spoke to the *Gold Coast Bulletin*, he met the issue head on, confirming the story and providing us with further details. He also put us in touch with a former classmate, Craig Jackson, who had since become a tutor at their old school. The following is a composite of their separate recollections of the incident.

The campsite, known as "Koonjewarre" and then newly established, was in grazing land right on the edge of dense forest. Thirty boys aged 12 to 13 and two teachers were lodged in cabins and it was from the window of one of those that the first sighting occurred. Bill and Craig were among the first to see it.

The creature was uphill and was at first lying on the ground. Eventually it stood up and moved slowly around. It was right out in an open, almost treeless area and no grass or underbrush obscured the view. Although about 400 metres away, the animal was so gigantic that the boys could see it clearly with the naked eye. It remained in the same spot for several minutes as they passed a pair of binoculars backwards and forwards. Each had time for a couple of good, long looks. It was 12:30 pm, the day was bright and sunny and through the binoculars the creature stood out in sharp detail.

"There was no doubt about it," Bill recalled, "it was like nothing we'd ever seen and it was *huge*." Its whole body, including its flat face, was covered with black or very dark

'Koonjewarre' campground. (Paul Cropper)

brown hair. It appeared to have no neck, as its head sat squarely on its shoulders.

"It was not really like a gorilla," said Craig. "In fact," he added with a laugh, "it was more like Chewbacca out of *Star Wars*, except that its hair was not so long and its body was much broader – it was very heavy around the shoulders. It looked rather slumped or hunched over. Its arms hung down past its knees and it took a couple of steps here and there with a sort of swaying, sideways movement. It seemed to be just looking around." It was standing right beside an excellent reference point, an isolated white-flowered bush. Finally, a couple of boys stepped outside the cabin. The ape-man spotted them immediately and ran for the tree line. By this time at least 20 of the 30 boys had observed it.

They told the teachers what they had seen and one of them, Kevin Brooks, an adventurous ex-soldier, decided to lead a party up the hill. Craig said the reaction of the campsite's caretaker was interesting: "When he heard what we'd seen his jaw dropped, he looked really frightened, and he urged us not to go." At the time Craig suspected the man had seen the creature himself and his suspicions have now been confirmed. The present caretaker recently told us that before his retirement his predecessor had spoken of encountering a yowie near the camp shortly after it was built.

Ignoring the warning, Mr. Brooks, Bill, Craig and two other boys advanced up the hill. To their amazement they found that the white bush, which had been waist high on the yowie, was four or five feet high. "The thing would have to have been eight feet [2.4 metres] tall at least." A trail of widely spaced imprints showed where it had bounded, slipping and sliding, across the hillside.

Craig admitted to being more than a little apprehensive as, armed with sticks, they entered the scrub, following a trail of pulverised underbrush. "A whole lot of saplings had been freshly twisted, shredded and broken off above head height. I started to think it might have become enraged." Eventually they came to a large impression in the ground where they assumed the creature slept. "Something very heavy had compressed the grass and twigs. They were packed down very tightly."

Meanwhile, the other boys back at the camp saw the monster emerge from the scrub near where the party had entered. It moved away along the tree line. Throughout the afternoon grunting noises were heard and the creature was glimpsed intermittently. "That night", Bill recalled, "it came back. There were the most incredible noises and crashing sounds all around the camp."

"It was no joke," Craig insists. "We were shit scared."

In the morning they again found footprints and also metre-high native shrubs that had been ripped right out of the ground. Both men insist no human, however strong, could have budged the bushes, which were strongly rooted in hard ground. "There were huge clods of dirt still attached to the roots and they'd been hurled all over the place." Because the ground was so hard, the creature's tracks, though quite visible, showed no detail.

Although the headmaster, Peter Rogers, later searched the area himself and did not dismiss the reports as imagination, the boys were ordered, on their return to Southport, to say nothing about the incident. An article about it in the school paper was also censored. Annoyed by this and feeling the story was too important to conceal, Bill risked expulsion by contacting the *Gold Coast Bulletin*. Anyone seriously interested in the yowie mystery should be grateful he did. Because of his independent spirit we now have an excellent record of one of Australia's best-ever multiple-witness yowie incidents.

Only a week before we interviewed them, Bill and Craig had attended their annual class reunion, where, as at every previous gathering, the yowie episode was discussed quite intensely. Both men remarked that all their former classmates seemed to need reassurance, on an annual basis, that they hadn't imagined the whole mind-blowing experience.

The scent of a yowie

Another frightening encounter occurred three months later, only eight kilometres north of "Koonjewarre".

Early one evening in January 1978, David Window (son of the ranger, Percy) and his girlfriend, drove their mate, Scott (surname withheld by request), to a house on Springbrook Road. They dropped him off, agreed to meet later, and drove away. Less than an hour later they returned, found the house deserted and, concerned for Scott's welfare, began searching. They eventually found him four kilometres away near Springbrook Homestead. He had run the entire way. Breathlessly, he poured out his story.

Shortly after entering the house, he'd heard something brushing against a window and, on looking out, was shocked to find a huge ape-like creature staring back. It was black, about two metres tall, with an egg-shaped head and a small "screwed-up" nose. Its deep-set eyes had a strange, glazed, "porcelain" appearance.

Scott's yowie sketch.

Weirdly, despite his own great fear and confusion, Scott had the strong impression that the creature was sad, lonely or a little lost. But if he'd had the slightest notion of inviting the meek man-ape in for a cup of tea, such feelings vanished when it moved to the open doorway, bringing with it a choking miasma of B.O. Later, groping for words to describe the stench, Scott said it was like "a badly-kept public lavatory".

Almost retching, he hurled a chair at the monster, which limped into the surrounding forest while he fled the house in the opposite direction. David had no doubt that his obviously terrified mate was telling the truth. When they returned to the house and found several large tracks in the lawn and garden beds, his credibility was further enhanced.

Word of the incident spread and within days Howard Smith, owner of Natureland Zoo at Kirra, arrived, interviewed Scott, and made plaster casts of one or more of the tracks.

As we have been unable to track down Mr. Smith's casts, David, now General Manager of the Forest of Dreams Glow-worm Research Centre, recently sketched the footprints for us. Although 22 years have elapsed, he says that, so intense was his interest at the time, his sketch is likely to be broadly correct. There were several imprints of both the left and the right foot, and the tracks were large, but not huge: about 10-12 inches long and 6 inches wide. With five toes, one of which was distinctly thumblike, they were more like gorilla tracks than human footprints. He is quite sure that, as in his sketch, the heel was very broad and rather squared-off.

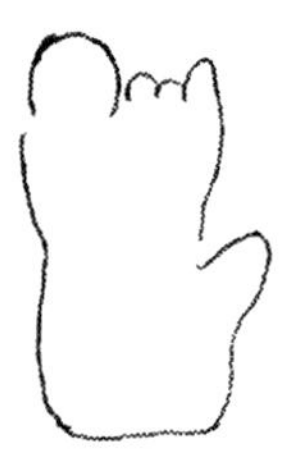

After the intense "flap" of 1977-78, yowie activity, or at least reported activity, in the Springbrook area subsided. For the next 20 years reports were infrequent and lacking in detail. In 1998, however, the giants returned.

Early that year, after hearing "blood-curdling screams" and finding large footprints on his Springbrook property, Gary Maguire contacted yowie hunters Dean Harrison and Tim the Yowie Man. Throughout the winter of 1998 Dean led several expeditions to the area, found a lot of interesting evidence and had several close encounters with what seemed to be a rather inquisitive, if not playful, yowie. In some places near the rainforest they found "up to 100 sticks standing upright in locations where there were no trees for them to have fallen from. We would knock them down, only to find them standing again some hours later. Another time we found over a dozen branches neatly stacked 'tepee' style against our cabin door."

One team member, Steve Bott, got reasonably close to the animal as it squatted on a track in front of him in the dusk. Its head and hunched shoulders, he said later, "resembled three bowling balls". Unfortunately, as Steve had a shotgun in his hands instead of a camera, he didn't photograph the creature. Fortunately, on the other hand, he wasn't tempted to shoot it.

On one rainy winter's night the team, having hiked through the jungle for hours, staggered back to their vehicles only to get both of them bogged on a muddy hillside. By the time they got them unstuck, they were physically and emotionally exhausted, but one task still needed to be done.

Dean drew the short straw: "I had to return to the bottom of the hill to retrieve some ropes, but I'd got only about halfway down when the guys radioed to say we had company. By that time I was in no shape for further physical activity, but as I made my way down I noticed that I was indeed being 'shadowed'. It was large, heavy, and walking through the bush to my right, obviously on two legs. It arrived at an intersection of tracks before I did and stood still in the bushes. As I arrived I was overwhelmed by a thick odour. It resembled urination and defecation and was awfully strong. Although aware of being watched, I didn't even have the energy to pick up a rock and throw it. Normally, being alone down there with something like that would have raised some fear, but I was just thoroughly disgruntled. I didn't have enough energy. I just looked in his direction and said, 'Not now – go away'. I limped back to the group. The creature didn't follow."

One morning the men woke at 6 am to find footprints only metres from where they lay in the muddy ground. "It was rather unnerving seeing them – it had basically walked around us between 3 am (when we fell asleep, exhausted) and 6 am." The tracks were bipedal and very large, and the team followed them for about a kilometre before losing them in the jungle.

On another occasion, Gary Maguire and Andre Clayden, manager of Springbrook

Homestead Restaurant, made a plaster cast of an unusually clear footprint. Andre recalls that there were nine tracks in all. "It had gone through muddy ground and most of the tracks weren't perfect, but we made a cast of the best one. Normally, maybe, I would have measured the distance between tracks, instep to instep or something, but it was raining and not ideal conditions. But it was a very large step: it seemed a very tall animal."

The Springbrook cast is one of the clearest we have seen. Five-toed and measuring 33 cm by 16 cm [13 x 6.5 inches], it looks somewhat like a huge, flat, misshapen human foot; in fact it looks very much like many North American bigfoot/sasquatch tracks. [It does not, however, even faintly resemble David Window's careful sketch of the tracks found four kilometres away in 1978. We will address that glaring anomaly in a later chapter.]

In 1999 the cast was sent to Dr. Jeff Meldrum of Idaho State University, who was eager to scan it for dermal ridges, such as those visible in the foot and toe prints of humans. When a plaster cast of a supposed bigfoot/sasquatch track was analysed by the same team of experts a few years earlier, they claimed to have discerned such dermal ridges. This time, however, the result was disappointing. Andre explained why: "They couldn't get any dermal ridges off the plaster cast because it had been handled by too many people and the fine detail had worn off." He laughed and shook his head ruefully. "We'd used it as a bit of a draw-card for the Homestead (Springbrook Homestead Restaurant) and about 10,000 people had already handled it!"

Recently, thanks to Dean and to naturalist Gary Opit, we heard about another sighting that occurred at virtually the same time as the great Springbrook yowie "flap" of 1977-78. An interesting aspect of this incident is that it allegedly took place about 15 kilometres

Andre Clayden and the Springbrook footprint cast. (Tony Healy)

east of the mountains, right on the edge of suburbia, on land that is today almost entirely built over.

Paul Cronk, now an environmental scientist in his mid thirties, grew up in Tweed Heads. As a boy, his best mate was Mark Gill, who lived on the edge of an extensive area of swampy bushland to the south and east of the Tweed Billabong Caravan Park. About 75 percent of the area is now covered by suburban Tweed Heads South and by the sprawling Tweed Heads golf course. The remaining one-square-kilometre segment is bordered to the north by Ukerebagh Passage and to the east by the Tweed River. Together with scrubby Ukerebagh Island, it is now designated Ukerebagh Nature Reserve.

One day in October 1977, when they were both 12 years old, Paul and Mark were exploring the swampy scrub when they heard a series of snarls, growls and grunts that sounded like the vocalisations of a very large cat. Shortly thereafter, they followed the sound of something large as it crashed through the scrub, crested a low ridge and descended into a small, grassy valley.

"In the bottom of the valley", Paul recalled, "was a large weeping willow being shaken violently by something obviously very large and strong. As the tree shook, a loud, repetitive, aggravated bellowing – like a distressed bull – emanated. We took off running for our lives ... we could hear the creature pursuing us, but it never came close enough for us to catch a glimpse of it. When we reached the safety of cleared land near Mark's house, we stopped and looked back. Behind us was a sparsely wooded ridge in partial shadow [as the sun was getting low], this is where we saw the large dark creature walk halfway down the slope. It stood upright and walked like a person, and had similar limb/body proportions to a human but may have been as tall as eight feet. [It] was . . . brown/black in colour and looked pretty similar to Chewbacca, the Wookie from *Star Wars*."

Both boys had often visited the Gold Coast Zoo where they had seen a Kodiak bear. The yowie was of comparable size and hairiness. As they watched, from a range of 70 to 100 metres, the huge creature

Paul Cronk. (Paul Cropper)

crouched down, supporting itself with one hand against a tree, "and just watched us watching it." After about 10 minutes, the boys ran to Mark's home.

Even though they had recently been plagued by a "mystery prowler", who had repeatedly invaded their yard and scratched on windows, Mark's parents refused to believe the boys' fantastic tale.

"We were determined to prove our story," Paul continued, "so we went back, but via a different route to try to ambush the thing and sneak a better look. We took Mark's Doberman dog and cane knives for protection. As we approached the valley with the willow tree, we witnessed yet another strange occurrence. Over the ridge, we could see the top of a tall silky oak tree. It was shaking and swaying. We could hear similar bellowing, but what was strange was that there were birds of at least two species swooping the tree. It was something like, maybe, two crows, one magpie and a hawk – about four or five birds.

"We assumed the creature was raiding a nest or something. Mark's dog took off over the ridge barking and attacking but within seconds returned, obviously very distressed, ran right past us with its stumpy tail between its legs, and jumped into a small billabong, repeatedly ducking its head under the water and pawing at its face, like it had been sprayed by a skunk or something. It was a trained guard dog and hated water with a passion. Needless to say, we ran home without looking back, the dog beat us home."

Although both Paul and Mark's parents remained sceptical, the boys were heartened to learn that two adult neighbours had also seen large hairy creatures. On the night of their first encounter, Mark's next-door neighbour had heard her horses acting up. Looking outside, she saw what she thought was a bear and later asked Mark's mother if such animals existed in Australia. The other witness was old "Spud" Murphy, the caretaker of an abandoned farm on the edge of the swampy scrub, who told the boys he'd "been seeing that thing for years".

Paul Cronk's sketch.

Within a few days of their own experiences, the boys also heard that (as Paul now recalls it) "a group of boy scouts" had just returned from the mountains, where they'd been terrorised by a hairy giant. This could only have been the "Koonjewarre" Campground incident of October 22-23, 1977.

We found it interesting that Paul, who had never seen a detailed account of the "Koonjewarre" incident, said his yowie resembled Chewbacca, the *Star Wars* Wookie. "Koonjewarre" eyewitness Craig Jackson had made exactly the same comparison. In fact, since the descriptions given by the two groups of witnesses are so very similar and because the creatures in both incidents were so aggressive, we think it just possible the same, rather cranky, yowie was involved in both cases. Two or three days seem to have elapsed between the events and the two locations are, after all, only about 25 kilometres apart – a very small step for a giant ape-man.

The way that Mark's Doberman threw itself into the billabong to wash its face brings to mind another remarkable incident that occurred only two months earlier but 100 kilometres to the west. On August 10, 1977, at the village of Woodenbong, a terrier was grabbed and then dropped by a yowie. Its owner, Jean Maloney, said that after the unwelcome hug the little dog's hair felt greasy and stank so badly she had to wash it in antiseptic. [Case 111]

As some readers may have noticed, another element of Paul's story tends to corroborate a much earlier Hairy Man report. His recollection of seeing birds swooping the yowie is highly reminiscent of H. J. McCooey's 1882 report [Case 25] which, we believe, Paul couldn't possibly have known about.

Even in 1977, the scrubby area where Paul and Mark encountered the yowie was not particularly extensive – perhaps five square kilometres. Paul doesn't think that such a large creature could have survived there continuously, but believes it could have made its way there occasionally from the mountains by utilising overgrown creek beds and forested reserves. Even so, it would have had to cross several roads, including the Pacific Highway.

Just recently, however, we have learned that some local Aborigines believe at least one Hairy Man *still* lurks in the vicinity of Tweed Heads. For this information, we are once again indebted to Kyle Slabb, already quoted in chapter one.

Kyle's people, the Goodjingburra, a sub-group of the Bundjalung, still occupy a portion of their ancestral land on Letitia Peninsula, directly across the river from where Paul and Mark encountered their yowie. During two lengthy interviews, Kyle, a National Parks and Wildlife officer in his late twenties, shared a considerable amount of Hairy Man lore. His stories gave us fascinating new insights into the Aboriginal view of the yowie, and also corroborated two elements of Paul Cronk's story.

Kyle, a quietly spoken, thoughtful man, began by telling us that his clan had known of the Hairy Man for a very long time. Several members of his parents' and grandparents' generations had told of encountering the creatures near their coastal settlement and also in the high country to the west. Although most sightings were of solitary creatures, Kyle's grandmother was once lucky enough to see a juvenile accompanied by its very protective mother. The elders also said that in the eastern reaches of the nearby McPherson Range there was a certain ledge from which, if they approached quietly, they could look down upon a whole community of Hairy Men.

Even more remarkable were their accounts of an apparently unique, amicable relationship that once existed between the Goodjingburra and one particular Hairy Man. When this relationship began is unclear, but it supposedly continued even after white people began to settle on the peninsula in the mid-1900s.

Kyle Slabb at the entrance to the Hairy Man's cave – now almost full of sand.
(Tony Healy)

Kyle was told that the yowie slept in a cave on the southern side of rocky Fingal Headland, where the lighthouse now stands. The cave is still readily accessible but because of massive changes brought about by sand mining it is now much smaller than it once was – its sandy floor having been raised by about 1.5 metres. Up until the early 1900s the Goodjingburra would travel up towards the mountains at different times of the year, "but instead of carting all their gear, fishing spears etc, they'd leave it here in one of the caves and the Hairy Man used to look after the stuff. I'm not sure if it could actually talk, but it could communicate. They said he was here for a long time and when most of the tribe was disbanded after European settlement – after the loggers and sand miners came in – he still stayed here. There was a dark old shed that was where the caravan park is now, and he moved into it [for a while] and then ... near the village, there's a lot of mangrove areas; there's a fair bit of scrub. There was evidence of him moving around here and there."

The marvellously located Letitia community consists of a community hall/church and about a dozen modern homes, several of which were built by Kyle's father. The layout is such that every dwelling backs onto trees or sand dunes.

"One of my uncles came home one night and the Hairy Man was in his kitchen. They all came in the front door and it was in the kitchen going through the food cupboards and ... he went out the back door and from the verandah into a tree – that was years ago. Some other people have seen them, mostly on the beach, and there have been plenty of footprints and hair. We've come across the hair [adhering to undergrowth] ever since we were kids. Long darkish-brown hair; we've found a fair few bits and they smell bad, sort of like lions in the circus – we all agreed on that when the circus came here. I've occasionally seen tracks on the beach. I saw a line of big tracks once, coming out of the dunes and back in. They were similar to a human's – five toes – but the toes don't come out clear."

Although he had heard about the Fingal Hairy Man all his life, when Kyle finally came face to face with it, he was still psychologically unprepared. It happened in 1991, as he and several friends walked down Letitia Road, only a hundred metres or so from his home. Kyle was 16 years old and it was a stretch of road he'd walked or driven virtually every day of his life.

"We walked past the Aboriginal cemetery and saw something down along the road. I remember noticing it. I thought it was just a bush, but there was not usually one in that spot. It was [dusk] – still a bit of light in the west. I forgot about it until we got quite close and then I walked over to see what it was … and he was just sitting there.

"I walked right up to him, right up to his feet – just one or two feet away. He was just sitting there, holding on to his knees. It wouldn't register – what I was looking at – until he looked up at me. He looked straight at me and I was frozen. The expression was really, just *blank* – blank as you could get – completely blank. There were a couple of other fellows with me, just looking at him and I was speechless … and I grabbed Ernie, who was next to me, and he just looked and caught his breath, and just *screamed*, and that started everyone to panic – everyone ran!

"[Sitting,] he was probably just up to my stomach; [standing,] he wouldn't have been taller than a good-size man. He was like a mixture [of human and animal]. A darkish-brown colour. It looked like a human face, but with long hair on every part except for around his eyes. He had pretty much just brown eyes; his jaw was small but stuck out more than a human's … the mouth looked really monkey-like. I can't remember the ears. His forehead was prominent, and the cheekbones. It had long hair, not really matted, but it was pretty rough, pretty long all over him. My first thought was 'it's a gorilla or a monkey or something', but when it looked up, it looked more human than anything.

"There's a track goes from where it was sitting right into the lagoon, and all the old people always tell kids, 'don't go down that track', but all our lives, we've always been down there. We all ran to my house and flew in and told Dad and the others and they said, 'Well, yeah, see: he *is* there!'"

Kyle's father had seen the same or a similar creature in virtually the same spot in 1959 or 1960. "When he was in primary school, he and my uncle were walking home for lunch and they were screaming – carrying on – and they heard this other scream coming from the bush like nothing they'd ever heard. And the Hairy Man came running out with a stick. Their screaming had annoyed him and they watched him coming and just took off. They were hysterical because my grandfather made them walk back up the same road to school after lunch."

Amazingly, considering the relentless approach of the Tweed Heads urban sprawl, many of Kyle's people believe the Hairy Man *still* lurks in the peninsula's scrub and dunes. His own most recent experience occurred

(Left) Kyle Slabb at the site of the yowie encounter. (Tony Healy)
(Right) Kyle Slabb's sketch.

one night in early 1999 as he and his wife, with two children in a pram and two large hunting dogs, walked down Letitia Road.

"At the same spot [as the 1991 incident], I could hear, coming through the bush, a snapping of branches – big branches cracking. Old Jake and Scar would usually chase anything but this time they just stopped dead and were growling and looking in, and they wouldn't run in. So I told my wife to walk back with the pram and I waited for them to get out of the road. You could hear the footsteps and the breaking branches, but it didn't come out ... and I dragged the dogs away."

In recent years, similar things have happened to several other people. Some say that as they walked at night, they could hear a large bipedal creature keeping pace with them just a few metres off the road – crunching through the tangled scrub, as no man would be able to do in broad daylight, let alone in total darkness.

Many readers, we know, will find some of the Goodjingburra yowie lore almost impossible to accept. Having spoken to Kyle and members of his family at length, however, and having found them to be open, generous-natured and sincere, we are not inclined to casually dismiss any of it. Almost every detail of Kyle's testimony – except for the unique story of the semi-domesticated yowie – is supported by similar testimony from other witnesses in various parts of Australia.

Undeniably, the notion of a laid-back, helpful yowie is, at first, a bit difficult to get one's mind around. On reflection, though, it doesn't seem completely implausible. Because a great deal of evidence suggests that yowies are reasonably intelligent, curious creatures – and rather man-like – it doesn't seem entirely out of the question that a solitary Hairy Man once forged a working relationship with a group of peaceful Aborigines.

Readers may find another detail of Goodjingburra yowie lore a bit problematical: some of Kyle's people believe the creature seen in recent years is the same individual that cooperated with earlier generations. If they are right, the Hairy Man in question must be at least 70 or 80 years old. Without dismissing that possibility out of hand, we think it more likely that several different creatures found their way to and from the peninsula over many years. The one that has been seen recently may have been marooned there, so to speak, by the rising tide of humanity between the coast and the mountains.

Kyle and some of his relations believe there are enough wallabies, shellfish and other bush tucker on the peninsula to support a large creature. Although we bow to their generations of local knowledge, we can't help feeling the poor old yowie must be hanging out for a decent meal most of the time. If it is indeed still there, the creature must, it seems, be confined to the approximately two square kilometres of scrub and dunes to the north of Fingal Head village.

Given that at least one third of the peninsula consists of a wide beach frequented by fisherman and that a small sand mining company occupies a small part of the northern tip, we don't imagine the lone yowie's life would be all beer and skittles. If, however, it could swim 100 metres at a stretch, it could cross the Tweed via Sandy Island and utilise the scrub-covered two square kilometres of Ukerebagh Nature Reserve – in the vicinity of Paul Cronk's 1977 yowie encounters. Throughout history, fugitive groups of humans, notably the remnants of California's Yahi tribe, have occasionally survived undetected for decades in comparably restricted areas, so what the Goodjingburra say about their last surviving Hairy Man just might be true.

Kyle Slabb shared a lot of very interesting stories with us, for which we are very grateful. He did not, however, share everything he knew. For cultural reasons, he explained, some aspects of his people's Hairy Man lore had to remain secret.

If any readers are inclined to dismiss the Goodjingburra stories as Aboriginal myth, they should remind themselves that the white boy, Paul Cronk, encountered his yowie less than three kilometres to the west. Paul had no knowledge of Goodjingburra yowie lore and Kyle had never heard of Paul's encounters.

The "Burleigh Bunyip"

We were so impressed with the testimony of Kyle and Paul that we looked again at an older report from a location 16 kilometres north of Tweed heads. The story had been gathering dust in our files for many years.

In early August 1973, 16-year-old Alan Livingstone of Southport, and his friends Brian Morey, 19, Ken Kilner, 19 and Gary Hoffchild, 15, claimed that they were chased by a "large, red furry creature" in West Burleigh Heads. Their adventure began as they were driving, at about 11 pm, along Tallebudgera Creek Road. Seeing what looked like a pair of glowing eyes in the scrub, they got out of the car to investigate: "but as soon as we got near the thing it came charging out towards us. It was about five foot eight inches tall, had two arms, walked on two legs like a man, and was covered in reddish fur. We didn't stay around too long to find out exactly what it was ... we took off like rabbits. As we ran back to the car, we could hear its padded feet thumping the ground behind us."

News of the encounter began to spread after the boys enquired at local zoos to see if a bear had escaped. Although the *Gold Coast Bulletin* mentioned that three similar sightings had been reported in the same week, it seems local and big-city journalists didn't know quite what to make of the story. Part of the problem was that the incident occurred at the tail end of the early modern era, when awareness of the yowie phenomenon was at an all-time low. As a consequence, the press labelled the creature the "Burleigh Bunyip" and cast around for an expert to debunk the story.

They didn't have to look far. The encounter had occurred just west of a fauna reserve run by one of Australia's best-known naturalists, David Fleay. Evidently even Mr. Fleay, who had led expeditions in search of the Tasmanian tiger in 1945 and '46, had never heard of the yowie. He offered the ludicrous suggestion that the boys had run in fear from a wild emu.

Because the encounter seemed to have occurred virtually in the back streets of Burleigh Heads, we too, had reservations about the story. After visiting the Gold Coast to document the stories of Paul Cronk and Kyle Slabb, however, we began to appreciate the area's unusual demography.

In the early 1970s the Gold Coast was probably the world's slimmest city. Although it stretched more than 40 kilometres north to south, the greater part of it extended inland just a few hundred metres. In those days, the area where the boys were monstered by the ape-man was beyond the urban strip, on the edge of some pretty rough scrub. It seems, in fact, that the region's unusual demography is the main reason why the Gold Coast hinterland has become one of Australia's hottest yowie hot spots. What was once just a string of tiny seaside villages is now Australia's most rapidly growing city. In recent decades the beach-hugging settlement has been expanding inexorably westwards, pushing into very rugged areas, like the foothills of the McPherson Ranges, that were previously, apparently, the exclusive haunt of the yowie.

In 1978, another almost-urban ape-man was sighted.

One Sunday afternoon in early August of that year, 13-year-old Shaun Cooper encountered an enormous hairy creature in bushland near his home at Nerang. He told Des Houghton of the *Gold Coast Bulletin* that he had been riding his bicycle at about 2.30 pm when he came across the animal stripping bark off a tree.

"Bark was falling down around its body, then suddenly it turned and looked at me, putting its arms by its side. It looked at me from about 50 yards away for no more than three seconds. I turned and went for my life. My dad wasn't home and my mum didn't want to go back and look for it."

The creature was about eight feet [2.4 metres] tall and covered in thick fur. Returning later with some friends, Shaun found several large footprints that went up a nearby hill and through a fence to a waterhole. *Gold Coast Bulletin* staff later found and photographed a tree with bark torn from the trunk, as well as several of the footprints. Clearly three-toed, the tracks were nothing like those found at nearby Springbrook in 1978 and 1998. They were, however, remarkably similar to a track which, as we will discuss later, was found in the Blue Mountains in 2001 [Cases 125 and 223]

Later that same month retired security officer Leonard Rye, his wife Nan and several others were travelling along Upper Coomera Valley Drive (about 8 km north-west of Nerang) after midnight when they noticed an absolutely gigantic, terrifying creature at the side of the road. Covered in dark brown hair, it stood between 10 and 14 feet tall. Crossing the road in two steps, it casually stepped over a four-foot high fence and walked into a paddock. It resembled a "huge, hairy gorilla", "a huge ape". [Case 126]

Shaun Cooper's sketch.

Diggers and *dulagarls*

Much of the Gold Coast hinterland is extremely rugged country. One indication of just *how* rugged is this: Australia's main jungle warfare training establishment is located at Canungra, just 25 kilometres west of Surfers Paradise. During the past 50 years, tens of thousands of camouflaged soldiers have crept through Canungra's gloomy rainforests, perfecting the arts of silent observation, evasion and ambush. If yowies truly exist all through the Gold Coast hinterland, then surely, one would expect, they would have been encountered at Canungra.

Recently, our resourceful colleague Dean Harrison has managed to establish that such is indeed the case: eyeball-to-eyeball encounters between diggers and *dulagarls* have occurred there on several, if not many, occasions over the last few decades. Because most of Dean's informants are still in the army and have given their information in confidence, we are at liberty to relate only a couple of stories.

Army widow Pauline Haimes said that while fighting a huge bush fire on the base in 1978 or '79 her late husband Max and another soldier saw two very tall, dark, grunting figures running *towards* the fire. The men called out a warning, but then realised the creatures were not human. "Everyone in the base seemed to acknowledge there was something there, and called them yowies", she added. "We used to hear strange howling noises at night – a high-pitched moaning."

At dusk on March 21, 1986, soldier Lester Davison watched a huge creature bound across a road about three kilometres from the base. It was 30 metres ahead, well illuminated by his car's headlights: " I had to brake so I wouldn't hit it. It took one, two steps on the asphalt; one more into the bush and it was gone. Big, fast steps. I'm 178 cm [5 ft 10 ins] and it would have towered over me. It was two and a half to three metres high, the size of a fully-grown cow standing up – just huge. It was hairy; shaggy, dark hair, maybe 10 to 20 cm [4 – 8 ins] long. It was leaning forward a bit from the waist. Long legged; I didn't notice its arms. Short neck, a funny shaped head, similar to an egg. All my hair stood up on end. I just drove on, dumbfounded. The infantry fellows in 10 RIC had seen quite a few, they say, in the bush in the Land Warfare Centre. Plenty of stories."

[For very close encounters involving military personnel in other parts of Queensland, see Cases 212, 219 and 265]

It is interesting to note that Gilston Road, where Aaron Carmichael collided with a yowie in March 2001, is less than five kilometres east of the Land Warfare Centre. Only a month later, in April 2001, another incident occurred just five kilometres north of the base.

"I'll swear upon the Holy Bible ..."

While jogging along Hartley Road on a bright moonlit night, 27-year-old David Holmdahl of North Tamborine noticed a huge, dark, seven or eight foot tall figure standing about five paces away, in a roadside clearing. At first he assumed it was a draught horse, but soon realised it was man-shaped and covered in short fur. In silhouette its head "looked like a hood, because it didn't really have any neck". It seemed to be about four feet across the shoulders and looked to weigh about 500 pounds.

On seeing him, it reeled backwards into the bushes, breaking branches and twigs as it went. It soon recovered, however, and ran through the scrub, parallel with him as he sprinted for his life along the road.

"It was playing with me, playing a game, whatever it was ... I could hear it and I could *feel* the ground thump; that's what freaked me out. This thing was making slower paces [than] my fast steps but was keeping up with me for about 100 metres ... then it just broke away, I think it had had enough ... then I just legged it, like I've never run before ... and got home and nearly collapsed."

David's mother heard him arrive: "We could hear his breath in the garage and went down and David couldn't speak ... he has an olive complexion, but was absolutely white. We could see the heart pumping under his shirt and my oldest son said, 'We'll have to give him mouth-to-mouth resuscitation'. But he calmed down, little by little. We could see the blood coming back to his face. It took maybe 20 minutes to be able to talk. He said, 'Oh, I pray to God, Mum!' That's the first words he said. And then, 'I saw this thing …'"

When David returned to the site three weeks later he discovered a couple of broken branches but no tracks. Behind the small clearing were avocado trees and kiwi fruit vines; David assumed the animal had been feeding from them. He is adamant that he didn't imagine the incident: "I don't take anything that influences or alters my state of mind. I'm into health and fitness and all my faculties are there, so I didn't imagine this. No. I'll swear upon the holy Bible, and I'm a Christian." [Case 242]

"I thought there was no escape … my life was about to end"

Frightening as it was, David's experience was far from unique. At least two other hair-raising pursuits have occurred in the Gold Coast hinterland. One, briefly mentioned earlier in this chapter, was the incident that inspired Dean Harrison to become involved in yowie research. It happened at Ormeau, about 20 kilometres north of Tamborine.

"At the end of 1996", he recalls, "I bought my dream home on acreage at Ormeau, which at that time was all bushland." He worked long hours as a sales manager and unwound after work by jogging along a narrow cleared strip that followed an overhead power line through the nearby bush.

"At about 11 o'clock one night in late July 1997, I was again heading down the mown area. The bush on either side was very thick and dark. To walk through that scrub at night without a torch would be a near impossibility. On this particular night I decided to make a quick phone call to my now wife Donna." While talking, he walked slowly along the edge of the scrub and stopped with his back to the tree line.

"So I'm standing there talking and I heard this thunderous crashing coming through the bush behind me, about 100 metres away, and I thought it must have been about three kids coming through, making a lot of noise – but what were they doing out at that time on a Monday night?"

Dean, a powerfully built boxing enthusiast, wasn't particularly worried: "I was talking away quite loudly. Then the noise stopped and I heard just a very loud crack about 75 metres behind me and then another and another, and I thought 'Oh – perhaps there's only *one* person out there'. I continued talking and made a point of not looking around. I kept facing the distant road – up on the road, about 80 metres away, there was a streetlight. This 'person' was slowly getting closer and closer, taking deliberate care not to make noise, and I then realized it was a *very* deliberate, stalking process. I could hear the placement of feet – very gently – and every time it made a noise it would stop for 10 seconds or so, and then I could hear the leaves of trees being parted as someone moved through.

"I said to Donna, 'I think some guy is sneaking up behind me', so I thought I would let him get close enough and then find out what his story was. I was very fit and feeling pretty 'bullet-proof" – I wasn't too worried.

"This thing came up to about 30 metres behind me and I was about 12 metres away from the bush line and suddenly I just had this almost indescribable chill which ran from my head to my toes. This unfamiliar and hugely terrifying sensation just overtook my entire body … I just stopped dead, stopped talking. I simply could not understand why

I suddenly felt as uneasy and vulnerable as I did, but I knew something was terribly wrong and it had something to do with whoever or *whatever* was behind me. I felt *really* unnerved. I didn't want to turn around and make myself obvious; that seemed as if it would have been a dangerous move. So I turned only my head and kind of strained my eyeballs around and saw the silhouette of a huge figure, about seven feet tall, slowly squat down behind the trees. Big wide shoulders. Right then I said: 'Donna, Donna, Donna ... I might have to do a runner here', and she began a barrage of questions. I said 'Just hang on a tick', then slowly lowered the phone.

"I looked up to the street light and planned it. I knew I had to physically force myself out of this paralysed state, so I literally counted: 'one, two, *three!*' Then forced my right foot forward. On my very first step, came a god almighty bellowing from behind me: 'WHAOOOR!' This thing just let out a *roar,* and this huge roar is something you'd never, ever, forget in your life! It's like a tiger, a bear – all these wild animals wrapped up in one – and it was so loud and aggressive. On that first roar it launched into action.

"I'm now running straight up this field, while it takes off through the trees to my left and it's just *tearing* everything down that's in its way, and on every heavy, pounding step there's this deep, guttural grunt, as if its diaphragm is bouncing around.

"It was big and it was heavy – you could hear it jumping over logs and the heavy footfalls as it ran, *boom, boom, boom,* still screaming, roaring and grunting the entire time. The raw aggression that came out was just phenomenal. I knew that this creature meant business, and without a word of exaggeration, if it got hold of me I wouldn't have had time to scream – it would have snapped my neck in a split second. You know how you can tell when someone's talking in your direction or facing the other way? Well, this thing kept its eyes on me the entire time, even while it was running and crashing through dark bush.

"So I'm running across this mown, flat strip and I'm no slouch as a runner. It was running parallel with me through the bush – and it ran about three to four times my speed, straight through this mess of trees with no trouble at all. It caught up with me in no time, still screaming – all the dogs in the area were going wild – and it comes right up beside me about 20 metres to my left and it's like it was verbally abusing me as it was running: it was right *at* me. I thought then there was no escape and my life was about to end. It was simply too big, fast, powerful and aggressive. I felt totally helpless.

"Then it overtook me, and I realised, 'This bastard's going to cut me off!' And sure enough, that was it, so I had to quickly change direction to my right, towards the street light and away from the bush line, and it came up and around and just lunged out of the bush – but luckily I was too far away for it to get a hold of me – and by this time I was looking back over my shoulder as I made my way to the street light. It returned to the bush line and squatted down behind the foliage, still watching as I explained to Donna, who was still on the phone.

"He was dark, about seven foot tall. It was very dark at the time, so it was mainly just a silhouette with just a hint of brown. Wide shoulders, thinner torso and hips. No glowing eyes or smell."

Any reader who is tempted to reject Dean's story as being simply too dramatic to be true should reflect on its similarity to David Holmdahl's experience at Tamborine. Another case that provides even better corroboration is well worth summarising here. Not only did the incident occur within 2.5 kilometres of Dean's encounter, but it seems quite likely it involved the same rather grumpy yowie.

The case is of special interest, also, because the witness, Jason Cole, had only recently returned to Australia, having spent most of his life in the United States, and had never even heard of the yowie phenomenon.

"I got the weird feeling it was hunting me"

The incident occurred one afternoon in late April 2003, as Jason was felling trees on extremely steep land off Plateau Road, Ormeau. At around 3:30 pm, as he turned off the chainsaw, he noticed a strange and frightening creature glaring at him from further down the slope. At first he could see only its hairy face as it crouched behind a bush. When it stood up, however, he was staggered to see it was "a weird-ass gorilla-man", towering to eight or nine feet and weighing perhaps 350 to 500 pounds.

It had a flat face, big eyes but no discernable neck or ears. Its arms reached to its knees and it was entirely covered in dark brown hair. "The only place it didn't have hair was around the eyes and nose." Just as unnerving as its huge size was its angry expression: "It looked a bit pissed off; it had, like, a mean look on its face."

Jason and the creature stared at each other for five or 10 seconds until it ran further downhill with a strange "crab-like running walk" and disappeared into the scrub. After a further 10 seconds a tremendous cacophony erupted, as if the huge animal "was just going nuts, like it was destroying something, like he was breaking up trees or picking up big rocks and throwing them – like how kids throw a tantrum. Then all of a sudden it got really quiet".

The incident had already been frightening enough, but as Jason quickly collected his gear and turned to retreat uphill, things got even scarier. The creature, he realised, had somehow circled around in an apparent attempt to cut him off. "I could hear it rustling and breaking branches and stuff. It had run – real quietly somehow – through the bush, up this 45-degree slope without me noticing. I didn't think that thing could run so fast up such a steep embankment … and I got the weird feeling it was hunting me."

"I went diagonally away from those bushes [where the creature appeared to be lurking] it was so steep I was holding onto bits of grass and roots to get up, and I had this weird feeling he was almost *behind* me. And I cut through into the clearing and ran back to my car."

The similarities between Jason's experience and that of Dean Harrison are inescapable. In fact, as we suggested earlier, it seems a reasonable bet that the same creature was involved in both cases.

Through sheer coincidence, a day after his encounter Jason happened to talk to a woman who lived on the edge of the scrub just around the corner from Dean's former residence in Ormeau. She said that someone or something had been thumping the walls of houses and running off, and that people had glimpsed someone looking in their windows. Everyone in the street thought a peeping Tom was responsible. As we will see later, yowies in another notorious hot spot have engaged in similar antisocial behaviour.

Dean also has met other Ormeau residents with stories to tell. A group of teenagers told him that they, too, had been accosted by a huge, hairy, roaring creature. It smashed their cubby house to bits and chased them out of the scrub. That occurred in late 1997, about six months after his own experience.

We could mention several other Gold Coast hinterland incidents, but we have probably made our point well enough. All manner of people – old, young, Aboriginal, non-Aboriginal, tourists, locals, a ranger and at least five people who'd previously never even heard of yowies – have reported seeing the same type of huge, shaggy monster, time and time again, right on the fringes of Australia's most famous beach resort.

The Gold Coast hinterland is certainly one of the hottest of yowie hot spots. There is, however, one that is even hotter.

Hottest: The Blue Mountains

There is little doubt that the hottest of all yowie hotspots is a huge, forest-covered tangle of towering cliffs, razor-backed ridges and deep, dark ravines just west of Sydney: the beautiful, beast-infested Blue Mountains.

The area's yowie-haunted reputation may be due partly to the fact that no other region has received so much attention from researchers. As mentioned earlier, P.J. Gresser collected Blue Mountains Aboriginal yowie lore in the 1950s and '60s, and veteran yowie hunter Rex Gilroy has lived in the area for about 40 years. Rex has collected a vast amount of local eyewitness testimony and has constantly promoted the area as a hotbed of yowie activity. After he first publicised the matter in the mid-1970s a few other enthusiasts began to visit the area and in recent years the number of people going there specifically to search for yowies has risen sharply. At present, as many as 30 individuals and half a dozen groups might be involved, to varying degrees. In recent years, too, several local residents other than Rex have become very productively involved in yowie research.

All these people have documented large numbers of sighting reports and altered public perception to the point where many local eyewitnesses who would previously have remained silent are now prepared to put their stories on record.

But the relatively large number of researchers who haunt the area is not the only reason for its reputation as "Yowie

Grose Valley, Blue Mountains. (Paul Cropper)

Central". The main reason is that the Blue Mountains, like the Gold Coast hinterland, is a place where a huge, impossibly rugged wilderness exists cheek-by-jowl with a large human population. This odd situation is, in fact, even more pronounced in the Blue Mountains than at the Gold Coast.

For Australia's first European colonists it was a case of "first the good news – now for the bad news". On sailing into Port Jackson in 1788, they discovered one of the world's biggest, safest and most scenic harbours. Soon, their first settlement, rum-soaked Sydney Town, was established right on the edge of that beautiful harbour. Off to the west they could discern an equally beautiful range of blue-tinged mountains. Magnificent. As the colony grew, however, they came to realise that the beautiful blue range formed a seemingly impenetrable wall.

At first the barrier was quite handy, as it discouraged disgruntled convicts who might otherwise have set out on foot for China (which some fondly believed was only a couple of hundred miles away). Soon, however, it became a frustrating impediment to the rapidly expanding grazing industry.

Many hardy souls tried to find a way through but whenever they ventured into gaps in the immense wall they found themselves mazed in deep, dark, leech-infested ravines. It wasn't until 1813, when Blaxland, Lawson and Wentworth thought to follow ridgelines instead of valleys, that a way was found through to the western plains.

The veritable goat track that was soon cut through the mountains by sweating, swearing convicts was dubbed the Great Western Highway. Improved slightly in recent years, it is still the main route through. Indeed, it is a measure of the ruggedness of the terrain that since 1813 only one other feasible route (now followed by the serpentine Bilpin Road) has been discovered.

Sydney now contains almost four million people and its western suburbs press right up against the mountains. At that point human settlement is abruptly squeezed down and funnelled westwards in an extremely narrow strip that clings to the ridgelines beside the highway.

Narrow, steep and serpentine, the Great Western Highway is distinctly *not* great to drive. Immediately beside it, hugging the same ridges, is an electric railway that enables residents in the eastern reaches of the mountains to commute to the city. As a result, some of the dozen or so villages and towns strung out through the mountains have become, to an extent, dormitory suburbs of Sydney.

Although dozens of yowie enthusiasts now visit the Blue Mountains, until about 1975 there was essentially just one, the aforementioned Rex Gilroy. Because our own interest in yowies was sparked by Rex's early writings, we engaged in a small amount of research there in the mid to late 1970s. Our relationship with Rex soon soured, however, and, in order to give him as wide a berth as possible, we decided to concentrate on other regions. Throughout the 1980s and early 1990s, therefore, we rarely ventured into the area.

Our attention was not drawn back to the mountains until 1993, when we received letters from a local resident, Neil Frost. He had an amazing story to tell. So amazing, in fact, that it seemed simply too good to be true. Neil had, he claimed, yowies in his back yard. The Hairy Men had been emerging from the forest on a regular basis for about 10 years, growling, roaring, damaging trees, leaving footprints, thumping walls, and even walking on his veranda. Many of his neighbours, he said, had also encountered the creatures. He felt he had a good chance of obtaining proof of their existence and invited us to visit.

All this was pretty mind-boggling, and even though Neil, when we phoned him, sounded perfectly normal, we found his tales of repeated yowie encounters – and even interaction with the creatures – almost unbelievable.

It may seem a little strange that people such as ourselves, who had travelled the world chasing sasquatches, yetis and lake monsters, and who were then in the process of writing a book about yowies, thylacines, marsupial tigers and bunyips, should harbour reservations about any strange animal story, no matter *how* strange. By the early 1990s though, we had grown used to thinking of yowie sightings as a once in a lifetime event for perhaps one lucky person in a hundred thousand. Also, despite being fairly familiar with the Blue Mountains region, we didn't fully appreciate the vast extent of its wilderness and the unique way it impinges on human settlement.

When one of us (Paul) visited Neil and his wife Sandy in late 1993 and camped below their house facing the scrub, he heard a loud, hollow thump of the kind the Frosts associated with yowies. Despite this, and despite the fact that Neil and Sandy were obviously sincere, intelligent people, he found it difficult to accept that a veritable riot of yowie activity could have been happening so close to Australia's largest city. He thought it just possible they'd been hoaxed but did not dismiss the case entirely and intended to revisit the site. At that time however, we were so preoccupied with the publication of our first book that we let the matter slip.

It wasn't until 1999, when we had time for lengthy chats with Neil, Sandy and their neighbours, that the penny finally dropped. The situation in their immediate neighbourhood was exactly as they'd always said it was: yowies and humans had been encountering each other there on a fairly regular basis for a very long time.

The Frosts' story is so remarkable that it is worth recounting in some detail. However, because it is so very remarkable and perhaps scarcely credible to some readers, we should emphasise at this point that Neil and Sandy are not ignorant or hysterical people. At university Neil majored in anthropology and in the 1970s took part in three expeditions to study tribal society in the New Guinea highlands and in the Trobriand Islands. Both he and Sandy are now high school teachers and are the authors of two books

The yowie-haunted Frost residence. (Paul Cropper)

about computing that are standard texts throughout the NSW school system. They are quiet, thoughtful, and as we shall see, both physically and morally courageous.

Their property, which they bought in 1983, is situated on the northern edge of one of the smaller mountain settlements (to discourage uninvited human visitors they have asked us not to reveal the precise location). Even before the first brick was laid, odd things began to happen.

"One day", Neil recalls, "just after the house site was cleared, we found a line of 20 or 30 massive footprints, going straight across newly-turned earth. I can't remember what shape they were because we simply didn't believe they were genuine; we thought someone was playing a joke – with a pair of rubber feet or something. They were deep, clearly defined, and were visible for six weeks. We looked at them but not very closely. I can't remember how many toes ... but they weren't human-like. They were fattish – not thin – they were big."

Although the Frosts and their neighbours have since experienced all sorts of apparent yowie activity and have even seen the creatures on occasion, they have never again, despite many hours of searching, managed to find tracks as clear as those first ones. Neil appreciates the irony: "It was a beautiful set of footprints – funny, isn't it?"

The young couple built the entire house themselves. They moved into it in 1985 and almost immediately became aware of something moving around outside at night. It sometimes sounded like a person running, which they found very puzzling, as the house was surrounded by bush on three sides. Anyone running through that scrub at night would soon scratch themselves to pieces. Neil went outside many times to look around, but for the first few years never saw a thing.

"Then sometimes in winter we'd sit on the veranda. It'd be misting rain, one or two degrees above freezing – and we'd see red glows in the bush. We thought for a long time it was kids having cigarettes and we thought, 'Gee, they're keen – it's so cold and wet.' They were, I'd say, 30 or 40 metres away and just looked like a red glow; at that stage I couldn't resolve two eyes – we weren't looking for two eyes."

Neil and Sandy's house is 110 metres from the road on a 1.5-acre [0.6 hectare] block. "There were no lights outside at all. These eyes were reflecting just the ambient light. Since then I've been really close to a yowie and seen how big their eyes are. They have to be, they're nocturnal animals. They dilate to a massive size. But anyway, when we initially saw the eye-shine we dismissed it altogether."

Then they began to hear thumps near the house at night: "They sounded rather like car doors being slammed so that's what we thought they were for a long while – even though we are a long way from the road and they sounded closer."

It took fully eight years for Neil and Sandy to accept that yowies might be involved, despite the fact that no human prowler could have been behind the disturbances, which occurred night after night, in all weather, in pitch darkness. Like everyone else in the mountains, they had heard of the Hairy Man, which they assumed was just a myth. Now, however, they began quietly quizzing their neighbours on the subject. A remarkable picture soon emerged.

As early as 1993 they managed to compile a list of more than 30 people who had encountered ape-like creatures within a two-kilometre radius. (The list has now grown to more than 60 people). Neil and Sandy kindly introduced us to several of those witnesses and we will describe their experiences later. At this point, however, because some readers may find some of what follows difficult to credit, we should look more closely at the unusual geography and demography of the Blue Mountains in general, and the Frost's local area in particular.

As previously mentioned, the Blue Mountains settlements basically cling to the ridges beside the Great Western Highway. Even the largest town, Katoomba, which occupies the broadest area of reasonably level ground, extends only three kilometres on either side the highway. Some of the other townships, particularly Bullaburra, Lawson, Hazelbrook, Woodford and Linden, are very narrow indeed. Although their town centres are on the highway, many of their residential streets are strung out along razorback spurs that run off either side of the main ridge. Virtually every house along those streets backs directly onto dense bush. The Katoomba 8930 – 1S and Springbrook 9030 – 4S Topographic and Orthophoto 1:25,000 maps show very clearly that, although the situation is most pronounced in the smaller settlements, many hundreds of houses on the edges of the larger towns also back directly onto wilderness.

So, as many residents of the Blue Mountains have been utterly gob-smacked to discover, the only thing separating their conventional suburban existence from the wild world of the yowie is a flimsy wooden fence.

In Neil and Sandy Frost's case, there isn't even a fence.

Neil points out that if a person (a very, very fit or very desperate person) hiked due north from their property he or she would

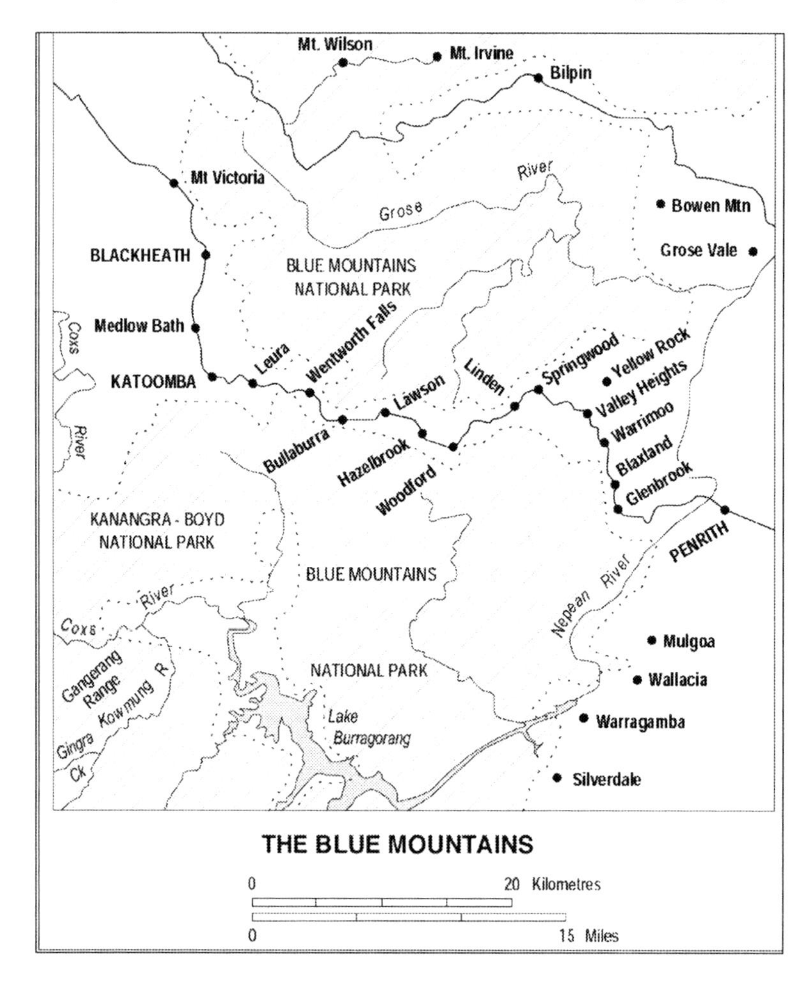

have to battle through 20 kilometres of dauntingly rugged wilderness before striking a road – the two-lane Bilpin Road. After crossing that, the intrepid tiger-walker would have to traverse another 120 kilometres [75 miles] of almost impossibly difficult terrain before reaching the minor road that connects Gulgong and Denman.

That enormous patch of scrub is penetrated by only a few fire trails and is, on average, 60 kilometres wide, which means Neil's hirsute nocturnal visitors have about 8,500 square kilometres [3,280 sq miles] of wilderness in which to hide. In fact, because the area in question is almost entirely covered by the Wollemi and Blue Mountains National Parks, our furry friends will probably be able to stomp around in there with complete impunity for centuries to come.

An indication of the ruggedness of the area is that in 1994 rock climbers discovered a grove of 23 rather alien-looking conifers deep in one of Wollemi National Park's innumerable ravines. It turned out that the trees, which grew up to 40 metres [132 feet] in height, were the only known survivors of a genus that had been thought extinct for 90 million years. As Carrick Chambers, director of Sydney's Royal Botanic Gardens, put it, the discovery of the Wollemi Pines was the botanical equivalent of "finding a small dinosaur alive on earth".

The previously mentioned estimate of 8,500 square kilometres of wilderness to the north is, in fact, very conservative: the 2,000 square kilometre Yengo National Park and the 694 square kilometre Goulburn River National Park, both separated from Wollemi National Park by only a single narrow road, are really part and parcel of the same huge wilderness. So this virtually trackless area is actually more than 11,000 square kilometres [4,247 sq miles] in extent.

And that, as Neil points out, is only the area to the north of the Great Western Highway.

To the south, an area of equally rugged country, which is occupied by the Kanangra Boyd and Nattai National Parks, the southern section of the Blue Mountains National Park, numerous smaller national parks and state recreational areas, adds another 3,500 square kilometres or so for the yowies to stomp around in.

So the wild country bordering the Great Western Highway just west of Sydney totals at least 14,500 square kilometres [5,600 sq miles]. It took us, as we have said, a long time to really get our minds around the enormity of that wilderness, and we suspect that most of our fellow Australians have never really reflected on the matter.

The sprawling city of Sydney, huge as it seems to us, covers an area of only about 1,840 square kilometres. The neighbouring wilderness covers *eight times* that area. Although it has been getting a little crowded lately, Sydney accommodates, in pretty reasonable comfort, approximately four million large, omnivorous primates. Is it entirely inconceivable, Neil asks, that a few hundred similar, if somewhat hairier, creatures could eke out a living in the huge, forest-shrouded wilderness right next door?

In the gully immediately behind the Frosts' house is a small tussocky swamp drained by a stream that eventually joins Grose River in the heart of the wilderness. Throughout the swamp and surrounds Neil discovered numerous pathways where an apparently very strong creature had pushed through the scrub, leaving large, indistinct footprints and, on either side, a trail of saplings broken off or twisted at about three or four feet above the ground. Night after night he strung lengths of black cotton across those tracks, at heights of five to seven feet, and night after night they were broken.

His enquiries turned up the fact that a local man, Roy Smallcoe, had told of repeated encounters with ape-men between the 1940s and 1990. He had often talked

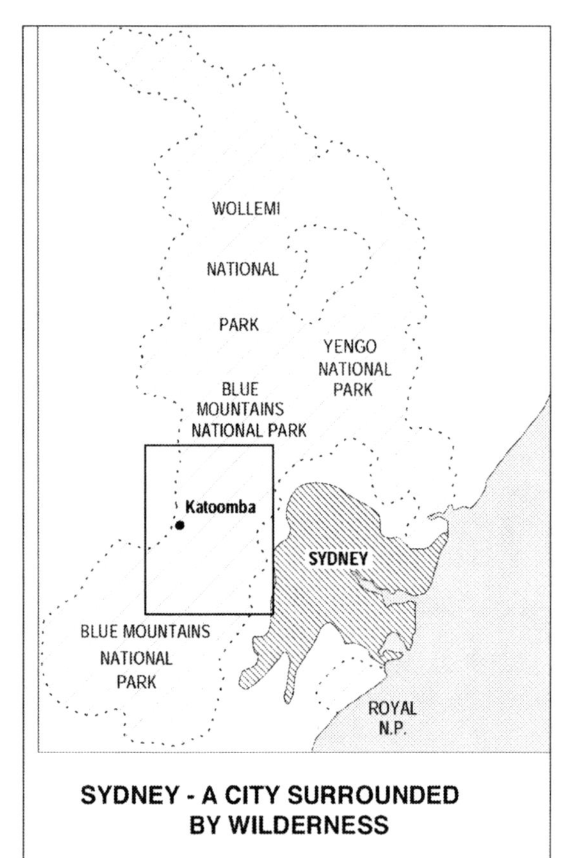

SYDNEY - A CITY SURROUNDED BY WILDERNESS

to other old residents about his experiences but, by an unfortunate twist of fate, died just before Neil could interview him. His friends, however, recalled that he'd said the creatures were the same shape, but larger than humans, that their eyes glowed red at night, and that they were very strong and very fast. He believed they used certain routes through the area and that they'd been forced to change their habits as new houses blocked their way. One misty evening while cutting firewood in the bush, he was approached by a particularly large yowie. Beating a hasty retreat, he left his chainsaw where it lay, and couldn't bring himself to return to the spot for several weeks.

Interestingly, Mr. Smallcoe saw the creatures most often in the swamp that is now just behind the Frost's property – a site that could be of special significance to yowies. On one occasion in 1993, Neil recalls, all the local dogs "went crazy, barked themselves hoarse all night; so we went straight down to the swamp in the morning and found a large area of ti tree and reeds flattened. It was about six metres by seven in extent – flattened, totally flattened. Unbelievable! The same thing happened almost exactly 12 months later."

He and Sandy set out food in the swamp and built a small cubby house for the kids that could double as a hide to observe the main "yowie feeding table" on the edge of the scrub. Food was often taken, but Murphy's Law – or yowie intuition – came into play whenever the lookout was manned, so the midnight snackers were never caught in the act. (The only person who had any success at all was yowie researcher Steve Rushton who later spent an entire night in the lookout. He glimpsed a pair of red eyes in the scrub behind the "feeding table" and heard a loud, hollow thump followed by the sound of something large running through the undergrowth.)

On rare occasions there seemed to be up to three creatures wandering around the property and adjoining blocks but there was usually only one. As the Frosts slowly became accustomed to its visits, it apparently got used to them.

Frequently, after about 1993, the creature would stand still, screened by only a few metres of scrub, as Neil, wearing dark clothes and soft footwear, moved slowly towards it. "In fact", he admits with a laugh, "I used to *talk* to it. Needless to say, I don't mention that very often! It was basically just to convey a non-threatening impression. I used to say things like 'How's the wife and kids? How's the hunting tonight?' A whole lot of stupid stuff. And it would let me approach – I could walk up to within 20 feet usually. On one occasion it talked back, saying 'mook, mook,

mook'. I replied and it repeated the sounds more loudly. I replied again, and it made the same vocalisations again at an even greater volume. A neighbour, Cheryl Price, was with me on that occasion. She commented, 'I think it likes you!' – and then asked to be escorted back to her house.

"I'd often hear it rocking – it would stand there and *rock*. I couldn't see it, but you could follow the shifting of the weight by the sound of the twigs crunching – first on the left, then on the right." Whenever he got too close or was carrying a shiny object like a camera flash or torch the creature would suddenly hurtle away through the scrub at breakneck speed.

Neil, logically enough, sought advice from the Aboriginal community. "Eventually a well-known tribal elder [name withheld by request] from the south coast phoned me a couple of times. He started describing our place without ever having been here. He said things like, 'How far from the swamp are you?' He knew what to ask, knew what he was talking about. He said Aborigines saw the creatures as 'the protectors of the environment'; they don't like trees being cut down or things being built. In fact, it *did* seem to act up when the neighbours put their fence in – and when we installed a high gate.

"He told me a lot of things, but mainly to watch out in terms of the kids – that there was an association with young children. That was a bit scary. I asked whether they would harm the kids, but he said 'No – they're just fascinated by them.' He asked if we'd lost any kids' clothing; I said 'No – but I've found some.' It turned out they belonged to a lady up the road. She said 'something big' had been in her house at two o'clock in the morning. She'd heard the steps creaking under the weight as it went up, walked around the kitchen and went out again."

Even for the unflappable Neil things were becoming a bit unsettling: " It really dawned on me then that we had a problem here. You see, our daughter Avril had apparently seen the yowie behind the house – in broad daylight, 11 am. She was then only two and a half years old, but spoke very well for her age. 'Daddy,' she said, 'there's a man down the back.' I immediately went down there with our Collie who sniffed around, bristled up, growled and ran back to the house. Later Avril said the 'man' had 'long, yucky hair – needed cutting'. So we banned the kids from going down the back and called the police."

Initially Neil complained only of a persistent prowler. Eventually, however, he revealed the true situation and was gratified to find that the police (after running a few character checks) took him seriously. Most Blue Mountains-based police are well aware of the yowie phenomenon and it seems almost certain that some have actually seen the creatures. In any case, local officers often staked out Neil's property and assisted him in evaluating evidence, some of which was photographic.

In mid-1993, he constructed an automatic camera trap: "It was an Olympus OM2n with a T20 flash, triggered by an infrared motion detector powered by a car battery. If anything intruded within 15 metres, it would take a photo – but it could take only one shot per night."

Before long the device succeeded in photographing what looked like a dark, non-human face peering out from behind a couple of saplings, about six feet above the ground. Murphy's Law, of course, ensured that the creature was at the extreme edge of the camera's effective range, so the image is far from clear, and Neil is the first to admit it would fail to impress anyone inclined to be sceptical. To be fully appreciated it needs to be viewed in colour and magnified several times, so we have not attempted to reproduce it here. But while it is murky, when viewed in optimal conditions the "face" has quite a startling impact. It is not the face of an ordinary Australian animal. As it also seems extremely unlikely, for a host of reasons, to be a man in a mask, we think it may well be, as Neil believes, the face of the legendary Hairy Man. Another consideration is that yowies have since been seen twice, in broad daylight, within just a few feet of where the camera was set up. [Cases 261 and 262]

Because the creature always gave the camera trap a very wide berth, Neil now believes, as do others who have attempted to use infrared devices to photograph yowies, that the creatures' eyes can detect IR light.

A second photo was also taken at night, "as I was stalking it, with a camera held over my head to clear the underbrush". This shot shows only two red eyes that, to the casual observer, appear rather small and distant. It did, however, interest the police, who took the camera, did some measurements and photographed a constable next to the saplings in the picture. They deduced that the creature was about 13 metres from the camera and that its glowing eyes were 6 feet 1 inch [185 cm] from the ground. "Of course", Neil points out, "the thing could have been crouching ... plus, say, another eight inches to the top of its head. So I'd say it was 6 foot 10 or 6 foot 11".

Bright eyes: Neil Frost's second photograph.

While there are often plenty of likely looking imprints in the scrubby swamp, its tussocky ground cover is not conducive to clear footprints. Away from the swamp, the generally dry, stony ground is equally unfavourable. Even so, as Neil is quick to admit, it seems very odd that clear footprints are almost never found. Part of the explanation could be that yowies are, by nature, extremely cautious. Neil once spread sand along part of a game trail in the swamp. He found that while wallabies and other animals walked straight across it, something much heavier repeatedly pushed through the scrub to one side, creating a detour. The creature(s) also gave a wide berth to trip wires Neil rigged up to a second camera trap.

The Frosts' best track find since 1984 occurred on February 22, 1995, when a line of footprints appeared on a patch of disturbed earth a few metres from the nearest neighbouring house. The neighbour had been awakened the preceding night by heavy footsteps and the sound of something scraping the bedroom wall. Interestingly, in view of what the Aboriginal elder said about the yowie's fascination with children, the neighbour's wife was at that time breast-feeding an infant.

There were eight tracks in all: "Left-right-left combinations; the left and right were mirror images of each other." Their size (11 x 7 inches) was what one might expect from a large, bipedal creature, but they were a very odd shape. They exhibited what looked like toes, and were distinctly non-human but were – it must be said – not a lot like sketches or casts of other supposed yowie tracks. It is interesting to note that, after Neil made a plaster cast of one track and placed it on the front veranda, his dog Bess growled and retreated from the object. It seems that, like Dave Reneke's dog at Kempsey [Case 203], she was reacting to odours in soil adhering to the plaster.

A plaster cast of 'Fatfoot's' track. (The heel has broken off.) (Tony Healy)

schoolteacher, but the two became firm friends. Soon they were not only waiting passively for the yowie but also chasing it, hell for leather, through the swamp. Neil was hoping to obtain clear photographic evidence; headstrong Ian was determined, if he couldn't wrestle the creature into submission, to at least rip out a handful of hair for analysis. "I was really inspired by Ian's courage", Neil recalls, "He would charge in there, full pelt. God knows what would have happened if he'd managed to grab it!"

They tried everything: stalking it, outflanking it, chasing it into the open – but it eluded them so easily that they became certain it could see perfectly in the dark. They were also convinced the creature actually enjoyed the contest. Its intelligence and sense of fun seemed quite evident. It would often use the sound of wind in the trees to conceal its footfalls, halting for minutes on end when the wind died. While being stalked it would occasionally throw a stone or branch some distance to baffle its pursuers. When pressed, however, it would use its tremendous strength and speed to charge away.

The men soon deduced that the mysterious, hollow "booming" sounds occurred when the creature wanted to attract their attention. It apparently created the sound by stamping its feet. They found they

As we have mentioned in relation to other cases, the question of exactly what a yowie's footprint should look like is not easy to answer. For the moment we will file Neil's footprints, along with several others, in the "Too Hard" basket, and look at them again later.

At about this time the Frost's nearest neighbour, Ian Price, a brawny, bearded, heavily tattooed biker, began to join Neil on his nightly vigils. Wildly extroverted, witty and impulsive, Ian seemed an unlikely sidekick for the quiet, thoughtful

Neighbourhood watch: Ian Price and Neil Frost. (Tony Healy)

could approximate the noise by hurling large flat rocks against the ground.

Once, while checking another area about 500 metres away, Ian was lucky enough to encounter a yowie that stepped out of the bush onto a fire trail. It was about 20 metres away, clearly silhouetted by bright moonlight and visible from the ankles up. It had long arms and legs and the by now familiar very short neck, and looked about the height and width of a normal doorway. It swiftly disappeared into thick scrub.

Over the years Neil, too, has glimpsed a few large silhouettes and experienced brief sightings of the creatures "from shoulder height and up, and a couple of times from bum up". A good sighting occurred in June 1993.

"One night some friends came for tea. They got to the door and said, 'Do you want to hear something weird? Something just followed us down the driveway!' I said, 'That's no problem – it's just the yowie'. And they said, 'You're kidding!' So after tea I took the other bloke down the back and, sure enough [we heard] the thing go up to the back of our neighbour's house where it would have been eating apples off his tree, sure as eggs. You could see the bloke up there, lit up in his kitchen window, washing his frying pan, clear as anything – it was quite comic. He dropped the pan and as soon as it hit the sink this thing took off.

"It ran down the block, slowed to a trot, then a walk, then stopped. It had found a tree with a currawong asleep in it, and it shook the tree. The bird let out a distressed call as the tree was shaken, came down through the branches flapping its wings and hit the ground. For a while, then, there was a deadly silence. Then the creature walked to the swamp and stopped again. I walked in a direct line to where I'd heard it... my mate was a few metres behind me. Not knowing how close I was, I lifted the torch up over my head and turned it on.

"It had been sitting down. It stood up. Literally six feet away from me on the other side of a bush. It was side-on and it was taller than me, maybe up to seven foot. It turned its head towards me, leaned forward over the bush – and *roared* right in my face. It was like a lion's roar – that's the closest thing I can compare it to. It was very deep and very loud. And instantaneously, up and down the valley, every dog went berserk.

"Despite the close proximity of the encounter and not wishing to play down its effect upon me, I cannot remember many specific details of the face and head. All I can remember with certainty are some general characteristics and a very few specific ones. In particular, I recall that the head was very large; at least 50 per cent larger than a human's. Surrounding the head was a thick layer of black hair, which may have added to the overall impression of size. The skin was black, which with hindsight implies the face was hairless. It had a nose. What the shape was, I couldn't say with certainty, but I think it was broad. The mouth was extremely large. I stood there and looked at it for ... moments seem to last an eternity ... probably a couple of seconds – and then it took off and *carved a path* through the bush.

"I had an unusual sensation, a kind of 'fight or flight' – and I felt 'fight' – so I chased it. I ran back towards my mate and shone the torch – a big, powerful quartz halogen lamp – on its head. Lit it up like Christmas. All we could see was its head towering over the bush as it ran, pushing things over. Every dog was going crazy. I remember seeing its head *snapping* around and the red eyes looking at me and it was *roaring* at the same time - *and* grunting. I think these things have got such a huge diaphragm and such a mass that the grunting is an involuntary thing once they get moving."

Neil's mate was so bowled over by the night's events that he subsequently rigged up a system of lights high in trees all around the property. At the flick of a switch they lit up the whole area. He and Neil then sat up night after night with a camera, but saw only distant eye-shine. Then Neil arrived home one day to find all the lights had been knocked down, apparently with a long, spear-like stick, which had been left poking into a tree.

Far from being frightened off by all the attention, the yowie, which the Frosts' had now dubbed "Fatfoot", began to venture closer and closer to the house. Neil had the impression the creature was playing a bit of a game. He had intruded into its swampy, forest-shrouded world; now it was intruding into his. It started to bash the side of the cubby house, making a loud and distinctive sound. "There was no doubt in my mind", Neil insists, "that it was the weatherboards being hit by a hand." Then it hit the house itself.

"We'd built a brick wall along the side, partly as a fire break but also to keep the damn thing away from the house. It began to thump the wall and when it did – that wall is close to 30 metres in length – the whole wall would vibrate and resonate."

The yowie's next move was downright unsettling: it ventured onto the veranda: "He's been on it five times, walking around, thumping his foot." Neil and Sandy reacted with commendable calmness. Refusing to panic, they left a sound-activated tape recorder outside and managed to tape noises that, though not of great value as evidence, do indeed sound like a large bipedal entity treading the boards. Once, when he heard the floorboards creaking, Neil opened the air-conditioner's exhaust vent to the veranda. "I put my ear to it and could actually hear him breathing. I went out and he took off straight across the driveway into the bush."

One night in late 1993 or early '94 a neighbour, nine-year-old Jeffrey, got a fairly good look at Fatfoot's face. It happened during a barbeque about 150 metres from the Frost property, when he took a flaming stick and went into the scrub exploring. As he rounded a stand of trees he was shocked to find a huge hairy creature, presumably

Fatfoot, standing just a few feet away. In contrast to its usual reaction on meeting humans, the yowie initially held its ground, as Jeffrey stared up at its face. Its head was big and its face, largely free of hair, was black and wrinkled, with deep folds, and looked almost emaciated. Surrounding the face was a thick layer of dark brown or black hair, with white or grey flashes at the side where ears could be expected to be located. After a moment, it turned and ran, as did the boy.

In the late 1990s, yowie hunter Dean Harrison, who'd had his hair-raising introduction to the phenomenon in southern Queensland, moved to Sydney. After contacting Neil and other informants, he began visiting the Blue Mountains three nights a week, staking out known sites and searching for evidence.

Before long, he experienced near Neil's property several encounters with a creature, which, though always just out of sight, appeared to be bipedal, very large, and very fast. Because it was non-aggressive and seemed to enjoy playing "hide-and-seek" for hours on end, he feels sure it was Fatfoot.

Dean often focussed on an area about a kilometre north of the Frosts' property and linked to it by a discernable game trail. One night he and two friends spent six hours there, playing cat and mouse with the creature. As usual, though its footsteps were clearly audible, it managed to stay just out of sight. Eventually, near exhaustion, the men slumped to the ground. As they sat quietly, gathering their strength, Dean realised that the yowie had also suddenly gone very quiet. Standing up, he held a spotlight out in front of him and switched it on. Luck was on his side: cutting straight through the gloom, the powerful beam hit Fatfoot square in the middle of his big hairy chest.

Apparently curious about what its pursuers were up to, the creature had crept much closer than usual. It now stood transfixed, just 15 metres downhill, and a metre or so from the closest tree. It was between six and seven feet tall and covered in dirty, matted, two-inch-long, reddish-brown hair. Its immense frame was as muscular as that of a weightlifter: "huge chest, shoulders like cannon balls". The skin of its face, where visible, appeared to be quite black. As Dean struggled to absorb the details, it "stepped very, very slowly behind the tree, then dropped to all fours and moved backwards into the undergrowth. There was no eye shine, no smell".

Although the creature's curiosity overrode its natural caution on that occasion, the incident nevertheless illustrates how elusive yowies normally are. Dean, who has led dozens of expeditions to many other hot spots, has never had another sighting anywhere near as good as that view of Fatfoot.

Neil and his neighbour Ian believe that Fatfoot's ability to afford the luxury of extended periods of "play" suggests that it and its hairy brethren enjoy an ample supply of food on the fringes of the mountain townships. "Most probably", Neil says, "the increased supply of food is closely linked with human settlement."

He points out that many of his neighbours, like hundreds of other residents of the mountains, have fruit trees overhanging their back fences. Many complain of fruit disappearing *en masse*, just as others complain of devastating raids on their vegetable gardens.

Yowies also seem partial to dog food. On several occasions during the early 1990s the Frosts noticed that Bess, though clearly hungry, was afraid to approach her food bowls, which were within a couple of metres of the scrub. The bowls, furthermore, kept disappearing. Thinking that a big, clever dog might have been carrying them off, Neil initially replaced the bowls with containers with sloping sides. Later, in frustration, he began using cheap, disposable containers. Meanwhile, Ian was experiencing similar problems: his guard dog Toby's food bowls often disappeared – from his fully fenced yard. Ian eventually chained a large double handled metal pot to a strong tree. That worked to an extent: although the pot was never completely carried off, it was often found empty, hanging over the fence by its chain. Many months later, on a neighbour's land about 50 metres to the north, Neil "...came upon a large collection of Bess's old feed bowls. They were concealed within a thick stand of trees and bushes."

Other pets may have more to worry about than stolen food: Neil has heard several stories of domestic cats vanishing without a trace. That, perhaps, might not seem so very strange, except that he once found large amounts of cat hair, arranged in a neat circle, in the yowie-haunted swamp.

A couple of yowie sightings, he says, have been specifically associated with raids on chicken coops. The scale of the destruction was beyond anything that could have been caused by quolls, goannas or foxes. Rabbit and guinea pig cages have also been torn open and their hapless occupants carried off. The bars of one cage, that would have defied the strength of any man, were spread wide apart.

Human settlement, Neil points out, has not only provided yowies with tasty domestic animals, it has also increased the availability of bush tucker. The local population of swamp wallabies, he says, "has multiplied during our time in the valley", due mainly to the acres of lush grass on hundreds of unfenced lawns and roadsides. The possum population, too, has exploded. Based on a rough count in his neighbourhood, Neil estimates there are as many as 800 possums per square kilometre throughout the settlements. Rich pickings for a strong, resourceful creature that can see in the dark and isn't particularly fussy about what it shoves down its throat. Neil suggests that, as a consequence, there are probably many more yowies near the settlements than there are in the heart of the wilderness.

Tree bites

One of the most problematical aspects of yowie behaviour is their supposed habit of biting into the trunks of trees to get at wood-boring grubs. Neil, Dean and several other researchers attribute this behaviour to yowies.

In the Blue Mountains the damaged trees are almost always Yellow Bloodwoods (*Corymbia exemia*). The Bloodwood's timber, though tough, is soft in comparison to other Australian hardwoods and is very susceptible to insect larvae infestations. The supposed yowie bites usually appear on young trees with diameters of four to eight inches [10-20 cm]. They certainly look impressive: usually between four and seven feet above the ground, they are quite uniform in size (about the size, one imagines, of a large yowie's mouth) and in some cases seem to display fang marks. In the centre of each bite there is a little hollow once occupied by an unfortunate witchetty grub.

There are, however, at least two good reasons for scepticism. Firstly, no one has ever reported seeing a yowie making such a bite. Secondly, and more importantly, Black Cockatoos have often been observed creating, with their amazingly strong beaks, similar tree damage as they forage for grubs.

Although he, like his colleagues, is well aware of this aspect of Black Cockatoo behaviour (and has, indeed, observed the feathered fiends ripping the daylights out of trees), Neil still maintains that many of the bites are made by yowies.

He reasons that:

• Many of the bites have been found in exactly the same areas where he and his neighbours have seen and heard yowies.

• Some of the bite marks have been made at night, when all good cockatoos are tucked up in bed. In March 1996, one very well defined bite mark was inflicted on a tree right next to the Frost's driveway between 8 pm and 4 am.

• In several of the neatest bites, the marks of what look like four large canine teeth are visible. These furrows are uniformly spaced: In many of the bites around Neil's property (the ones apparently made by the individual known as Fatfoot) the upper set are 60 mm apart and the lower set 55 mm. We observed and measured several of these "fang marks" ourselves, and found them impressive. It was difficult to imagine a cockatoo creating such identically spaced gouges on tree after tree.

• Cockatoo-inflicted tree damage is usually very messy; saplings are commonly ripped, shredded, and left riddled with bite marks of many different sizes and shapes.

• Whereas cockatoos frequently gnaw trees from their base right up to their higher branches, the large, neat "yowie bites" are almost always between three and seven feet six inches (1 – 2.3 m) from the ground. In fact, approximately 80 percent of the bites near Neil's property are six feet above the ground (give or take a few inches).

• Many of the bite marks have appeared alongside the game trails that Neil believes are used by yowies. Black cotton stretched at a height of six feet across those paths has been broken night after night. In early 2001 Neil and some friends used a Global Positioning System to establish the exact location of hundreds of apparently yowie-bitten trees. When transferred onto a map, the data revealed a linear distribution that, says Neil, "highlighted the creatures' habitual routes [between the swamp and other local hot spots]. There was a close and obvious association between the bites and other established indicators [of a yowie's presence] like twig snaps and larger branch breaks".

• Large imprints that may well be yowie tracks have been found under many of the damaged trees. Although most are ill-defined, we have been shown one or two that appear to display toe marks.

Neil has cut sections from several bitten trees and sent them to academics in Australia and overseas. He has also approached scientific organizations including the CSIRO. The boffins, however, ignore the fact that some of the bites were made at night and other inconvenient details and always attribute the damage to Black Cockatoos. "The preferred method of disposing with the evidence", Neil resignedly observes, "is to make it fit into the appropriate paradigm, no matter how badly."

Only one academic had enough scientific curiosity and self-confidence to visit the site and examine the bites. Dr. Helmut Loofs-Wissowa, now retired, was Reader in Asian History and Civilisations at the Australian National University. A close friend of "the father of cryptozoology", Dr. Bernard Heuvelmans, he has been actively involved in the field for many years. As well as researching the yowie phenomenon, the remarkably fit 78-year-old French Foreign Legion Indochina veteran has taken part in expeditions to Vietnam and Laos in search of hairy hominids rumoured to exist there. He agrees that the interpretation of the bites as being made by yowies makes eminent sense and should be retained as a working hypothesis.

Few people know the bush better than environmental consultant Gary Opit, who has spent nearly five decades studying wildlife in every corner of Australia, as well as parts of New Guinea, Malaysia and North America. After examining the relevant tree bites he declared, "I'm positive they're not made by Black cockies, which are always messy. Nor were they made by the Yellow-tailed Glider". He, too, comments on the "fang marks". They conform, he notes, to "a particular pattern: evenly-spaced parallel cuts across the wood. The marks look like they could have been made by two large canines".

Pat Ryan, a mountaineer, outdoor educator and lifelong Blue Mountains resident who assisted Neil in the GPS work, agrees. He has found many similar bite marks alongside well-trodden game trails in the heart of the wilderness. The grass below many of the bitten trees, he notes, is trodden down, as if by something quite heavy. He sometimes finds strange stick formations close by: usually crossed sticks piled up in the forks of trees; sometimes free standing pyramid-like structures. "I've seen teepee-like structures formed by the crown of small trees, like the larger banksias, pulled down around the trunk".

Our only Aboriginal reference to tree bites was passed on to us by "J" [Case 183] who was told by Aborigines near Hebel, Queensland, that the hairy giants "leave bite marks in trees at a height higher than a man".

Although it may be difficult to imagine any ape-like animal casually chomping through Bloodwood, we have one good report that illustrates the enormous power of the yowie's jaw. Stan Pappin, who observed a huge yowie at close range near Gootchie, Queensland, later examined a steer that had apparently been killed by the same animal. "It had eaten the rear half," he told

Tasty grub: one of many apparent yowie bites. (Tony Healy)

us, "bones and all. The leg bone was bitten clean through, so it must have had terrific jaw strength." [Case 172]

More yard invasions

Most Blue Mountains yowie reports, like stories from other areas, concern unexpected encounters by people who had never before, or since, had any contact with the creatures. We will relate several of their stories, some of which are extraordinarily interesting, towards the end of this chapter. At this point, however, we will describe the experiences of two other families whose properties, like that of the Frosts, have for some reason been visited time and again by the legendary Hairy Man.

Out of the Blue Labyrinth

Jerry and Sue O'Connor live in the same township as the Frosts. Their house, however, is on the extreme southern side of the settlement, on the edge of a vast maze of twisting, scrub-covered ravines known as the Blue Labyrinth.

Since moving into the area in September 1997, they have experienced a wide range of yowie activity strikingly similar to that experienced by the Frosts. The O'Connors' story is, in fact, even stranger in some ways than the Frosts'; so strange that many people will find it almost impossible to accept. As we will see, however, everything the O'Connors have reported, even the very oddest detail, has been mentioned at some time by other witnesses. We now know Jerry and Sue well. They are, like the Frosts, decent, honest and courageous people.

Some aspects of the activity around the O'Connor property are so remarkably similar to the Frosts' experiences that it is tempting to wonder if the same very curious yowie or the same group of yowies is/are involved at both locations. Despite being on the other side of the Great Western Highway, the O'Connors' house is less than two kilometres from the Frosts' and, as we will learn later, yowies have been seen crossing that road in the wee hours of the morning. Be that as it may, the yowie activity began almost as soon as Jerry and Sue moved to the area.

It began when they heard something that sounded the size of "a bull elephant" crashing through the surrounding scrub. Then, on several occasions, in the dead of the night, their screen door rattled, the lid of their power box slammed and someone or something pounded on the weatherboard walls. Prior to and during these disturbances, their cats invariably "Snarled, hissed, generally freaked out".

Next came the decidedly unpleasant sensation that something was stalking them as they walked, very early in the morning, along the badly lit, scrub-fringed road towards the Hazelbrook railway station.

Like the Frosts, the O'Connors have no back fence and, like Neil and Sandy, they first assumed they were the victims of a very persistent human prowler. That theory began to seem a little inadequate after November 1999. One evening that month Sue, while on the back verandah, noticed a movement behind a large bush in the backyard. Jerry stepped outside to investigate but was stopped dead in his tracks by "the most God-awful *roar*", similar to that of a lion but indefinably alien. It was immensely loud and utterly terrifying. Jerry, six foot three and solidly built, is not a timid man, but on that occasion, when getting ripped limb from limb seemed very definitely on the agenda, he decided to proceed no further.

It should be noted here that in the colonial and early modern eras, the native Gundungurra people and white pioneers of the Blue Mountains were sometimes thrown into utter panic by enormously loud, lion-like vocalisations like those described by the Frosts and O'Connors. According to M. Feld, writing in 1900, the Gundungurra attributed the sound to *gubba*, "a wild, hairy man with feet turned backwards". In *Cullenbenbong* (1940), the renowned bushman Bernard O'Reilly said the vocalisations had the quality of infecting all who heard them, even "men who were hard to scare … whose lives were wedded to the lonely bush", with "unreasoning terror". Dogs, too, "never failed to show a complete panic [and] the most staid horses invariably bolted." [21]

Although it was now apparent their nocturnal visitor was some kind of wild creature, the O'Connors were very reluctant to believe it might be the "mythical" yowie – until they actually saw it.

The first sighting occurred in August 2000, when Sue awoke at 2 am, looked up at the small, high window to the left of their bed, and saw a huge creature staring back at her. It had a human-sized head, but because the head was set low into a pair of absolutely huge shoulders it looked disproportionately small. Thanks to a nearby streetlight and a full moon, she could see the creature had a slim nose, very wide mouth and a rounded clump of tan-coloured hair on top of its head. She had absolutely no doubt: whatever the peeping prowler was, it was not human.

In the morning Jerry went outside with a tape measure. To peer through the window as it did, the creature had to be over eight feet tall. It was then that he and Sue finally accepted that their mysterious visitor was indeed a yowie. The realisation knocked Jerry, a down-to-earth veteran of 20 years Navy service, for a six. "My whole life spun on its axis," he said, "it changed my whole belief system."

Because of the unusually bright moon on that first occasion, Sue's initial sighting is still their best, but since then both she and Jerry have seen the creature, or at least its silhouette, at the window on six other occasions.

Completely unaware that similar events had been occurring for years only two kilometres away, Jerry began researching the subject on the Internet and soon contacted

Sue O'Connor's sketch.

Dean Harrison at Australian Yowie Research. Dean quickly put him in touch with Neil Frost who suggested he and Sue search the surrounding scrub for tree bites like those that had appeared around his own property. They soon found 30 of them.

Neil and Dean visited the site and confirmed that the apparent bite marks, like those at Neil's place, were almost all four to seven feet above the ground in young bloodwood trees (and the occasional mountain ash). In the middle of each bite mark was a small hollow that had contained a wood-boring grub. In several of the bite marks grooves were visible that looked as if they'd been made by the upper and lower canines of a very large animal. As the presumed upper canines were 80 millimetres [3.2 in] apart, it appeared the creature was considerably larger than the one that haunted the Frost's property.

Like the Frosts, the O'Connors have observed Black Cockatoos feeding and are convinced they aren't responsible for all of the tree damage. They, too, are sure that at least some of the trees have been bitten during the hours of darkness. They have found apparent footprints around the base of several damaged saplings. Though frustratingly indistinct, the 16-inch (40 cm) imprints were too large to be those of a man.

After a couple of nocturnal visits, with no sign of aggression from the creature or creatures, Jerry and Sue began leaving out food – "peanut butter rolls, apples, bananas; anything we thought a yowie might fancy"– in buckets suspended seven or eight feet above the ground. Although the food disappeared, it was possible that agile and ingenious possums had taken it. So Jerry and Sue took the additional step of stringing lengths of black cotton at heights of six feet or more between the trees. The food was always taken, the thread frequently broken.

Jerry O'Connor indicates the height of the nocturnal visitor. (Tony Healy)

Other broken threads indicated that to get to their bedroom window the creature normally walked along the side of the house via the backyard. For several weeks they hid four surveillance cameras, loaned to them by Dean, at various key points around the yard. The whole scene was "illuminated" by infrared light, which is invisible to the human eye, while the cameras operated continuously at the rate of one frame per minute.

During those weeks the yowie approached only once. Choosing the only night when a narrow quadrant was not covered by the cameras and lights, it reached the bedroom window at 2:15 am and was seen peering in as before. To get there it must have circled right around the front of the house where it would have been visible from across the street had any neighbour been awake.

This incident seemed to confirm what Neil, Dean and others who have tried photographing yowies with infrared devices had long suspected: that the creatures' eyes can actually detect IR light. Either that, or, as Jerry remarked in frustration and not entirely in jest, "the buggers can read your mind!"

Midnight creeper

Although the camera-shy critter easily avoided being photographed, Jerry and Sue did manage to capture some of the its shenanigans on audiotape.

On October 25, 2000, they suspended a feed pot by fishing line from a branch that hung over a six metre [20 ft] cliff, approximately 500 metres down a ridgeline behind their house. "It is", Jerry explained, "the only access point from the valley floor for about a kilometre or so – the rest being vertical rock face." Just below this little cliff was the place where local man Brad Croft had seen a yowie only 12 months earlier [Case 232]. Further down is a small swamp. Intriguingly, a well-worn game trail runs up from the swamp to the cliff and then proceeds directly to the O'Connors' backyard.

"At dusk on the 25th", Jerry continued, "I hid Neil Frost's voice-activated tape recorder inside a hollow stump, which I thoroughly camouflaged with branches, leaf litter, etc. I placed food in the pot, which was so high I could hardly reach it, and we left for the night." Throughout these proceedings Sue noted, "our cats were behaving very strangely, making odd meows and staring at the foliage below." They refused to walk any further down the track, something they'd been comfortable doing on the preceding two days.

What follows is Jerry's interpretation of the sounds that were recorded that night (an interpretation that we, having heard the tape, consider entirely reasonable).

25 Oct

20:10: Sounds of us leaving the area.
Between 20:10 and approximately 23:15: The sound of bipedal footsteps entering the area (crunching leaf litter and twigs).
Between 23:15 and 23:45: The distant sound of the first coal train of the night (a useful time check).

26 Oct

Approximately 01:00: The sound of the second regular coal train.
Between 01:00 and 06:00 the recorder was activated about six times:

* "Yowie" discovers tape recorder. Sound of camouflage branches, etc. being pulled away. The sound clearly demonstrates the dexterity of the fingers as it examines the recorder. Plastic bag that the recorder is inside is ripped by what appeared to be a claw or sharp fingernail.
* Tape activates a couple of times. No sound detected. Then the unmistakable sound of the feed pot being "belted" with what sounds like a stick.
* Sound of birds, plus a plane or helicopter.

06:00: Sound of my footsteps as I arrive at cliff (dawn was at 05:45).

On arrival at the site, I could not see any sign of the feed pot, also all of the branches, leaves, grass etc were scattered everywhere, as if someone/something had furiously searched for what we imagine was the sound of the tape recorder activating. I finally found the feed pot about 30 or 40 feet away, down below the cliff. It had a big split in the side. All the fresh rolls were missing, plus all the

apples. Interestingly, the two stale rolls were untouched.
We feel sure that what was recorded was a yowie, for these reasons:

1. The area in question is practically in our backyard, and, since settling here in 1997 we have never seen *anyone* in the area. We have only about 12 houses in the entire street, mainly elderly retirees. For the most part, the street is dense bushland.

2. No one apart from Sue knew where I had hidden the recorder. It was hidden so well that you could stand right beside it in broad daylight and not see it. I also covered the recording light with thick masking tape. *The sound of the recorder activating was inaudible to the human ear.*

3. While setting up the recorder the following evening, Sue positively spotted a reddish-brown, hair-covered, man-like creature dart between trees in the valley below. It was hunched over, with both arms dangling out in front, and moved with lightning speed. [Jerry has also experienced a brief daylight sighting of the head, shoulders and upper back of a hairy biped disappearing into the scrub behind their house.]

4. In addition to the seven occasions when we have seen the head and shoulders of creatures at our window, we have now had three or four other encounters near our house. We have heard howls or shrieks emanating from this valley, and have found many large footprints.

5. If the visitor taped on the night of October 25 was a human, don't you think he would have kept the brand new recorder he found out there in the bush? Instead, it was uncovered and left in place inside the stump.

6. Strange that only the fresh food was consumed. I don't know of any possums that are so fussy.

7. Neil and myself tried to replicate the effect of the pot being knocked down by a possum. We dropped the pot from the same height seven or eight times, but it never landed any more than 10 feet away.

8. At the time, no one apart from Dean Harrison and the Frosts knew of our interest in yowies. We had kept our encounters to ourselves out of fear of ridicule.

9. Finally, we both swear that neither of us has perpetrated a hoax of any kind, and are quite prepared to take lie detector tests on this, or any other statement we have made concerning this subject.

The forest floor in the vicinity of Jerry and Sue's block is rough, hard and not particularly conducive to footprint discoveries. They have, however, as already mentioned, found many partial tracks. The best of the partial tracks, though incomplete in the heel, displayed five toes and was larger than Jerry's booted foot. The only complete track appeared in a patch of waterlogged turf beside their house. Strangely, however, that footprint – which is 35 cm [14 inches] long, was clearly *three-toed.*

Sometime after that discovery, a police officer (who wishes to remain anonymous) found a single track in the bush below the house. Although the cast shows only the heel and ball of the foot (there was a rock under the toes), it matches the heel and ball – and a distinctive "V" crease in the middle of the ball – of the O'Connors' three-toed track. We will discuss these puzzling tracks further in chapter six.

Jerry and Sue have now been receiving visits from what they believe are three yowies for more than five years. They have become used to the strange visitations and they cite several instances of apparent communication. Sometimes the communication is quite basic. Late one night, when the cats suddenly bristled and a neighbour's dog began barking, Jerry impulsively pounded on the bedroom wall and yelled, "How are ya goin' mate?" Two nights later a tattoo of knocks came by way of reply.

More recently he has interacted with the creature(s) via patterns and structures he makes with poles up to 15 feet (5 metres) long. He creates, say, a star pattern or a tepee-like structure and frequently finds the patterns have been altered overnight. Sometimes the sticks have been used to gouge six-foot long, three-inch-deep ruts in the ground. He feels the yowie(s) enjoy the game.

Strange: the five-toed track. (Jerry O'Connor)

By any standards, what we have written so far about the O'Connors' experiences is pretty strange. Some other elements of their story, however, are stranger still. So strange, in fact, that we will put them aside for the moment and return to them in chapter seven.

Stranger: the three-toed track. (Paul Cropper)

Although we may find the mindset of dyed-in-the-wool sceptics exasperating, they might still argue that in the absence of irrefutable photographic or other evidence it is impossible to rule out the possibility that the Frosts and O'Connors were the victims of hoaxers. We believe, however, that most readers, even mildly sceptical ones, will acknowledge that much of the phenomena that has been reported time and again by both families – the roars, the tree bites, the uncanny evasion of infrared light, etc. – very strongly indicates non-human activity.

It may be difficult to believe that huge hairy ape-men are invading backyards virtually on the outskirts of Sydney, but when one knows and trusts the principal witnesses as we do, when one hears supporting evidence from other local witnesses, and when one takes into account the length of time that these events have been going on – 20 years for the Frosts, eight for the O'Connors – it is even more difficult to believe that human hoaxers could have been responsible.

Neil Frost thinks that over the past hundred years or so, many other Blue Mountains residents must have experienced yowie "yard invasions". Privacy concerns and fear of ridicule have, he believes, caused nearly all of those stories to be "buried". He was, however, able to put us in touch with a third family that has experienced repeated yowie visitations and was willing to talk about them.

The tree ripper of Yellow Rock

Hardworking, prosperous, good-humoured and hospitable, Lynn and Gordon Pendlebury live with their four children in a large modern residence on several acres of land on Purvines Road, Yellow Rock. The property is at the extreme eastern end of the Blue Mountains and only three kilometres from the Nepean River, which now delineates the western edge of Sydney.

Although they have lived there since 1980, Lynn and Gordon didn't experience anything out of the ordinary until 1995. At that time their daughter Zoe, then five or six years old, began waking them at night to say that something was knocking on the outer wall of her room.

On checking outside they found, to their dismay, a broad patch of flattened grass below her window, with a discernable trail winding away into the scrub. There, amid the trees, they found a spot "where something had been lying in the grass. And then we found a pile of manure – it looked like dog manure to start with, but when you looked at it closely it was all vegetation – grass, leaves, things like that – most unusual."

At about the same time something began attacking trees all over the property. Mature stringy barks bore the brunt of the assault, and virtually every tree of that type had its outer bark lifted, shredded and torn away from its trunk from a couple of feet above the ground to a height of about twelve feet. On some nights as many as a dozen trees were ripped, until more than a hundred were effected, and, because the inner skin of newly-damaged stringy barks is blood-red, the results were as spectacular as they were unsettling: "They were ripped to pieces

Lynn Pendlebury with one of the damaged trees. (Tony Healy)

– some looked like they'd been shot with a cannon!"

Many smaller trees, mainly angophoras, were also damaged, but less dramatically: they exhibited odd, "V", and occasionally "X" – shaped scratches which, though deep, were no longer than about five inches (13 cm). Because of their uniform shape, because they were all within five feet of the ground and because there was no evidence of wood-boring larvae, they didn't appear to be the work of possums or Black Cockatoos. (Among the many trees on the Pendlebury property, there are very few bloodwoods – the type that has been bitten in such a distinctive manner elsewhere.)

Amazingly, the area most often targeted was what is effectively the Pendleburys' front yard – a park-like space of about one acre, devoid of underbrush, overlooked by several windows and containing only well-spaced mature trees. Despite hours of scanning with night binoculars, and despite rushing outside whenever their dogs barked, the Pendleburys could never catch whoever or whatever was tormenting them. And the tree-ripping continued – right up to within twelve paces of their front door.

Once, at 2 am, when the dogs kicked up a hell of a fuss, Lynn took the night glasses and walked out onto the front driveway. "Then all of a sudden I heard an almighty roar – like a tiger's or a lion's – and I ran back inside!"

She had the impression the sound came from the vicinity of the fowl yard, which is near the edge of the scrub about 50 metres from the side of the house. Although bark had been ripped from trees all around the pen, and although the perpetrator would presumably possess the strength to tear chicken wire, none of the fowls has ever been harmed.

While none of the Pendleburys' pets or livestock has sustained any unusual injuries, a dog belonging to their next-door neighbour, film critic Bill Collins, was almost killed in a mysterious attack. The dog is a knee-high, poodle-like Bichon Frise. Mr. Collins' wife Joan told us they found it one morning lying, terribly injured, under some chairs beside the house. It had severe head wounds, a dislocated leg and had been partly disembowelled: "It was terribly mauled around its little anus."

The vet, who had to remove one of the poor creature's eyes, couldn't say what had inflicted the wounds. Mr. and Mrs. Collins, not inclined to speculate about yowies, think a large dog or perhaps a rock wallaby was responsible. Joan Collins does, nevertheless, believe something rather strange is going on. Like the Pendleburys, she and Bill have had stringy barks spectacularly but inexplicably damaged, and she cheerfully admits that, during the time of the most intense tree ripping, she'd been afraid to venture outside at night.

Meanwhile, back at the Pendleburys', Lynn and Gordon were making renewed efforts to identify their mysterious visitor. Courtesy of a friend who owns a large security firm, they installed a huge roof-mounted 500-watt illuminator that saturated the entire front paddock with infrared light. Next to it on the roof was a permanently running video camera that was hooked up to a recorder and a monitor in the living room.

No prizes for guessing what happened next. As had happened at the Frost and O'Connor properties, the midnight creeper, apparently able to detect the infrared light, avoided it like the plague. To Lynn and Gordon's frustration, the whatever-it-was continued to rip trees – but only to the side and back of the house.

Only one unusual image was recorded: eye-shine from an unidentified creature that moved slowly across an area right at the edge of the light's coverage. Although the eyes did not appear to be particularly large, Gordon asked Neil Frost to help him establish exactly where the creature had been.

When Neil, equipped with a two-way radio and guided by Gordon at the monitor, located the spot, they found the animal had been a lot further away than it had seemed. They also found that, since it had been traversing a low-lying area below the road embankment, it must have been about as tall as a man.

While conceding that the video was not clear enough to prove anything, Neil thinks that the eye shine probably *was* that of a yowie. He believes it was creeping along below the bank, popping up occasionally to steal a glimpse at the (to its eyes) brightly lit front yard.

Finally, in June 2000, after five years of fruitless night vigils, Lynn actually saw what was presumably their strange, nocturnal visitor. Murphy's Law, of course, always comes into play at times like that, so when she encountered the creature she wasn't actually looking for it and had no camera at hand. It happened as she walked her dogs down the road, about 200 metres from her front gate.

"It was dawn, just barely getting light, and this creature ran from the right hand side of the road [the Pendleburys' side] only about 15, maybe 20 metres ahead of me. It looked like an overgrown monkey on two legs and it was letting out the weirdest *screams* – and the dogs just chased it!

"It kept making that noise – a very high-pitched squeal – as it ran down into the gully. It was maybe four to five feet tall and had long arms which were sort of swinging, and it was black. It *did* look like it had a tail to me ... but being such dim light ... It was perfectly upright, very straight – monkeys seem to bend when they run, don't they? It was a most unusual-looking thing."

They didn't know it until recently, but the Pendleburys aren't the only family in their neighbourhood to have been visited by yowies. Some years back Neil interviewed a woman who told of a huge hairy creature that had frequently peered through high windows, broken tree branches and thrown rocks at her property about three kilometres to the west – throughout the early 1980s. As if engaged in some migratory or seasonal pattern, it appeared only in the month of November. On one occasion it ventured underneath her high-set house and pounded on the supporting stumps, sending shudders through the whole structure.

The Frost, O'Connor and Pendlebury "yard invasion" stories are, for a host of reasons, some of the most interesting in our files, and have all involved sightings of creatures that can only have been the mysterious yowie. As luck would have it, however, none of them has involved a sighting of long duration, in good light, unimpeded by tree trunks or foliage.

Before we conclude this chapter, therefore, it might be a good idea to record the testimony of some other Blue Mountains residents who have been fortunate enough (although at least one man might quibble with the word "fortunate") to obtain a clear look at the strange creatures.

The road runner

In September 2002, a Springwood resident, who asked to be identified only as Susan, told us of a remarkable yowie encounter that she and her father had experienced, in almost perfect viewing conditions, two years earlier.

"It was when I was 19 years old, in about May 2000. We had a property up at Cowra and were going up there to do a bit of work. So at about 4 am we were heading west on the [Great Western] highway, and as we got to Woodford – near the turn-off to the left [Park Road] towards the Rural

Fire Brigade building – we saw this really huge thing running down towards us, down the grassy strip on the [opposite] side of the road. We thought it was just a person, and it looked like it was in trouble. It was running, stopping, and looking around and we thought it might have been someone who was hurt or was being chased. It was loping, but … like when you've got a sore foot and are all off-balance … it's hard to describe.

"We were in a Toyota Hi Lux utility, we used to go shooting all the time, so we had powerful lights, a spotlight and everything. There was no traffic around, so we moved into the middle of the road and stopped. It was about 40 metres away. It stopped and we put the spotlight on and it sort of looked into the light, and you could see it was absolutely huge – it wasn't a person.

"It was probably one or two feet taller than a big man, really big – like a fully-grown cow standing on its hind legs – that's the height it would be. [But] very slender – like an adult male, with no hips but these massive shoulders that stuck right out from its neck … the shoulders were a *lot* wider than the rest of the body [which was] just straight. The hands were almost down to its knees – really, really long arms - really big hands, massive feet.

"It seemed to be covered in hair: reddy, browny … matted, dirty-looking. Out at Cowra we see wild goats all the time. It was that sort of hair. You couldn't see any genitalia or anything like that. There was virtually no neck … [the head was] like a human's, maybe a bit longer; the face was virtually human except that the jaw was really strong. A really strong face, a normal nose, and the eyes were set a long way back. It was hairy all over except for the face … hardly any hair on its face."

The amazing creature stood there "like it was stunned, and we slowly went onto the median strip … and it turned around and ran down a side street on the [north] side of the road [probably Woodbury Street] and we drove across the median strip and followed it. It would run and stop and face us. And the closer we got, it would slowly back away, and as soon as we would turn the spotlight off, it would run, and when we turned it on again it would turn and face us. [They sometimes got within] 20 or 30 metres – if we got any closer it would run. It didn't swing its arms as a human would. They were "just hanging down by its side, and the way it loped along, it looked heavy, as if it was hard to lift its feet up".

Woodbury Street has houses on both sides and stretches north for 750 metres before terminating at the edge of the forest.

"It was running down the middle of the road, or swaying all over the place, running off to the side. It just looked really confused and out of place, really disorientated. It would run and stop and weave … but always moving towards the end of the road. And when it got closer and could see the bush it just bolted. There's a bush track at the end of the street but it didn't follow that; it just went straight into the trees.

"We put the spotlight on, and you could see the tops of the trees swaying – and it wasn't a windy night – and hear crashing noises. There were a couple of dogs going pretty crazy. We didn't follow it because it was the middle of the night and we were scared, but three days later we walked down there and there were six or seven trees – not huge trees, but big enough that a person couldn't push them over – just *snapped* [at the height of] about a metre or a bit higher."

Susan's testimony lends additional credence to what the Frosts and the O'Connors have been saying for many years about the area's strange nightlife. Only one detail is puzzling: although they had the creature transfixed by bright lights on several occasions, Susan and her father noticed no eye-shine. Although that seems at odds with the Frosts' experience, we don't doubt Susan's honesty. Hers, in fact, is not the only nocturnal encounter in which eye-shine was strangely absent or not noticed.

The hairy fellow of Fairy Dell

Mary Camden is a well-spoken, physically fit schoolteacher in her early sixties. Prior to settling in Springwood in 2001, she had heard little, if anything, about the yowie phenomenon. Until she saw one herself, she hadn't the slightest idea that gorilla-like creatures had been reported anywhere in Australia – let alone within a couple of hundred metres of her own house.

Her sighting occurred on the southern side of Springwood in mid December 2001, just before the massive bushfires of that year.

"It was during my regular morning walk at about 6 am [broad daylight at that time of year]. I went down the western side of Magdala Creek to Picnic Point with the idea of crossing it at the bottom and coming back up to Fairy Dell, which connects up to Springwood Avenue. Just after crossing the creek I saw something on the track ahead.

"At first glance I thought it was a kangaroo, but soon saw it wasn't. It was heavy, black, solid, about my height [163 cm or 5 ft 4 ins]. I weigh 60 kilos [132 lbs] but it must have been a *lot* heavier. I could see it from about the chest up. It was standing in the middle of the path, perhaps 20 or 30 metres away, looking in my direction, obviously startled. I assume it had been drinking at the creek and had to cross my path to get away to safety.

"It was upright and I was very surprised and very puzzled, because it had an almost human stance, but no neck, really solid. My first impression was [that it's upper body was] 'rounded', that it was upright, but that the head and shoulders merged – if that makes sense. My impression was that it was covered in long black hair – shaggy hair. I particularly noticed there was no neck. It would have had a neck, of course, but you know, when you hunch your shoulders up? And presumably … the hair would sort of fill in the space between the head and shoulders. I got an impression of a face, that it was startled and looking in my direction, that's really about all. I felt it had a flatter face than a gorilla or a human, but I didn't see it in profile. It's like nothing I've seen before, and if you've got nothing to liken it to … .

"And I was really *interested*, but it immediately turned to go up the hill through all the undergrowth. The noise it made was quite incredible. It was like a 100 kilo man just belting his way through the forest; certainly not elegant and quite amusing in a way, because it didn't seem to be using any track of any kind – just crashing, really, up through the undergrowth. And that was that.

"I told one or two people about it, but you know the kind of reception you get. They think: 'She'll be talking about UFOs next!' So I didn't elaborate to anyone, but I was very puzzled and very interested because I assumed it must be a primate – a pre-Aboriginal primate – in Australia. Then, about a year later, I was really quite amazed to see the ABC's *Catalyst* program [about Neil and Sandy Frost and their yowie experiences] and I thought, 'I've seen one of those!'"

Three on the road

Our next witness, Faulconbridge resident Justin Garlick, could hardly be more different from the erudite, ladylike Mary Camden.

When we interviewed him 17 days after his yowie encounter, Justin was 20 years old. A cheerful, easy-going, knockabout young fellow, he was, like most blokes his age, interested mainly in girls, cars and socialising. We have learned, however, that while he may be a bit rough around the edges, Justin is a very public-spirited young guy. During the time we knew him he performed a remarkably selfless act for the good of his community.

We found him to be likeable, amusing and transparently honest. We were also struck by the fact that although he is not particularly highly educated, he is, in his own very colourful, individualistic way, quite eloquent. In the hope that readers will find this decent young fellow's words as convincing as we did, we present them here, with a minimum of editing. The incident occurred on the night of Monday August 26, 2002.

Justin Garlick.
(Justin Garlick)

"It was about 9:30 or 10 pm. I had a fight with my missus [girlfriend] and thought, 'Oh, f*** it – I'm just going to go for a drive'. And I'm just driving along and thought 'I'll just go down here.' [Down Tableland Road, which runs south from Wentworth Falls.]

"We'd been down there before, trying to freak ourselves out – because there's that old quarantine station down there [the decommissioned Queen Victoria Memorial Hospital]. It was reasonably thick bush and the night was overcast – no moon or stars, pretty pitch black. And I was going heaps slow, just putting along, and then this big, black, tall, really *built* thing came out, and for a second I thought, 'Oh yeah – some idiot playing a joke in a suit'. And then I looked closer and you could just tell it wasn't someone in a suit.

"I just jumped on the brakes and *shat* myself ... just looking at it, saying, 'What the f*** is that?' And then, from my right, two other ones came out and they were even bigger: probably seven, eight foot. They were real big, their backs were hunched over and they were *built* – like real 'tank'. They didn't have, like, a beer belly ... they were real chunky things. I couldn't really tell whether they had fur or hair; they were just a straight black. And I just looked, and I wasn't calm at the time – I was really peaking out – and they just *looked.*

"So one of them was out on the road and the other two were sort of still half in the bush. And they were all looking at each other like they were plotting something, like, they sort of half pointed at the car. [The arms] were pretty long ... slightly longer than a person's. One big one pointed, then the other big one and then the first [slightly smaller] one. When they were pointing, they didn't have their arms at full stretch, [only about] 45 degrees. I didn't see fingers or anything, [but] it just looked like they were pointing. And I was just freaking out by this stage, and I just buried my head, like, between my knees and, like, *shat myself.*

Tablelands Road. (Paul Cropper)

"I looked up again and they started walking towards the car, and I thought my number was up ... I thought they were gonna smash through the car and drag me out and kick the shit out of me. Oh shit yeah – they could rip your bloody head off! And that's when I full-on, like, peaked out. I just buried my head and was just yelling at the top of my voice, 'Just *f*** off!*' Yelling *anything*. I was peaking out, I was full gone into shock, I reckon. And I had my head down for a while and I looked up and they were walking around the car, pretty close, like, say, one and a half metres away. And I was, like, shitting myself, hardcore, by this stage.

"But they just kept walking around. They'd sort of do a U-turn round the back of the car – because they wouldn't walk in front of the lights. But one walked around the right side and jumped the headlights with its back facing the front of the car. It's only a little Mitsubishi Lancer and the lights are no good. I've frigged around with them a bit, and they're pointing right down in front of the car. Another thing I noticed: it smelled like I was behind a garbage truck – that sort of smell.

"Anyway, they just kept circling around, like just a U shape – from the passenger door to the driver's door and back." One of the animals reached towards the front passenger side window: "it sort of put its hand up. It might have seen a reflection of itself – that's what my brother suggested later. [The hand] looked like a big sort of plate, like an oval. I couldn't see how many fingers because I was sweating, I had, like, water in my eyes and I was almost crying my eyes out and I was, like, full-on shitting. I was just *shitting bricks.*"

Perhaps because they seemed to be carefully avoiding the headlight beams, Justin didn't see any significant reflection from the yowies' eyes. He did, however, actually see their eyes at times when they were close to the driver's side window: "They had real big eyes. They looked like just big, black, shiny eyes. I could just see these two big, round ... you know, just the shine on each edge of the eye. One came up to my window and ... it was pretty friggin' tall ... and the seat in my car is low and I could just see the shine on his eyes looking down. I didn't really want to look at its face! By that time I had heaps of water in my eyes, and I was blinking and sort of trying to look, but I didn't really want to make eye contact unless he put his fist through the window! And then he walked back.

"I've got air horns in my car and I thought, 'Oh yeah – that'll scare them off!' And when I beeped my horn, WHAAAP! There was one of them on the left and I just seen ... a massive stride – *one* – into the bush ... and it was gone! I assume the others followed – I didn't really see. It just took one big bound and, DOOSH! It was off. Pretty freaky.

"And that was it – I was just sitting there, counting my lucky stars.

"I just sat there, recouped for a sec, dropped the clutch and drove off, turned around where I could and drove back out really quick. When I first met my girlfriend I was in hysterics. I just jumped out of the car and just hugged her. I was a mess."

One of the things we liked about Justin was his willingness to admit to being scared into a state of near-hysteria by his close encounter. Actually, we consider him a very gutsy guy. He emerged from his horrendous encounter with his body, mind and sense of humour intact – which is more, perhaps, than many others would have managed to do.

Two dog night

Justin didn't know it, but he wasn't the only person to have encountered yowies on Tableland Road.

In 1996 or '97, shortly before the Queen Victoria Memorial Hospital was closed, "B.J." lived and worked there as a male nurse. He told Neil Frost that the institution, located deep in the scrub over two kilometres from the nearest settlement, had unusual problems, such as semi-feral dogs that used to dig up waste at the hospital tip.

B.J. and his colleagues shot several of the scavengers and on one occasion, using a backhoe, buried two carcasses quite deeply. The following dawn revealed an unsettling sight: something large had dug down, creating a big burrow, and eaten them both.

One night shortly thereafter, he encountered the dog-muncher face-to-face near the residential quarters. It was a huge, hairy yowie, apparently rummaging through a garbage bin. It ran off, but 15 minutes later, B.J. and his wife saw what appeared to be the same creature on the other side of the building.

The hairy giants still roam

The Blue Mountains area is unquestionably Australia's premier yowie hot spot. Although many people have helped us with our investigations there, none have helped more than Neil and Sandy Frost. One day, when other commitments allow him the time, Neil will write and publish the full story of his family's unique 20-year association with the hairy giants. Until then, we hope that our summary of the situation will suffice.

Every member of his family has now experienced at least one sighting. Apart from his face-to-face encounter of June 18, 1993, Neil has seen glowing eyes and partial silhouettes on several occasions, as has Sandy. On May 9, 2003, 11-year-old Drew had his first sighting.

It happened at dusk a couple of hundred metres from the family home. While walking through the scrub, he heard a noise and suddenly noticed a huge creature just a few metres away. It was standing with its back against a tree and seemed to be watching his sister Avril, who was some distance ahead.

As soon as he caught sight of it, the creature turned and looked at him.

Like Justin Garlick, Drew instinctively avoided eye contact with the towering animal. Lowering his gaze, he beat a hasty retreat to the nearby road. Consequently, although he was left with the general impression of a huge, hair-covered primate, his only specific memories involve the creature's legs. They were, he recalls, very solid and covered with hair that looked like it was "teased out" from the knees down. Neil suggested later that the "teased" effect might have been created by the creature wading through low undergrowth.

At 9:10 am on Sunday May 23, 2004, 13-year-old Avril saw a similar creature right next to the family home. Just as she lifted the lid of an outside rubbish bin, she caught sight of a dark figure standing, almost unobscured by underbrush, just 20 metres away. It was covered with jet-black hair and was standing in a strange posture with its arms up, "as if it was surrendering, except that its hands were together above its head". Weirdly, it was swaying from side to side, and she had the impression it was pretending to be a tree (nearby saplings were swaying in the breeze). It was about seven feet tall but didn't appear to be massively built. Its head was so hairy and black that despite the bright, sunny conditions she couldn't make out any details of its face. Interestingly, it was standing only a few feet from the spot where Neil's camera trap photographed the dark "face" in 1993.

As the realisation of what she was looking at struck home, Avril dropped the rubbish bin lid. The creature charged off towards the swamp; she ran screaming into the house, where Neil was having breakfast.

Within 20 seconds he was outside looking for the creature, and within seven minutes neighbour Mike Williams was on site, video camera running. Together they scoured the scrub, but although they heard slow footsteps moving away to the north and although Neil's dog barked dementedly, they saw nothing. The Hairy Man, as always, simply faded into the vast Blue Mountains wilderness.

Chapter 5

Littlefoot — The Junjudee

"It was awful. The face was small and drawn back like that of an ape."

For a couple of weeks in early 1979, the citizens of Charters Towers were in an uproar – and in the national news – over reports of "little hairy ape men" supposedly attacking teenagers on a hill overlooking the old north Queensland mining town. The encounters had apparently been occurring since about August or September of the previous year, but it wasn't until six months later that the story hit the headlines.

It had all the makings of a good schlock-horror film.

Terrified teens tell

Late one night in early March, the sedate routine at Charters Towers Police Station was shattered when a breathless young man burst though the door to report that one of his mates had gone missing in strange circumstances on Towers Hill, a popular lovers' rendezvous on the outskirts of town.

Nineteen-year-old apprentice baker Michael Mangan was, according to police, ashen-faced as he poured out the story. Earlier that night, as on other occasions, he had persuaded a group of friends to join him in searching the rugged hillside for mysterious "little hairy men", but somehow one of the boys had become lost. They could clearly hear his terrified screams but, try as they might, could not locate him in the inky darkness.

As the whole area was riddled with dangerous old mine shafts, the police responded immediately and within minutes were at the base of the hill. No sooner had they begun the ascent, however, than they met the missing youth running frantically down the road in the opposite direction.

Safely back at the station, the badly shaken lad told of being attacked by one of the hairy midgets. He had, he said, fought it off with a rock. During the interview the officer in charge, Sergeant Gill Engler, noticed blood on the uninjured boy's leg – evidently not his own. When the gob-smacked gendarmes quizzed Michael Mangan again, he supported his mate's story, saying the incident was the culmination of many sightings and searches for the little ape-men.

It had all started one night about six months earlier, as he was parked on the hill with his girlfriend: "I looked across to the passenger side and saw a black, hairy face at the window. It was awful. The face was small and drawn back like that of an ape. We both screamed because we got such a shock. I started the car to get out of the place and this thing raised his hand and smashed the passenger side window." The creature was about one metre tall and covered in black hair.

When the story first came our way, in the early days of our yowie research, we didn't quite know what to make of it. At that stage, we had collected quite a few reports of hairy ape-men, but they were all said to be large animals, with an average height, based upon eyewitness estimates, of over seven feet (2.13 m). If the three-foot-tall Towers Hill creatures were yowies, they must surely have been extremely young ones – mere ankle-biters. If that was the case, how was it that throughout their months of searching, during which they'd seen several little ones, Michael and his friends had never encountered Mama and Papa yowie?

The Charters Towers story seemed to be a one-off and we filed it, initially, in the "Too Hard" basket. It wasn't long before we had to retrieve it, however, because soon afterwards we began to hear bits and pieces of Aboriginal lore that seemed to refer to a widespread belief in the existence of similar tiny hairy men in various parts of Australia.

Michael Mangan at Towers Hill, where he encountered 'little hairy ape-men'. (Northern Miner)

The little creatures were known by many names, including *dinderi*, *kuritjah*, *magulid*, *net-net*, *nimminge*, *nimbunj*, *njmbin*, *waaki*, *wadagadarn*, *waligada*, *waladhegahra*, *winambuu* and several variations on the word *junjudee*. Some northern NSW Aborigines commonly referred to the little creatures as "brown jacks".

According to a 1977 edition of *The Richmond River Historical Society Bulletin*, the hippie Mecca of Nimbin was named after small hairy creatures that Aborigines said lived in the area. They were described as "sort of hobbits". When he was a child, Bundjalung elder Gerry Bostock was told that the name of a neighbouring town, Mullumbimby, also means "little hairy man". [22]

Although it was some years before we interviewed people who claimed personal encounters with junjudees, we soon met Aborigines who told of sightings by relatives or friends. Disconcertingly, however, our informants didn't appear to be speaking of an entirely uncouth race of wild little monkey-men. They sometimes attributed to them strange behaviours and semi-magical qualities that were reminiscent of the fairy lore of Britain and other places.

The Ualarai people of central northern NSW, for instance, believe that the *winembu* will sometimes persuade a human being to follow them home. The person's spirit is then somehow stolen, but when the victim returns to human society he or she finds it impossible to describe the ordeal. Among the tribes of the Lower Clarence Valley in northern NSW the little hairy men are known as the *nimminge*. In 1991 a Bundjalung elder, Ron Heron, recalled a story that had distinct echoes of the European fairy tale "Billy Goat's Gruff": "[When] I was 16 or 17, Frank Randall, a friend of my father's, told me of a little hairy man living at Ashby. Some nights, when Frank was walking home from the ferry … he would come to a small wooden bridge where there would be a little hairy man waiting for him. Frank would have to wrestle with this man before he could go across the bridge. He said this would happen as many as 10 times each year. Since then … I have heard similar stories from other older people."[23]

Well, this was very confusing, not to mention inconvenient. We were just beginning to assemble enough data to make a reasonable case for the existence of the big hairy yowies, but now had to deal with distracting tales of hairy midgets. Magical hairy midgets at that! For some time, we tried to accommodate the Aboriginal reports, which were, at that stage, all second or third hand, by filing them as native folklore. The Charters Towers reports we consigned again to the limbo of the "Too Hard" basket.

That was the way things stood in the early 1990s when we were working on our earlier book, *Out of the Shadows – Mystery Animals of Australia*. As a result, although we wrote a lengthy chapter about the yowie, we dismissed the matter of the tiny junjudee in just a few paragraphs. Since then, however, we have collected much more Aboriginal junjudee lore, as well as eyewitness reports from both Aborigines and non-Aborigines. We would like to be able to say that, armed with this new information, we now know exactly what the little hairy men are and exactly how they relate to the yowie phenomenon. We'd be less than honest, however, if we made such a bold claim. Frankly, we are almost as baffled by the junjudees today as we were when we first heard of them 23 years ago.

Over the years we have, of course, toyed with various theories, but the pesky little creatures, like the elves and fairies of Europe, have always managed to avoid being pinned down. At times, like many of our colleagues, we have favoured what seems to be the most logical explanation: that the junjudees are simply juvenile yowies. That assumption, however, flies in the face of the apparently unanimous belief among knowledgeable Aborigines that such is not the case.

The magical aspects of junjudee lore have made us, at times, strongly inclined to dismiss the whole phenomenon as native myth. Whenever we drifted too far in that direction, however, we were dragged back to "reality" by eyewitness reports by both Aboriginal and white Australians, who like Michael Mangan, seemed to be describing encounters with very solid, very real little creatures. Perhaps, at this point, it would be best for us to present various representative items from our junjudee file, so that readers can attempt to make up their own minds.

In January 1977, 86-year-old Henry Methven told Patricia Riggs of the *Macleay Argus* about a little creature he'd seen while hunting near Jervis Bay in about 1901. Having become separated from his companions, the then 10-year-old Henry returned alone to a temporary camp.

"I was stripping off my shirt and when I looked around, the Hairy Man was standing right behind me. He was only about ... two or three foot … a handsome little fellow… he had a long straight nose and he was the colour of a real full blood … dark and coppery … everything about the little bloke … seemed to be human."

The creature was strongly built with a short neck. There was hair on the back of its hands. On its head, the hair was about two or three inches long and "a bit smoky looking, a bit grey." Its body hair was different: "darkish brown". That was as much detail as the startled boy could absorb: " I took off into thc bush and got stung with stinging nettles.

"The next day we tracked him. He had feet like a human's … five toes." On a nearby ridge, they found evidence that the little creature lived with others in a small cave and dined on shellfish. Henry said all the tribal elders knew about the creatures. They called them *wallathegah*. They were said to be harmless, but to have a great fondness for honey. Earlier that day, Henry's party had harvested honey from a native beehive and he had carried it back to camp in a coolamon. The elders said, "He could smell the honey and he followed you along."

Folklorist Aldo Massola, author of *Bunjil's Cave*, heard similar stories of little hairy men, known locally as *net-nets*, from Aborigines at Lake Condah, Victoria, in the 1950s and early '60s. As well as being hairy and very small, *net-nets* were said to have claws instead of finger and toenails. They were mischievous but harmless and were believed to live in natural hollows among jumbled heaps of boulders.

Andrew Arden told of encountering one in about 1932, while hunting with his wife in the Stony Rises near the lake. He had just shot a rabbit when "one of the little people" suddenly appeared, seized the carcass, and ran away over the rocks. Mr. Arden gave chase but soon lost sight of the light-fingered Lilliputian.

Remarkable as they most certainly are, Mr. Methven and Mr. Arden's stories seem straightforward reports of encounters with some sort of apparently real, if extremely rare, animal or hominid. Would that life were so simple, Lord. Many, if not most, junjudee stories are (to the "Western" mind, at least) not quite so down to earth.

Frank Povah, a researcher, lecturer and writer of mixed Aboriginal and European ancestry, collected many stories about the little hairy men and included them in his fascinating book, *You Kids Count Your Shadows*. Most of the tales were told by Wiradjuri people, whose country covers about 87,000 square kilometres [33,600 sq miles] of central NSW.

Several of the stories contain rather magical details, but while Frank sees the little hairy men as "indigenous fairies ... [the] Aboriginal equivalent ... of white Australian folklore", he is open to the possibility that they also have some kind of objective reality. His informants provided fairly uniform descriptions of the little beings, which they knew by various names, most commonly *yuurii*:

"A little man about so high – a metre, say. Small, real small..."

"Covered in hair with long nails and big teeth."

"*Yuuriwinaa* means hairy woman ... about three feet [in height], a bit more. Real hairy ... teeth like a greyhound, big fangs ..."

A couple of informants said males sometimes sported long beards, and foul body odour was occasionally mentioned: "Real stinkin", "real smelly". They lived in mountain caves, in holes in the ground and in the gidji scrub. Like Henry Methven, the Wiradjuri said the little men had feet that were quite human-like, but whereas he had seen five-toed tracks, they insisted the creatures had only four. Testimony about the creatures' behaviour also contained minor anomalies. They were said to be quite harmless, but, paradoxically, most people seemed to be afraid of them. Some parents used tales of the little men to scare children away from dangerous locations, but others "always used to say to the kids, don't be frightened of them".

In a twist reminiscent of European fairy lore, the *winambuu* sometimes punished human troublemakers. One "little feller" challenged a drunken bully who swung a punch, missed, hit a tree and broke his arm: "They get them type of people, see". They were said to be guardians of certain places and would punish anyone who harmed certain animals, such as the mopoke – a "ghost bird".

Another remarkable attribute seemed to place the little creatures firmly in the realm of fairy lore: *they could speak,* although they would normally converse only with old initiated

Tony Healy (left) with a junjudee-sized friend. (Paul Cropper)

"clever men" in "the lingo" [the Wiradjuri language]. At least some Wiradjuri believed the "little fellers" possessed considerable occult knowledge, "same as the high initiated people". Aborigines in some other parts of the country have similar beliefs.

In 2002 an Aboriginal elder from the NSW south coast told Gary Opit of tribal lore that emphasised the junjudees' supernatural nature and also their connection with children. He said that an ancient initiation ceremony involved children smearing their bodies with blood and ochre and then bathing in waterholes near Mumbulla Mountain. His people believed the little Hairy Men were then spontaneously created out of the blood and ochre as it flowed down the Murrah River. He had seen the little creatures himself.

Henry Buchanan, a Kumbaingeri man of Nambucca Heads, NSW, believed the little hairy man had a material form – at least sometimes – but was also a kind of magical elf/benevolent spirit. In 1976 he told a *Macleay Argus* correspondent, Sue Horton, "The Hairy Man is just a little mite, like a little monk [little monkey]. If you catch them, they are as good as the Lord. They do things for you."

He claimed that they emerged from holes at Middle Head and that he had seen one there, but had been unable to cry out: "I couldn't make a sound." On being asked about the feasibility of trapping one, Mr. Buchanan's reply was, to the "Western" mind, rather confusing: "You can't catch him. No ... He's a spirit, but he's a live thing too. As soon as you catch him, he goes into your blood and his spirit goes into you. He's with you all the time. Any people get sick ... like might be dying, they ... send for you ... and you say 'Listen, I want you to go to this place' and this [little] fellow knows where to go. He ... fixes it up. They call him the little brown jack."

After such a mind-bogglingly strange but apparently sincerely-told story, it would not be surprising if many readers now feel that the entire junjudee phenomenon should be written off as a colourful, widespread Aboriginal myth. But if it is only a myth, how do we account for the numerous sightings by non-Aborigines?

In mid-1997, while driving along the Mount Lindsay Highway, in northern NSW, Mark Pope, of Bexhill NSW, encountered what may well have been a junjudee: "It was just on daybreak, I still had the headlights on. I was heading down to Tooloom; heading south. There's a State Forest there; it has massive white gums in it. I think it's before Woodenbong. I came around a corner; I was a bit tired, and there was a combination of my headlights and enough [natural] light to see – just. There was something on the other side of the road. I was about to enter a left hand bend, and this thing looked like it had just crossed the road before I'd got there. It was about to go into bushes on the other side, which was up a slight embankment. And it looked for all the world like it had heard me, stopped, looked over its shoulder, and was looking to see what I was doing.

"If I had to say it looked like anything, I'd say a chimpanzee. As to whether it was a chimp, I'd say no, but something in the same line; I can't quite say what ... As much as I could tell, it was covered in hair. It was quite dark, dark brown or black. Its face ... I can't remember it very clearly, except that it seemed fairly flat.

"It wasn't very big: about the height of a guidepost [about one metre]. It was standing on two legs [but] was sort of leaning forward. One of its legs was down the bank ... and straight ... the other was further up. One arm was down on its forward leg or on the ground. Its head was turned towards me, shoulders slightly towards me. When it decided to move, it took off in a hell of a hurry and used arms, legs and everything to claw its way up the bank – and then it was gone." [Case 221]

Two other non-Aboriginal eyewitnesses were Mr. and Mrs. Roy Locke of Theodore, Queensland. In early October 1979, as they were driving west on the Wide Bay Highway, they saw, just before dusk, a one-metre-tall hairy animal standing beside the road about 20 kilometres north-east of Murgon. Mrs. Locke told the *South Burnett Times* that the creature had broad shoulders and stood looking at them as they drove past. [Case 140]

The Cherbourg Aboriginal Reserve is situated only six kilometres south of Murgon. When told of the Lockes' experience, Les Stewart, chairman of Cherbourg's Aboriginal Council, said "there is a small man called *junjurrie* who was seen here as recently as eight years ago. He was about a metre tall and used to play with the children in the old hospital. Several adults claimed to have seen him when they heard the children laughing at night."

In the early 1980s, Alice, a member of Queensland's Waka Waka tribe, told us that while she was growing up at Cherbourg in the 1960s it was commonly believed the little people, which she knew as *junjudees*, were in the area and that they supposedly kept a protective eye on sick children. Her father spoke of occasional encounters with them, saying that once, as he walked along a darkened road, a little Hairy Man emerged from the shadows, took his hand, and walked beside him for a while. Alice recalled her father saying something to the effect that the hand felt soft and strange.

One hundred and forty kilometres to the north-east, at Scrub Hill Farm, near Hervey Bay, members of the Butchella tribe are also familiar with the little creatures, which they call *jongari*.

In February 2000, Mally Clarke of the Korrawinga Aboriginal Corporation told the *Fraser Coast Chronicle* there had been many sightings at the farm over the years. "One Islander fellow who stayed here once got a real fright", she said. "He was out chasing cows one night and came back shaking because he had seen a little hairy man. We all laughed because we knew what it was." There were three different sightings in 1996, one witness telling Hervey Bay Police that little hairy hominids had run across a road into the farm. "Years ago," added Ms. Clarke, "they would play with the kids. I saw one once and thought it was a monkey because it was so hairy."

In February 2000, after small, furry, unidentified animals were seen beside the Burnett River at Gayndah, 120 kilometres west of Hervey Bay, a local Aboriginal man, Sam Hill, suggested the creatures were not feral bears, as assumed by non-Aboriginal witnesses, but *jongaris*.

After we rushed to Gayndah, however, we found that all the eyewitnesses were adamant that, even though the creatures occasionally stood, and even walked, on their hind legs, they weren't at all ape-like. They really did closely resemble bears. It turned out that that semi-mythical ingredient of many cryptozoological mysteries – the crash of a circus truck – really *had* occurred near Gayndah in February or March 1959. Brett Green, who lived in Gayndah at the time, inspected the site the morning after the accident. Two vehicles, in fact, had slid off a dirt road in a storm, down the jungle-covered side of Binjour Plateau. Two lions escaped but were quickly recaptured, but several monkeys and three small bears, possibly Himalayan bears, North American Black bears or Malayan Sun bears, escaped into the scrub. There were two females and one male, and sightings have been reported from the plateau and surrounds ever since. [24]

Although the presence in the area of feral bears complicated the situation, we found Sam Hill's statements about the *jongari* very interesting. The little creatures, had, he said, lived beside his people, the Waka Waka, from as far back as anyone could remember. In recent times, however, most of them had moved away because of drastic changes to their environment brought about by European settlement: "when they blew up the mountain near here to build the railway ... a lot of them ran out of the hills". The normally inoffensive little creatures could, in fact, react violently to destruction of the natural environment: Sam said his grandfather had been attacked by one while ringbarking trees.

When we met him a few weeks later, Sam's father Rodney Hill made it clear he disapproved of Sam talking to the media. His people, he explained, normally never mention the *jongari* to outsiders.

Because of this tradition of reticence, Mr. Hill senior, though by nature a kind and generous man, was reluctant to discuss the matter further. However, when he realised we were genuinely interested in the phenomenon and had travelled far to learn about it, he did share a few details, while emphasising that a great deal more must always remain secret.

The word, "jongari," used by the media, was not, he said, quite correct. His people's term for the little people would be more correctly rendered as *jungurrie*, although *junjudee* was an acceptable variation. The *jungurri* had always been, and still were, guardians of his people, and he concurred with Alice's statement that the little people keep a particularly close, protective eye on sick children. If, however, they are ridiculed or even talked about too loosely, they are liable to punish the talkative person or members of his family with illness. Then only traditional rites can cure the afflicted – not Western medicine. But now, he lamented, all the Waka Waka "clever men" had gone: the last of them had passed away with his father's generation. That was partly why the situation at Gayndah – with stories about the little people being bandied about, sometimes jocularly, in the media – worried him so much. If, as a result, sickness was visited on his people, there was no one left to cure them.

The whole episode saddened him. Their secret *jungurri* lore was one of the few things the Waka Waka had left – just about everything else had been taken. He was particularly upset when some insensitive fool of a businessman hatched a plan to sell "jongari" T-shirts in Gayndah. Fortunately the scheme had come to nothing. Mr. Hill's final remark on the T-shirt episode illustrates how deeply his people feel about the *jungurri*: "This is a religious matter. Selling those things – it would be like someone walking around with Jesus Christ on his shirt!"

While there may be, at present, up to two dozen fairly active yowie investigators in Australia, we know of only one person, Grahame Walsh, who has concentrated his efforts on the mystery of the tiny junjudee for any length of time. No dreamy, yurt-dwelling, faery-lore enthusiast, Mr. Walsh, a former National Parks and Wildlife officer, writer and photographer, is one of Australia's greatest authorities on Aboriginal rock art.

Carnarvon National Park, where he conducted most of his investigations, is situated about 350 kilometres north-west of Gayndah. An immensely rugged area, it is cut through by twisting chasms, crystal clear creeks and white sandstone cliffs up to 300 metres in height. It is famous not only for its rugged beauty, but also for its abundance of ancient Aboriginal rock art – the kind of art that Mr. Walsh has presented so magnificently in his various books.

During his years with National Parks and Wildlife at Carnarvon, Mr. Walsh heard many references to junjudees. He interviewed several eyewitnesses, all of whom described the creatures as being hair-covered, ape-like and about three feet tall. Most also mentioned the creatures' terrible smell. There were seasoned bushmen, he says, who would not camp in certain areas for fear of the little creatures.

"There were a lot of reports [up to about the mid 1970s], but people don't get out on their properties [on horseback] the way they used to. Nowadays people go in a vehicle."

One witness, timber man Graham Griggs, was kept awake by junjudees that leapt around in the shadows on the edge of his campsite and repeatedly jumped between his tent and the fire, leaving many tracks and scaring him so much that he abandoned the site altogether. Another timber getter, Leo Denton, of Injune, found tiny tracks and heard cries "like chooks cackling". His wife Joy also saw fresh tracks "like a kid's bare feet" in remote bush locations.

Interestingly enough, another local person compared junjudee vocalisations to those of birds. Retired timber worker Paddy O'Connor told journalist John Pinkney that he once encountered two of the little creatures while camped in the vicinity of Carnarvon Gorge. They gave off an absolutely nauseating odour and were, he said, "pointing at my billycan. [They] seemed to be exchanging comments about it. I wasn't

Grahame Walsh. (Courier Mail)

in much doubt they were using some type of language. It was a kind of chirping, but seemed to have a shape to it." Although it was just on dawn, and there wasn't enough light for him to discern the colour of their fur, Mr. O'Connor noticed that the creatures' eyes, like those of yowies, seemed to shine: "their reddish eyes were very visible". [Case 197]

Grahame Walsh finds the eyewitness testimony and Aboriginal lore quite compelling, and considers the Carnarvon area rugged enough, yet productive enough, to conceal and sustain a small population of creatures of the presumed size of junjudees. Although he has yet to experience an actual sighting, he has, at least, seen apparent junjudee tracks. Although they were in a very remote location near the headwaters of the Maranoa River, where no sane person would wander barefoot and alone, they were similar to those of a five-year-old child. "I followed them up a hill, but then I lost them."

Mr. Walsh doesn't pretend to know exactly what the mysterious creatures are, but is determined to keep looking until he finds out. "People will think I'm mad", he says, "but ... I'll just keep looking." [25]

Whatever junjudees may be, they are certainly widely distributed. Gary Opit recently sent us some material collected by Les Holland of Tully, in tropical North Queensland.

Nathan Moilan, whose parents are Aboriginal and Indian, told Mr. Holland that his father, a timber worker, often spoke of seeing little Hairy Men in the Kirrama Range behind Tully. The sightings supposedly occurred between 1990 and '91, just before rainforest logging was halted in the area. Mr. Moilan senior said the creatures regularly appeared beside the gravel road at around 10 pm and stood watching as he drove the last truckload of timber down and out of the mountains. Although he never actually stopped, he always felt secure enough, locked in the cabin of his truck, to slow down and have a good look.

Nathan's father had another remarkable story to tell. He said that one night, when he and his uncle were sharing a three-room hut in the mountains, a little Hairy Man attacked his uncle as he lay on his bed. Hearing desperate cries for help, Nathan's father rushed in and together he and his uncle wrestled with the very powerful little creature. Just as they began to overpower it, it broke free, jumped out the window and fled into the night.

This rare account of junjudee aggression is, as readers may have noticed, very reminiscent of George Gray's ordeal as described in chapter three. This is interesting, but also a bit confusing, because we had always assumed the creature that attacked Mr. Gray was not a junjudee but a small yowie. It would be strange if two different kinds of creature engaged in the same odd behaviour.

This brings us back to the most important question about the junjudees: are they merely juvenile yowies, or are they an entirely different animal?

Perhaps it should be mentioned before we proceed further that although most Aborigines east of the Great Dividing Range seem to know of both the yowie and the junjudee, there are some groups to the west of the mountains, notably the Wiradjuri, who seem to know of only the little Hairy Man. Among Aborigines who know of both creatures, however, opinions seem to be unanimous: the little Hairy Men are definitely a different breed from the "big fellas". Gary Opit points out, also, that none of the presumed junjudees seen by either Aborigines or non-Aborigines have been accompanied by full-sized yowie "parents". This, to Gary, strongly suggests the little creatures are, indeed, a separate species.

Because we would like to be able to prove the existence of the giant yowies, or at least persuade the reader that it is possible they exist, the junjudee has always been, to us, a very inconvenient little creature. It is difficult enough to believe that one sort of wild, hairy hominid could be running around, undiscovered, in the Australian bush, but the suggestion that there could be *two* entirely different types out there seems way over the top. The junjudee's reputed magical powers and fairy-like behaviour adds an extra, mind-bending measure of confusion.

Names of the beast

While it is impossible, in the absence of physical remains, to say for sure whether junjudees and yowies are separate species, a comparison of all known terms for both creatures seems to confirm, at least, that many Aborigines have long *believed* the two types are completely different.

There is, with only one exception, no similarity between the terms. The exception is the Wiradjuri term *yuurii*, which is pronounced "yawree" as in "story". Perhaps it is coincidental, but this word sounds very much like *yourie,* a variant of yowie, which was used on parts of the NSW south coast. It could be argued that "Hairy Man" is another exception to the rule, because Aborigines often use that term for both the yowie and the junjudee. They do, however, usually differentiate the two by saying "the big fella" or "the little fella".

But whatever you call them, *winambuu*, *yuurii*, *dinderi*, and whatever they actually are – monkey, man or nature spirit – junjudees obviously have a direct bearing on the yowie mystery. If this book is to be a comprehensive and honest analysis of the yowie phenomenon, we can't simply ignore the little creatures, inconvenient as they may be.

By now most readers are probably almost as confused as we are, so at this point it might be a good idea to summarise the reported similarities and differences between junjudees and yowies as follows:

Physical characteristics

Similarities:

- Both types are said to resemble hair-covered ape-like men or man-like apes.
- Neither have tails.
- Both are mainly bipedal.

• In the case of both creatures, there is confusion over the number of toes. Both are sometimes said to have four toes and sometimes five.
• Both often exhibit eye-shine – usually red – at night.
• Both are sometimes said to have sharp teeth and long, strong fingernails or claws.
• Both are sometimes said to smell very bad.
• The ears of both are rarely noticed or mentioned.
• The genitalia of both types are rarely noticed or mentioned, but most creatures are assumed by witnesses to be male.
• Both are often said to have black or brown skin.

Differences:

Yowie	Junjudee
Said to have virtually no neck .	Witnesses have not commented on the neck
Has extremely long arms.	Arms are apparently not noticeably long
Is often said to slouch.	Stands up Straight
Average height estimate is about 7.5 feet.(2.3m)	Average height about 3 feet. (1m)
Rarely, if ever said to have beards.	Occasionally said to sport long beards

Behaviour

Similarities:

• Both types are, to put it mildly, extremely elusive.
• Both leave far fewer tracks than might reasonably be expected.
• Both are omnivorous.
• Both are said to be fond of honey.
• Both are more active by night than by day.
• Neither uses fire.
• Neither uses tools, although both occasionally throw stones.
• Both are said to be curious about human children.
• Both sometimes enter houses.

Differences:

Yowie	Junjudee
Sometimes said to be dangerous to man.	Said to be fairly harmless.
Are solitary Creatures.	Aborigines say they live in groups.
Not noted for living in caves or other shelters.	Said to live in caves or holes in the ground.
Rarely reported very far west of the Great Dividing Range.	Frequently seen a long way west of the divide.

One difference between yowies and junjudees is that while Aborigines attribute many semi-magical qualities – talking, curing the sick, teleporting etc – to junjudees, they attribute less in the way of supernatural powers to yowies.

In reference to the junjudees' magical attributes, Kyle Slabb, of the Goodjingburra, says, "We're brought up with that awareness of the spiritual side of life. To us, that's part of whole life, part of our culture." As for the big hairy yowies, "They're not a spirit – you can touch them like you can touch a man".

Mr. Moilan and Mr. Gray both said the little hairy men were extremely difficult to wrestle with physically, and the same is certainly true figuratively. As we admitted earlier, the mischievous little junjudees have had us pretty well baffled for the last 23 years. Despite the Aboriginal lore and the sighting reports from non-Aborigines, there is not a skerrick of physical evidence to indicate such creatures ever existed in Australia.

In 2004, however, relatively recent remains of junjudee-like creatures, *Homo floresiensis*, were discovered on the Indonesian island of Flores – less than 700 kilometres from the Australian coast. Mike Morwood, of the University of New England, leader of the party that found the remains, thinks it quite possible that the ancestors of *H. floresiensis* went on to establish themselves in Australia.

We'll return to the junjudee and to the amazing *Homo floresiensis* in chapter seven.

Chapter 6

Summarising the Evidence

Aboriginal lore and eyewitness testimony

When one considers that most indigenous knowledge has been passed down, by word of mouth, over many generations, it isn't surprising that Aboriginal yowie lore contains some inconsistencies. Inevitably, because of the rather mystical worldview of many Aborigines, their lore also contains some semi-magical, folkloric elements. That having been said, we believe that the bulk of Aboriginal lore provides good corroboration for the testimony of non-Aboriginal eyewitnesses.

Non-Aboriginal testimony has been strong and consistent since the colonial era. One thing that lends it credibility is the fact that so many of the cases (115) involve more than a single witness. In 57 cases, two witnesses were present; in 29 cases three; in 10 cases, between four and six; in one case seven and in 16 cases "several". In two cases as many as 20 people supposedly viewed the creature. More than 100 sightings occurred in broad daylight and many encounters were at extremely close range.

Another consideration is the high quality of many eyewitnesses. In addition to scores of farmers, loggers, hunters, bushwalkers and other knowledgeable outdoorsmen, our files contain reports from three rangers, three surveyors, six soldiers (including members of the SAS Regiment), a zoologist and an environmental scientist.

Our witnesses are people of all ages and many different ethnicities. Interestingly, several of them insist that prior to their own encounter they were totally unaware of the yowie legend. Greek-born Kos Guines, Katrina Tucker (born and raised in Scotland), American resident Jason Cole and the "N" family (migrants from the Middle East) are among those who had never heard of the hairy giants before seeing one.

Aboriginal lore and non-Aboriginal eyewitness testimony have provided us with a pretty reasonable idea of what the "average" yowie looks like.

Physical characteristics

Yowies resemble huge ape-like men or man-like apes. They are frequently likened to long-legged gorillas. A full-grown adult is seven and a half to eight feet tall [2.3 – 2.44 m] and very heavily built; covered from head to foot in dark hair; its dome-shaped head may seem small in comparison to its very wide, but rounded, shoulders; skin is brown to black; eyes large and deep-set; ears small, set close to side of head; nose flat; mouth wide, lips thin; teeth large and fearsome; upper canines sometimes protrude over the lower lip; neck extremely short and thick; arms very long and muscular; hands roughly human-like with very strong nails or claws; legs as long, proportionately, as those of a human.

It is useful to examine some of those characteristics in greater detail.

Height

About 160 eyewitnesses have attempted to estimate the creature's height. While some younger people made their estimates in metres, we found that most people of all ages still seem to think in terms of feet and inches. We therefore converted everything to feet and inches and rounded off each estimate to the nearest foot.

The results were as follows:

Est. height in feet:	*No. of witnesses:*
2	1
3	5
4	9
5	17
6	34
7	45
8	24
9	11
10	7
11	2
12	4
13	1

These figures yield an average height of 6 foot 10½ inches [2.1m]. But because the smaller creatures were almost certainly juveniles (or even junjudees) it seems fair to assume the average mature yowie is considerably taller. So we did a second calculation excluding every estimate of five feet and less. That calculation suggests the average mature yowie stands about seven and a half feet [2.3 m].

It seems reasonable, in fact, to assume the average adult is even taller than that. While almost half of our witnesses didn't attempt a precise estimate of height, it is impossible not to notice, when sifting through their reports, the great number of instances where terms such as "very tall", "immense" and "much too tall to be a man" were used. If, as seems reasonable, we can assume those witnesses meant the creatures

If it had longer legs, a large male gorilla like this would be almost as tall and heavy as a full-grown yowie. (Courtesy of John Green)

were taller than, say, six and a half feet, then the height of the average adult yowie would be bumped up to something approaching eight feet [2.44 m].

Colour of hair

Hair colour is specifically mentioned in 111 cases:

Colour	*Number of reports*
Black	42
Brown or dark brown	27
Light brown / tan / fawn	13
Reddish-brown	10
"Reddish" and red	4
Grey	12
White	3

It is clear that the vast majority of yowies are brown to black in colour. In addition to the witnesses who specifically described them as black or some shade of brown, many others referred to the animals as "dark". In many other cases, where the creatures were said to resemble apes or gorillas, the implication is that they were dark haired.

"Jim", who repeatedly encountered yowies at a NSW south coast location, suggested that they might moult at certain times of the year. The hair of what he assumed were the same two or three creatures seemed to be brown in summer and grey in winter. [Case 187]

Face

While some witnesses are convinced the creatures are more ape-like than man-like, others are left with the opposite impression.

This dichotomy is most obvious when it comes to the creature's face. Despite its piercing, deep-set eyes and its "most horrible mouth … ornamented with … two large canine teeth", Charles Harper thought the face of the creature he encountered was "very human". Susan "X" was left with the same impression: "the face was virtually human". On the other hand, George Osborne, H.J. McCooey, Percy Window and many others noticed only bestial characteristics.

Many witnesses, in any case, have said that the face was so obscured by hair that they could hardly see any features at all. The obscuring hair sometimes hangs down from the top or sides of the head, and sometimes actually grows on the face. The facial hair is no longer than hair on the rest of the body, and is almost never referred to as a beard. It sometimes covers virtually the entire face, including the forehead and nose, leaving only the lips and a small area around the eyes exposed.

People who have encountered creatures with relatively hairless faces often say the brow is heavy (and sometimes receding) the eyes are deep-set, and the nose and face are flat. Almost everyone who has specifically mentioned the chin has said that it is not particularly pronounced.

Facial expression

Like Kyle Slabb, Sue O'Connor, who also observed a yowie at very close range, said it "was totally poker-faced – no hostility or otherwise was shown". The yowie met by Gary Jones and party also "had no expression – just blank". [Cases 179, 185 and 223]

Others have reported a range of expressions: "Looking … frightened"; "its eyes and mouth were in motion, after the fashion of a monkey"; "blinking its eyes and distorting its visage"; "grimacing" face and "piercing" eyes; "the eyes were glaring"; "penetrating" eyes; "tired eyes"; had a "shocked" look on its face; "a really angry expression"; "it stared as if it hated us"; "ugly and angry"; "a bit pissed-off; it had, like, a mean look on its face". Scott "X" thought the

Springbrook yowie looked sad, lonely, or a little lost. [Case 121]

We're still waiting for someone to report a smiling yowie.

Eyes

The eyes are often said to be set deeply under a heavy brow. Several people, such as Charles Harper, Percy Window and Justin Garlick, who have seen yowies at close range, say the eyes are quite large. H.J. McCooey, however, thought his creature's eyes looked "small and restless."

Eye colour – daytime

Most people who refer to the colour of the eyes say they are black, brown or just plain "dark". The only other colour mentioned is yellow. [Cases 30, 68 and 123]

The notion of an ape-like creature with big yellow eyes may not be as strange as it at first seems. Many gorillas have light brown, honey-coloured eyes, and some Madagascan lemurs – nocturnal animals – have large, glaring eyes with striking yellow irises. The question of why some yowies *don't* have yellow eyes remains, of course, unanswered.

Old red eyes is back: eye colour – night

There are several accounts of yowies' eyes lighting up at night when close to camp fires or caught in the beams of headlights. While that seems reasonable enough, other people have noticed eye-shine in situations where there is virtually no artificial light for the creatures' eyes *to* reflect. Even odder, many people have shone very bright lights on the creatures without producing any eye-shine at all.

Yowie eye-shine is usually bright red. Geoff Nelson's comment that the eyes were "vivid, red and glowing, like two flashlights shining back at you" is fairly typical. In a statement that contained an interesting similarity to *mooluwonk* lore, another witness described eyes that were "a deep, dark red, but with a brightness to it ... like a bright glow". [Case 156]

The next most frequently mentioned colour is yellow. Neil Frost was told a yowie that lurked around a property at Linden many years ago was known locally as "old yellow eyes". Neil has observed yowies' eye-shine changing from yellow to red and back again, as has a Sydney policeman who wishes to remain anonymous. Professional shooter Steve Croft has also observed the phenomenon and says that the creature's eyes glow red when it is facing both the light and the observer, but change to yellow as it begins to look away.

One witness recalled seeing green eye-shine. Another was certain the eyes were "a bright, iridescent greenish-blue". [Cases 153 and 258]

According to lower Murray River Aborigines, the *mooluwonk* has "dark red eyes". We assume they mean that the eyes reflect red at night. Percy Mumbulla and Guboo Ted Thomas said the *doolagarl's* eyes are red. Some contemporary Aborigines refer to the Hairy Man, colloquially, as "red-eye". In one old story the *thoolagal's* shining eyes were mistaken for stars.

Where's Willie?

Although nearly all witnesses assume the creatures they see are males, penises are almost never noticed. As the hair that covers its body is often said to be only two or three inches long it could be that the yowie's penis, like that of the male gorilla, is very small in proportion to the rest of its body.

Another possibility is that the organ, like that of many other animals, can withdraw almost completely inside the body. That would, of course, be very handy for any creature that habitually charges through thorny underbrush with no pants on. It might also explain why, on very rare occasions, sizable penises *are* observed.

When the yowie dropped her terrified little dog and stood up, Mrs. Jean Maloney [Case 111] couldn't help noticing that it most definitely had a penis. It was "quite large: maybe nine inches long" and seemed very similar to an uncircumcised human penis. A human-like scrotum was also evident. In that case the penis may have been all the more evident because that particular yowie, though very hairy on its head, shoulders, legs and arms, had only sparse hair on its chest and abdomen.

Hairy ladies

Female yowies have been reported only five times, and only three of those reports refer to a specific female characteristic: George Birch said that one of the creatures he saw "appeared to have breasts" and Malcolm "X" said the "hairy lady" of Four Mile Creek had "sagging breasts". At Springbrook, a woman spoke of seeing a yowie with "long, pendulous breasts". [Cases 85, 123 and 162]

Because it seems highly unlikely that males are encountered 50 times more often than females, it is tempting to assume that female's breasts are usually very small, like those of female gorillas. Perhaps the breasts increase in size to the point of being noticeable only when females are nursing their young. It seems reasonable that nursing females would avoid places where they might be seen and harassed.

Small animals of indeterminate sex are seen occasionally. Although they are sometimes assumed to be juvenile yowies, they are almost never seen in the company of full-grown creatures. Some of them may be junjudees.

Feet of clay?

Thanks to Aboriginal lore and non-Aboriginal eyewitness testimony, we have a pretty good composite picture of what the "average" yowie looks like from the top of its head down to its ankles. When it comes to anything below ankle level, however, the data is anything but consistent.

By far the most difficult thing to establish about the yowie is the shape of its feet. From the early colonial days to the present the question of exactly what they look like has been a frustrating mystery within a mystery.

Ten years ago, in *Out of the Shadows*, we observed that "everything about the yowie's appearance has become more or less known, and amounts to a more or less coherent picture – but only down to ankle level – where contradictions, uncertainty and confusion set in". At that time, when we had fewer than 90 reports to work with, we hoped that, "as the years go by more and better tracks will be found, photographed and cast, and the vexed matter of foot shape will be solved".

Dream on, yowie hunters. As the years have gone by, more tracks *have* been found, photographed and cast, but far from clarifying the situation, they have made it even more confusing.

Aboriginal lore is vague about the shape of the feet and non-Aboriginal eyewitness descriptions are contradictory. It is undeniable, also, that discoveries of possible yowie footprints are exceedingly rare. Sceptics might well say *suspiciously* rare. Whereas the tracks of most large wild animals are seen much more frequently than the animals themselves, actual sightings of yowies are reported about nine times more frequently than discoveries of supposed tracks (more than 300 sighting vs. only about 35 recorded track finds).

Surprisingly, although many have said that they are similar in shape, but larger than those of a human, Aborigines have rarely described yowie feet in detail. Although most imply the creatures have five toes, not all subscribe to that school of thought. The Yalanji people of Cape York, for instance, believe that Turramulli has, or had, only three clawed toes on each foot. Even stranger, Aborigines of the Great Victoria Desert identified the weird *two-toed* tracks found by Peter Muir as those of the *tjangara* or Spinifex Man. [Case 87]

As noted in chapter one, Aborigines of the colonial era occasionally mentioned a very bizarre belief: that the hairy giants' feet are turned backwards, so that their tracks confuse anyone trying to follow.

Non-Aboriginal testimony is equally vague and contradictory about the shape of the feet. Descriptions and sketches, even photographs and casts, of supposed yowie tracks display little uniformity, either in their general shape or in the number of toes. While the five-toed variety are most commonly reported, four-toed and three-toed tracks are found almost as frequently. To make matters worse, several of the weirdest-looking tracks were discovered immediately after credible witnesses reported actually sighting yowies.

Observations of yowies' feet

Although many eyewitnesses get the impression that yowies' feet are rather similar to those of humans, very few have had the presence of mind, during yowie encounters, to carefully observe them. In many cases, too, long grass or undergrowth has screened the feet and lower legs from view.

We know of only eight witnesses who have attempted to describe the feet in any detail. Unfortunately, although there is no reason to suppose any of them were anything less than truthful, their descriptions display little uniformity.

* Asterisks indicate first-hand reports.

Case 6: "Starfish-like feet"
Case 14*: "About 18 inches [46cm] long, and shaped like an iguana, with long toes."
Case 20: "Three claws on each foot."
Case 21: "One of its feet resembled the hoof of a horse and the other was club-shaped."
Case 33: As it ran, it "tore up the dust with its nails".
Case 35*: "The metatarsal bones were very short, much shorter than in the genus homo, but the phalanges were extremely long, indicating great grasping power by the feet." (The main part of the foot was short but the toes were very long).
Case 219*: The feet "were great big fat things. Huge ... the big toe was ... two inches wide and splayed out – way bigger than the other toes [but] the second toe was much longer than the first ... gnarled nails. The side of its foot, too, was really padded".
Case 226*: The creature's feet seemed different to those of a man. The heels looked odd, but they couldn't say exactly why.

Sometimes the only consistent thing about descriptions of yowie feet or footprints seems to be their inconsistency. In its march towards respectability the creature is hampered rather than helped by its mighty "plates o' meat".

Matching tracks

On the positive side, we know of two instances where very similar footprints have been cast or photographed in widely separated locations. In 2001, in a yowie hot spot near Kilkiven, Queensland, Dean Harrison found a line of large tracks and made a cast of one of them. Although the toes came out very badly, the cast is strikingly similar in the heel, instep and ball to the cast of a five-toed track made in 1998 in Springbrook National Park, 250 kilometres to the south. The two

A yowie's feet, as sketched by Richard Easton. (Case 219)

casts have so many points of similarity, in fact, that it seems certain they were made by the same type of creature.

In the other instance, a cast made on the yowie-plagued O'Connor property in the Blue Mountains in 2001 was found to be remarkably similar to a track photographed after Shaun Cooper's yowie sighting at Nerang, Queensland in 1978. Inconveniently, however, the O'Connor and Cooper tracks, though very similar to each other, are both clearly three-toed, and therefore nothing at all like the Kilkiven and Springbrook tracks. This glaring anomaly will be addressed in chapter seven.

The extreme scarcity of uniformly shaped, plausible-looking tracks is, not unreasonably, seen by sceptics as proof that the yowie has feet of clay. All serious researchers concede it is a major problem and some have attempted to explain it. We will examine their theories later.

Yowie behaviour

The creature is seen most often between about 1 pm and 3 am and least often between 4 am and 10 am; is probably most active at night; possesses very good night vision; is afraid of fire; shuns bright lights; is solitary; bipedal; occasionally quadrupedal; occasionally sways or waddles sideways in a crab-like manner; can move amazingly fast; occasionally stamps feet, creating a booming sound; stalks and covertly observes people; occasionally (in 9.5 percent of cases) emits a devastatingly foul odour; occasionally climbs trees; breaks and twists saplings; bites

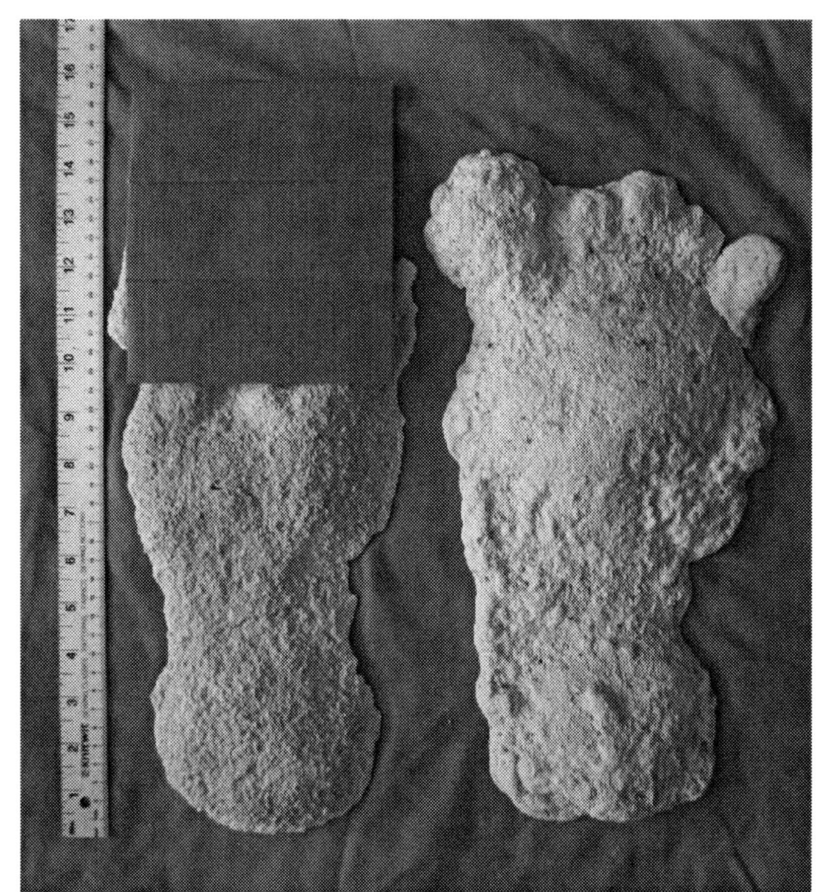

(Left)
Though incomplete, the Kilkiven cast (left) is strikingly similar to the Springbrook track (right).
(Paul Cropper)

(Below)
Left track – Nerang, Qld 1978 (Case 125). Right track – Blue Mountains, NSW 2001 (Case 223).

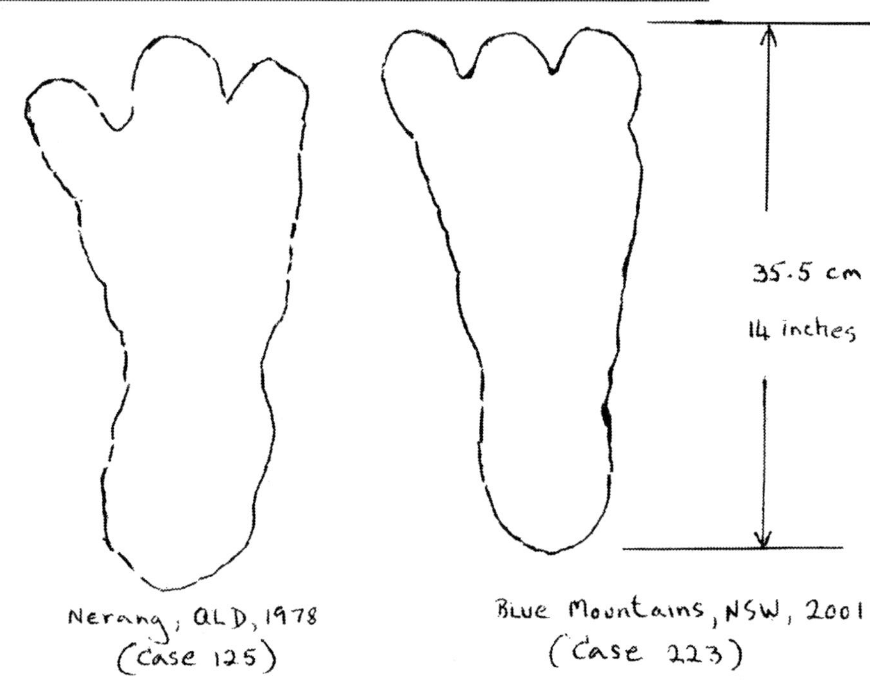

saplings; tears bark from trees; occasionally constructs teepee-like structures, presumably as territorial markers; occasionally builds nests on the ground; occasionally raids campsites; occasionally approaches and even enters buildings; swims; appears to be less active in summer than in other seasons; may visit certain locations at particular times of the year; may occasionally use sticks for hunting and foraging; throws stones, sometimes to disorientate prey or pursuers; occasionally engages in terrifying displays of aggression, uprooting shrubs, thumping trees, houses and cars; sometimes chases, but very rarely hurts people; occasionally tears the heads off smaller animals; emits a wide variety of bestial noises and occasionally more sophisticated vocalisations.

Night stalkers

Eighty-seven witnesses specified the actual time of day or night that their sighting occurred. (See graph below.)

Another 116 witnesses specified only "day", "night", "dusk", "dawn", "early afternoon", "late afternoon" or "early morning". Because there is only a slight difference between the number of "day" cases (35) and "night" cases (38), there is little point in adding them to the graph. The great disparity between mentions of "dusk" (19) and "dawn (5), and between "afternoon" (18) and "morning" (1), on the other hand, is statistically significant.

If we take "dawn" as meaning 5-7 am, "dusk" as meaning 5-7 pm, "afternoon" as meaning noon to 4pm and arbitrarily allocate the single "morning" report (the witness actually specified "early morning") to 7 am, and add those statistics, we get a slightly different picture. (See graph next page.)

This graph shows that only slightly more yowies are encountered during the day (7 am to 6 pm inclusive) than during the hours of darkness (7 pm to 6 am). It is evident, also, that the frequency of sightings rises sharply around noon and stays high (apart from a strange drop to almost zero at 8 pm – dinnertime for humans?) right through to about 3 am, after which it drops down to a very low level at dawn. The fact that there are few yowie sightings at dawn might simply

Sightings By Time Of Day

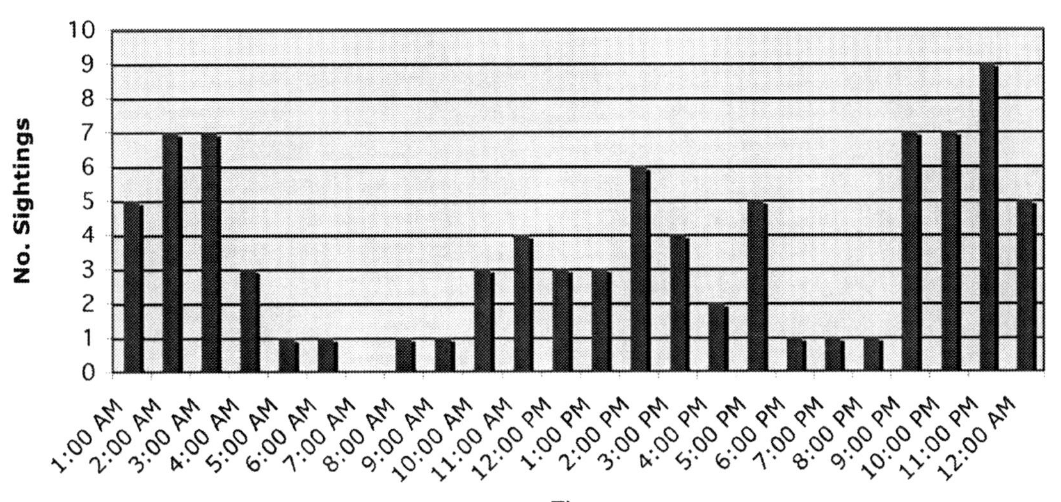

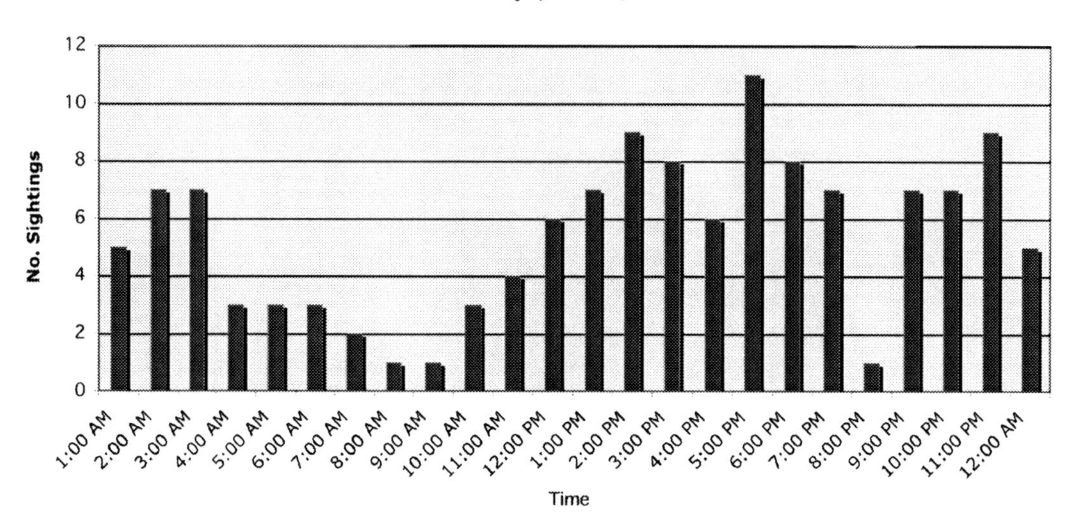

reflect the fact that few Australians these days are such early risers, but the similarly low number of reports from 7 am to 10 am is more interesting, and suggests yowies might grab a few hours sleep during mid-morning.

While few yowies are seen just before dawn, it is worth bearing in mind that *very* few humans are abroad at that time. Virtually no one walks through the scrub at that hour, and it would probably be safe to say that very few – certainly less than 5 percent – regularly drive country roads between 4 am and 5 am.

So between 4 am and 5 am, when less than 5 percent of the human population are abroad to see them, the frequency of yowie sightings remains at about 40 percent of the 4 pm to 5 pm level. All this suggests that what Aborigines have long said about the hairy giants is correct; they are predominantly night stalkers.

It occurred to us, while crunching the above numbers, that the ratio of daytime to night-time sightings might have been different in the days before motorcars were common and when electric light was not universally available. Prior to, say, 1940. Sure enough, on looking again at the early reports, we found that during the years 1847 to 1940 yowies were reported almost twice as often during the day as during the night (26 day vs. 14 night).

The fact that yowies' eyes have been seen to reflect light strongly suggests they are primarily nocturnal. Artificial light, in fact, seems to irritate and disorientate the creatures. They have been seen to shade their eyes against torchlight, and when Charles Harper's chainman threw dry leaves on the fire "the creature stood erect, as if the firelight had paralysed him".

Events at three sites in the Blue Mountains [Cases 160, 205 and 223] strongly suggest yowies can detect infrared light – a handy talent for a nocturnal animal.

Neil Frost, Dean Harrison and others who have seen yowies run at breakneck speed through thick scrub in the dead of the night have no doubt: the creatures can see in the dark.

Seasonal activity

One hundred and thirty one eyewitnesses specified the month of the year in which their sighting occurred. Surprisingly, when that data is plotted onto a graph, we see that considerably fewer yowies are reported in summer than in the other seasons. This is remarkable because, not only are the days much longer in summer, but a vast number of people take their holidays at that time of year and "go bush". One would have expected many more sightings to be reported in summer than in any other season.

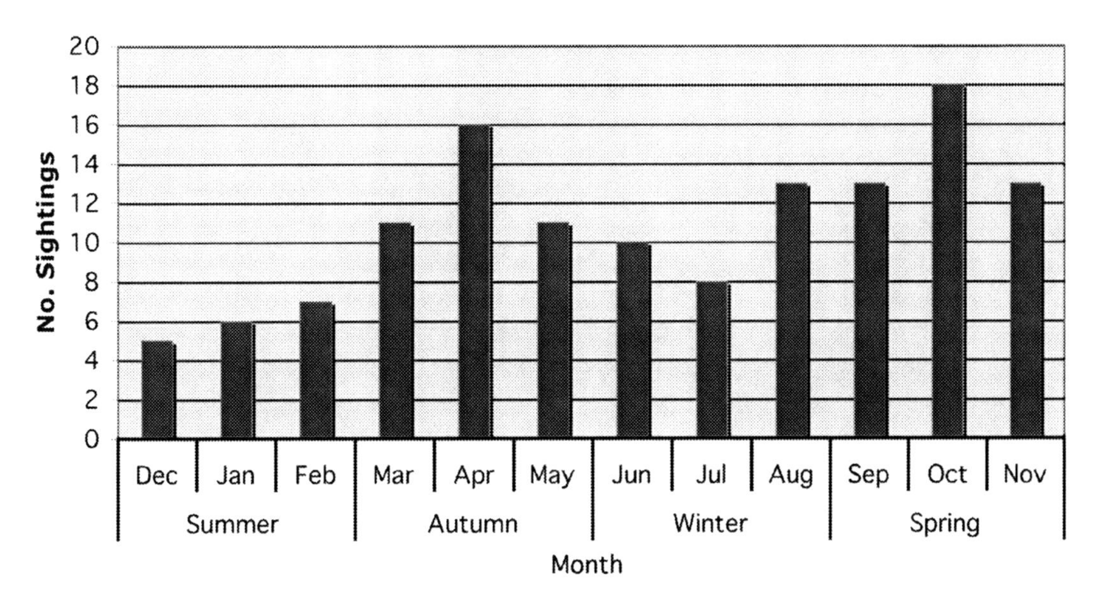

Seasonal migration

In one case – a series of "yard invasions" at Springwood – the creature always visited in November. In another, the "hairy lady" of Four Mile Creek – or at least her odour – was always noticed in winter. [Cases 187 and 162] Other strong indications of seasonal migration, discovered by yowie hunter Paul Compton, will be discussed later.

Diet

Yowies are omnivorous, opportunistic hunters and scavengers. They have been seen eating, or have been reasonably suspected of eating, the following: a steer, wallabies, kangaroos, possums, dogs, cats, birds, meat scraps, eggs, fish, garbage, dog food, potatoes, carrots, tomatoes, fruit of the lilli pilli, bananas, apples, plums, bread, peanut butter, honey, leaves, bark, beetles, crickets, grubs, termites, ti tree pods, bracken fern and saltbush berries.

We have two reports of yowies rejecting food. At the O'Connors yowie "feeding station", the creatures sometimes took fresh bread rolls and hurled stale ones to the ground. In Royal National Park a yowie apparently ate steak, bread, a tomato and a banana, but ignored lettuce and snow peas. On one occasion, it toyed with the proffered food in a gruesome manner: beheading and eviscerating a chicken, but leaving the body uneaten. [Cases 223 and 233]

It is interesting to note that although bananas aren't native to Australia, yowies, or at least some yowies, seem to have worked out how to peel them. [Case 233]

Aborigines say yowies are omnivorous, sometimes adding that they are very partial to honey. Some, particularly those interviewed during the colonial era, said the creatures were man-eaters. Percy Mumbulla said they are tantalised by the smell of cooking fish.

The scent of a yowie

Few large omnivores are particularly fragrant, but the adjectives people use when struggling to describe the smell of a yowie make it clear that, when it comes to body odour, the Aussie ape is in a class of its own.

About 9.5 percent of witnesses speak of unpleasant odours, and the mildest of their comments compares the smell to the strong, offensive odour of a ferret. Most use more colourful terms: "like a badly-kept public lavatory"; "like a badly-kept hen house"; "like a cave full of bats"; "urine"; "vomit"; "a filthy stench"; "like I was behind a garbage truck"; "like a dead rat"; "terrible ... something like a pig ... sour, stale ... like something rotten"; "like rotting flesh"; "like rotten meat in a garbage bag"; "like rotten meat but worse, more primal, somehow scary and sickening."

Offensive as they might be, odours like excreta and rotten flesh don't seem entirely inappropriate for a huge, shaggy, unwashed omnivore. More difficult to rationalise are references to yowies that had about them a strange, "burnt" odour. Geoff Nelson, for instance, was struck by a "strong, acrid, electrical smell, like burnt bakelite – like when you blow up an old radio – a sulphury stink"; Alwyn Richards was assaulted by a "terrible burning smell" and "Sue" was revolted by a smell "like a burnt mattress". [Cases 94, 108 and 225]

The stench of a really smelly yowie can be absolutely overwhelming. The particularly foul Springbrook creature caused Percy Window to vomit on the spot, and the Acacia Springs yowie left Katrina Tucker gagging. [Cases 123 and 220]

Let's face it: yowies stink. But if they are so terribly smelly, why do most (90.5 percent) of witnesses – even many who encounter the creatures at close range – detect no odour at all? Are only some of the creatures smelly, or are *all* of them smelly – but only some of the time?

It has been suggested that, like some other animals, adult yowies might exude strong odours in mating season. Dean Harrison, who encountered a yowie that smelled like cow manure, heard another theory. Local graziers told him dogs that rolled in manure seemed to attract fewer ticks. Perhaps the yowies had learned the same trick. That theory, of course, would only apply in cattle country.

Another suggestion is that the yowie, like an enormous antipodean skunk, can release a stupefying odour at will. Although that theory may sound pretty wild, it corresponds with Aboriginal *mumuga* lore, and the experience of one of our best witnesses seems to support it. At Springbrook, Percy Window and the yowie stood facing each other for some minutes until "it suddenly gave off a foul smell that made me vomit". While the ranger was occupied decorating the daisies, the pungent pongid "made off sideways and disappeared". Paul Cronk's dog – which ran from a yowie, jumped into a pond and frantically washed its face – may have been another victim of the creature's chemical weapons arsenal. [Cases 123 and 114]

The yowie's scent often lingers for a remarkably long time. When he returned to Scotts Creek three days after his encounter, Geoff Nelson was struck by a "strong, acrid ... sulphury stink" that seemed to permeate everything in the vicinity. A similar stubborn stench was apparent after the Woodenbong yowie groped Jean Maloney's terrier.

The prize for the weirdest smell of all must go to the "hairy lady" of Four Mile

Creek, Victoria. That particular creature left, on several occasions, a lingering odour described as "something like chicken broth". [Case 162]

The big tramp

The fact that the same obscure, offbeat details of yowie behaviour crop up in eyewitness accounts, in widely separated areas, many years apart, strongly suggests they are real animals. Six sets of witnesses, for instance, mentioned the yowie's employment of a strange, sideways movement:

"A sort of forward and sideways movement ... a bit like Charlie Chaplin used to walk – sort of a waddle, but not exactly ... because there was a sort of stamp or stomp at the same time"; "[it] walked to one side in a crab-like style ... a swaying, sideways movement"; a "crab-like running walk"; a strange, swaying, side-to-side motion; "made off sideways"; "walked funny: a bit of a waddle, from side to side ...". [Cases 59, 115, 220, 256 123 and 219] Significantly, that unlikely-sounding behaviour is also mentioned in Aboriginal lore. Percy Mumbulla and Guboo Ted Thomas once stated that the *dulagarl* "walks from side to side".

Curious about children

One rather unsettling aspect of Aboriginal lore is the belief that yowies are curious about human children. Although sinister motives are not generally involved, there are some stories of children being carried off. [e.g. Case 38].

While several dramatic yowie episodes *have* involved families with young children, our files clearly indicate yowies are just as interested in adult humans as they are in children.

Peeping pongids

Yowies have approached young lovers on a few occasions [e.g. Cases 99, 127 and 183]. While the number of such cases may not be statistically significant, the behaviour might still be worth noting.

Although the *dulugar* in Big Charley's story [Case 43] supposedly wanted to force its loathsome attentions onto a human female, we can find no specific reference in Aboriginal lore to voyeuristic yowies.

Aggression

While yowies usually seem peaceful enough, there are plenty of instances of aggressive behaviour, from mildly menacing growls to terrifying fits of branch ripping, car rocking and wall thumping. People have had large rocks thrown in their direction. Several have been chased.

Survivors of such chases are convinced the creatures meant to tear them limb from limb. It is reasonable to wonder, though, whether the yowies, so immensely strong and fast, really wanted to catch them. Perhaps their terrifying displays, like those of gorillas, are mainly bluff. On the other hand, yowies reportedly chased horsemen, with murderous intent, on two occasions near Byron Bay. A horse was mortally wounded and a rider injured. In 1932, a Victorian man had his shirt torn to ribbons. [Cases 218, 256, 40 and 58]

Some yowies seem to find motor vehicles irritating. Cars and trucks have occasionally been thumped and damaged. [Cases 127, 136, 241 and 135]

It seems that yowies thump the walls of newly built houses because they are angry at having their territory encroached upon. But if the thumping is meant to drive the occupants away, it generally fails to do so. In every case we know of, the residents, though jittery, have stayed put. Why then don't the immensely strong animals proceed to bash the doors down or lob great rocks through the windows? Perhaps, like their practice of chasing, but never catching, fleeing humans, this restraint indicates they aren't as aggressive as they seem to be.

Even in cases where yowies actually grabbed people, there is reason to doubt they meant serious harm. Although the smallish yowie that tried to drag George Grey out of his house could have bitten and scratched him, it confined itself to wrestling. Similarly, the "big, hairy monster" that grabbed Mrs. Nott could have strangled her on the spot, but it fled when she screamed. [Cases 83 and 70]

We have heard two stories about police supposedly retrieving, from yowie hot spots, human bodies that had been torn to pieces. One story was at best, second or third-hand, and contained that staple of shonky fakelore, the "high-level cover-up", but the other seemed a bit more substantial.

In the 1990s Neil Frost interviewed a Blue Mountains-based lawyer who'd made a grisly discovery about 20 years earlier. When he was 16 years old he found the decapitated body of a man in the bush behind Warrimoo Bush Fire Brigade station. The head was found later, 50 feet away. The police, he told Neil, were baffled: the man had apparently been killed and beheaded on the spot, but they couldn't determine how (the head had seemingly been *pulled* from the body) or why it had been done. Interestingly, one of Neil's students, who knew nothing of that story, told him he'd seen a yowie in the same area in 1997. [Case 268]

Rumbles in the jungle: vocalisations

Many people have heard loud, frightening vocalisations in yowie hot spots. Many have actually observed the creatures vocalising.

The terms most frequently employed by people struggling to describe the often very frightening vocalisations are "screaming" and "screeching". Terms like "growling/snarling"; "moaning/groaning"; "grunting"; "bellowing" and "roaring" are also frequently used.

Very similar descriptions of vocalisations have come from witnesses in different eras and in widely separated regions. Stella Donahue, in Tasmania, and Paddy O'Connor, in Queensland, for instance, both said the Hairy Men made strange bird-like noises. [Cases 173 and 197] That very odd detail is also an obscure element of NSW south coast Aboriginal lore.

Very few people are better qualified to assess animal calls than naturalist Gary Opit. On June 21, 1978, on Tamborine mountain, he heard "a very powerful, continuously repeated roaring, bellowing call ... a deep-throated, booming 'yee-yee-yee-yee' that continued without a break for five minutes. I could clearly hear the calls being pumped out of a massive chest, and they sounded more like the call of a big primate than anything else. It was much more powerful than the roaring-grunting of a koala or even the bellowing of cattle".

On June 1, 1996, from his house on the slopes of the Koonyum Range, in northern NSW, he heard powerful bark-like calls rending the air. "[They] were mostly in a series of three, making a sound like 'arroo-ARROO- arroo'... between the sets of three barks ... a disturbingly strange, soft, gurgling call. 'gu-gu-gu-gu', could be heard. It continued ... for five minutes. They had a primate feel to them."

In the morning he found three toe prints in a creek bank: "Each was about the same size as a human big toe, but each slightly smaller in size towards the right". Only two months earlier a smallish yowie was seen within 200 metres of Gary's house. Another appeared on an adjoining property in 2002. [Cases 210 and 272]

In the Blue Mountains, outdoor educator Pat Ryan and several of his associates have heard whooping calls they feel sure are produced by yowies. Others have reported more sophisticated vocalisations. Pat has talked to reliable people who have heard, while in the proximity of yowies, murmurings reminiscent of some Aboriginal dialects. Near yowie-plagued Ormeau, researcher Dave Glen heard crudely rhythmic mutterings a little like rap singing.

Jenny "X" [Case 258] wrote a particularly good description of fairly sophisticated yowie vocalisations:

"The voice quality was so foreign that I knew it wasn't human and my heart started pounding. The voice(s) was very deep and had a resonant quality suggesting a very large chest area. At times the voice(s) had a somewhat steely quality, but the talker(s) sounded as if he/she/they were trying to keep the volume down, so there was a breathy/half whisper quality too. He/she/they used mostly monosyllables consisting of a consonant followed by a vowel. The one I remember well was "jy" (rhymes with "my"). There were at least four different monosyllables of the same consonant-vowel pattern. They were uttered alone, not strung together as phrases or sentences ... it sounded as if [one creature] was commanding or instructing another. There were other sounds, all guttural – a deep sound a bit like throat clearing and a deep, oscillating sound in the throat."

One night in the middle of the Pilliga Scrub a large, bipedal creature approached an Australian Yowie Research campsite. The animal remained out of sight but Dean Harrison got to within a few metres of it, whereupon it angrily "told him off". "It sounded a bit like language", Dean recalls, "something like you would expect to hear if a cave man or Neanderthal was trying to express itself, if you can imagine that."

No furry bodies

The most difficult thing for yowie enthusiasts to explain is the undeniable fact that not a single carcass – or even a small part of one – has ever been dragged in from the bush and placed on a laboratory bench for all to see. Yowies' hulking bodies, it seems, are as elusive in death as they are in life.

There are a few old Aboriginal stories of yowies slain, and a few cases in which non-Aboriginal people claimed to have killed or wounded them.

Yowies shot, speared and stoned

We know of two legends concerning large-scale warfare between yowies and Aborigines in centuries past. [See chapters one and seven] If those stories are true, then many yowies must have been slain by heroes whose names are long forgotten.

While those tales were handed down by word of mouth through untold generations, a first-hand account of yowie-slaying was provided by "Black Harry" Williams. As mentioned in chapter one, he claimed that in 1847, when he was about 10 years old, he saw a large group of Aborigines spear a Hairy Man to death. It was like a black man but covered all over with grey hair. Two warriors dragged it downhill by its ankles. Mr. Williams was a highly respected man, and we like to think his story has a distinct ring of truth. On the other hand he was describing, in 1903, an event that occurred about 56 years earlier. Intertribal skirmishes were still occurring during the 1840s. Sceptics might suggest he simply misremembered one such skirmish. [Case 2]

Stories by non-Aborigines are equally problematical. Although we were convinced of the sincerity of Jim Banks and Kos Guines, who told us how they'd shot and wounded yowies in 1971 and 1977 [Cases 90 and 116] we have yet to find a really convincing account of one actually being killed.

A rock in the head

On October 28, 1893, the *Goulburn Evening Penny Post* reported that Arthur Marrin encountered a strange animal a couple of miles south of Captains Flat. Covered with tan coloured hair and six or seven feet tall, it chased his dog out of the scrub. When it moved towards him with its arms extended, Marrin threw a stone, hit it on the temple, brought it to the ground and finished it off with the butt end of his whip.

He allegedly took the carcass back to Braidwood where staff of the *Braidwood Dispatch* examined it. In a suspiciously brief description they said its torso was four feet [1.2 m] long and its head eleven inches [28 cm] wide at the forehead. Its face was "very much like that of a polar bear". It weighed more than seven stone [98 lbs or 45 kg]. "Its forearms were strong, with great paws that would be capable of giving a terrible grip". [Case 32]

Anyone who has seen a bear or a gorilla at close quarters must find the idea of a man slaying a comparable creature with a stone to be nearly incredible. Because of odd references to "giant wombats" in a follow-up article and some laboured attempts at humour, we are pretty sure the story was a load of claptrap.

Monkey business

Because of the rather off-hand manner in which he told the story, and his failure to describe the carcass, we also harbour doubts about Arthur Bicknell's supposed slaying of the monkey-like "wood devil" in north Queensland in the late 1880s. [Case 31]

The lady vanishes

That leaves us with Tom Chapman's story of having shot dead a female "ape-creature" near Watsons Creek NSW in 1923. He supposedly left the carcass where it lay and fled. Two questions, of course, come to mind. If the animal was dead, why did he flee? And why wasn't the carcass retrieved for proper examination?

The story seems rather flimsy and is, in any case, fourth-hand. Mr. Chapman supposedly told it to a friend, Henry O'Dell, who passed it on to a Tamworth newspaper, from whence it found its way to the *Psychic Australian* in August 1977. [Case 45]

We know of four instances where yowie carcasses or parts thereof have supposedly been discovered.

Walla heck was it?

The story of the carcass supposedly discovered in the Walla Walla Scrub in 1876 contains several tantalising details. Local settlers believed the scrub was inhabited by "the hairy man of the wood", and said that horses and cattle shunned the area. Apart from the weird shape of the feet – possibly gnawed by scavengers – the description given by the unnamed sawyers tallies with descriptions of live yowies. The "down side" is that the story was effectively third-hand before it saw the light of day in the *Australian Town and Country Journal.* [Case 21]

Where's the cave, man?

Much as we like H.J. McCooey's story of seeing an "Australian ape" in 1882, we find his assertion that "the skeleton of an ape, 4 ft in length, may be seen at any time in a cave 14 miles from Batemans Bay" difficult to accept. McCooey clearly knew the importance of such a relic. His failure to produce the bones speaks for itself. [Case 25]

Off with its head

Although the next story has a ring of truth, it must also take the cake as the most frustrating yowie story of recent years. In 2004, Carol Schaeffer of Yowrie, NSW, told us that her late Uncle Neville often spoke of a huge skull he found in the bush in the 1930s or '40s. Carol found the story so interesting that she asked him to repeat it many times before his death in 1989. The skull, he always insisted, was twice the size of a human's. He thought it looked like that of a "prehistoric man", so he carried it back to his property near Kybean. On arriving home he put it on top of a fence post while attending to some chores.

No prizes for guessing what happened next. Cows nudged the post, toppled the skull and trampled it to pieces. Nice one, Cosmic Prankster.

There's no arm in it

In 1979, *The South East Magazine* quoted yowie hunter Rex Gilroy as saying that during the preceding year, at a southern NSW location that he was not at liberty to reveal, skeletal remains including a huge non-human arm, hand and finger bones, were found on a snow-covered hillside. He was quoted as saying that the bones, which were still attached to long brownish hair, "confirm that the length of a full-grown yowie's arms does indeed reach down toward the knees and that the animals do reach a height of 12 ft." We heard nothing more of this potentially earth-shattering discovery until 2001, when Rex disavowed the story, saying that he had been misquoted.

One would expect that during the course of 60,000 years Aborigines would have discovered yowie remains on many occasions. Although most tribes moved around a lot and travelled very light, it still seems strange that not a single yowie relic – not a skull, not even a tooth – seems to have been preserved by any Aboriginal group anywhere.

So apart from the skeletons known as the Kow Swamp People [see following chapter] no remains of any possible relevance to the yowie phenomenon have been discovered in Australia.

Camera-shy critters

As far as we know, no one has ever taken a clear photograph of a yowie. The only two pictures worth considering are those that were taken by Neil Frost in the Blue Mountains. Although very murky, one of them shows what looks like a dark, non-human face peering around a tree; the other shows only a pair of glowing eyes. Neil cheerfully acknowledges that neither picture is clear enough to convince anyone who is inclined to be sceptical. He is, however, quite convinced that they really do depict, however imperfectly, the legendary Hairy Man. Given that two good sightings of yowies have occurred within just a few feet of where both shots were taken [Cases 261 and 262], we believe he is right. [See chapter four]

Like most of our colleagues, we have many pictures of bitten, broken and shredded trees and large, generally rather indistinct, footprints. We also have scores of photos of bemused witnesses pointing into empty paddocks, saying things like, "It was right over there and it was seven feet tall, so help me." Such photos might excite the likes of us, but they leave sceptics completely unmoved. There is one yowie-related photograph, however, that any fair-minded person will find harder to dismiss: the picture of the dog that was supposedly grabbed by the yowie at Woodenbong in August 1977.

Jean Maloney was within six paces of the huge, shaggy ape-man as it clutched the screaming dog to its chest, fully illuminated by a powerful yard light. Her husband and a neighbour heard it, and Gary Buchanan of the Lismore *Northern Star* photographed its tracks. Another neighbour, Thelma Crewe, saw two similar creatures in her own yard nine months earlier, at very close range in good light. Is it reasonable to suppose all those people were hallucinating or lying? Gary Buchanan's photograph shows quite

Jean Maloney's dog, showing the wounds inflicted by the yowie. (Lismore Northern Star)

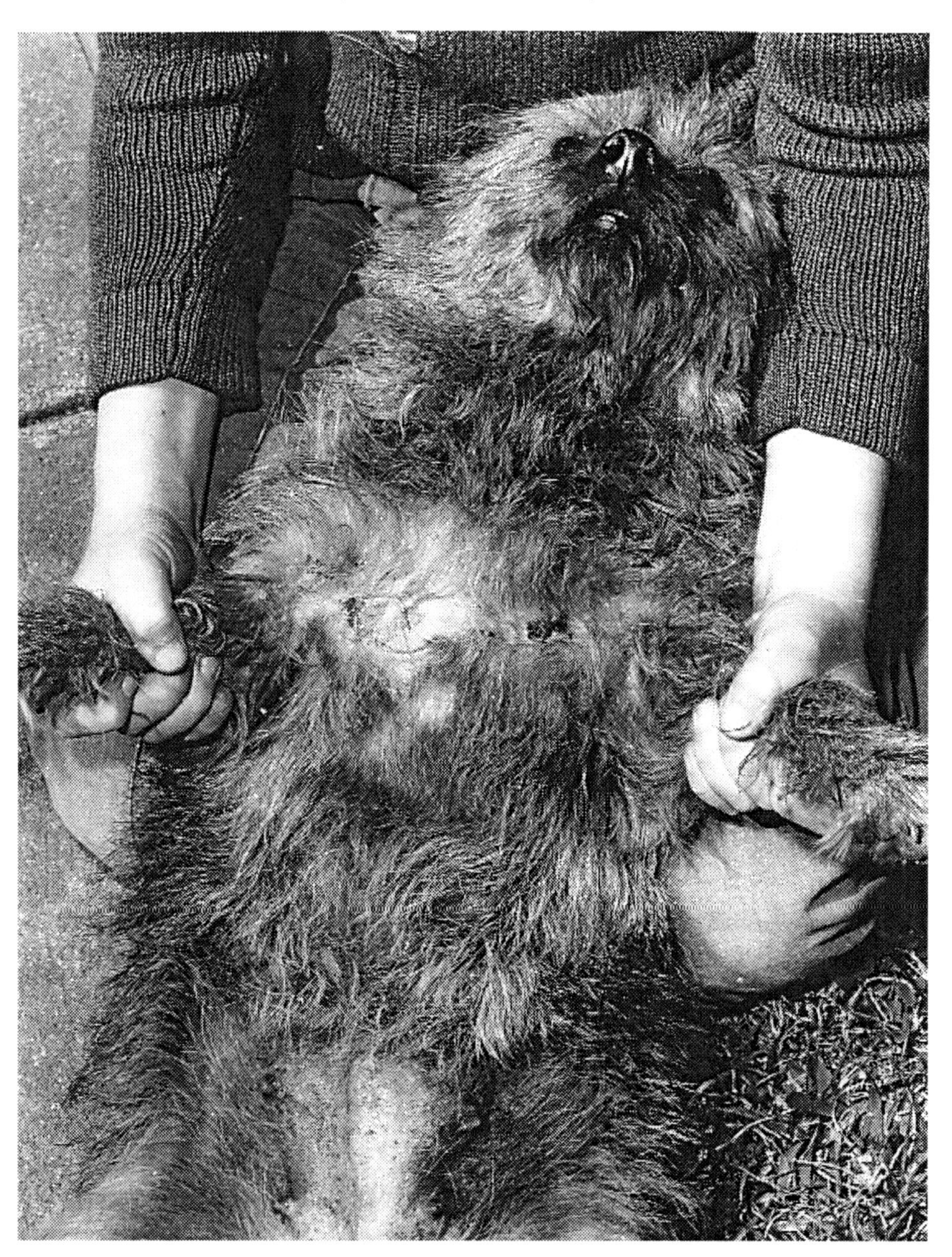

clearly that the dog was wounded by something. The poor animal stank so badly afterwards that it had to be washed in antiseptic. It never recovered from its ordeal and died a short time later. [Cases 105 and 111]

Paul Compton, of Glen Innes, has taken several other very interesting yowie-related photographs. (See pages 154-155.)

Tangible evidence

During and after yowie sightings, people have reported car windows smashed, doors scratched, trees bitten and broken, strong bushes uprooted, heavy logs moved, heavy water tanks pushed over, horses injured, steers, goats, wallabies and possums killed and eaten. While such damage might leave sceptics unimpressed, eyewitnesses and yowie researchers interpret it as tangible evidence of the creature's existence.

Gary Opit has found physical traces that, he believes, could only have been left by a yowie. In October 2003, in Jimna State Forest, he examined trees that had been damaged in a way he'd never seen before in his 50 years as a naturalist. Over the course of two days, in an area notorious for yowie sightings, he discovered over 500 Green Wattle trees (*Acacia irrorata)* from which wood-boring grubs (larvae of longicorn beetles) had been cleverly extracted. The trees, about ten to 20 centimetres [four to eight inches] in diameter had all been damaged in exactly the same way: at heights of between half a metre and one metre above the ground, they had been torn or ripped as if by tremendously strong finger or thumbnails.

Every rip or tear "consisted of a strip of bark torn vertically down the trunk and measuring 12 to 35 cm in length and 5 to 6 cm wide. There were never any claw, teeth or beak bite marks on the smooth, soft bark. However there were almost always a series of thin, straight indentations 1.5 cm in length vertically along one or both sides of the tree tear. Sometimes there were identical indentations directly above and below the tree tear. These appeared to be the impression of a thumbnail and it appeared as if the unknown animal was pressing its nail into the bark to determine the location of its prey.

"The strip of bark was always left hanging and then the same thumbnail appeared to have been used to tear out 2 or 3 pieces of the underlying timber. These were usually 12 to 15 cm long, 3 to 4 cm wide and 1 to 2 cm thick and were lying directly beneath the tree tear where they had fallen. This always exposed the 5 to 7 mm wide tunnel of the wood-boring larvae that was aligned vertically within the tree tear.

"Only a 5 or 6 cm length of tunnel was exposed, always near the base of the tree tear and only the top quarter of the tunnel was removed when the elongated chip of wood had been torn out. There was never any damage to the tunnel such as a cockatoo would create to enlarge the hole ... Instead, the animal must have inserted an elongated fingernail and hooked the larvae cleanly out of its resting place.

"I was particularly impressed by the efficiency used in the predation ... with every attempt to locate and extract the grub exactly the same and every attempt successful. There was no sign of tool use and the bark and timber was too strong to have been ripped open by human fingernails. I could think of no known animal that could predate on grubs in such a way."

Yowie researchers Nigel and Jeanie Francis, who guided Gary to the site and who sometimes camped there, said that, as fresh tree rips were discovered each morning, the grub hunter seemed to operate at night. Once, while spotlighting, they disturbed a

Gary Opit. (Tony Healy)

bipedal, black-furred, human-sized animal that quickly moved away.

Big jobs

It seems likely that two massive droppings – one found by Katrina Tucker in the Northern Territory and one by "Jim" in south-east NSW – were left by the hairy giants.

Mrs. Tucker made her fragrant discovery at the spot where she encountered a huge, screaming ape a few days earlier. Jim found the other mighty nugget in an area where he frequently saw yowies, at the base of a sapling that had been snapped and twisted about six feet from the ground. Both droppings were more like those of a human than anything else, but were about four or five times larger, and infinitely more smelly.

Neither Mrs. Tucker nor Jim attempted to preserve the droppings for analysis, and if the steamy calling cards were as massive and foul as they say they were, we don't blame them for a minute. [Cases 187 and 220]

Although Lynn and Gordon Pendlebury [Case 205] took apart droppings found on their yowie-plagued property (finding only vegetable matter), they didn't preserve any of them.

Paul Compton is the only person we know of who has both analysed and preserved an apparent yowie dropping. When he found it, in an area where he and others have seen yowies and discovered other evidence [see Case 279], the massive object was still hot and steaming. It bore no resemblance to the scat of any known Australian animal and, since it was 18 inches [45 cm] long and approximately eight inches [20 cm] in circumference, it was difficult to imagine that it had been left by a human being. Its content, too, suggested a non-human animal.

Whew! Katrina Tucker indicates the size of the yowie dropping. (Tony Healy)

"I dried it out", Paul wrote, "and the break up of its contents was approx 50% leaves/bark, 30% black beetle/crickets, 10% white ground grubs, 5% tea tree pods, 5% bracken fern (this is poisonous to livestock, so he must be immune). There was definitely no bones or hair evident in the scat."

The dropping also stank abominably: two years after its discovery, its dried out remains still smell terrible.

Yowie nests

After they saw and tracked a yowie in the Gangerang Range, "M" and his brother found "a superb sleep site, constructed of grasses, leaves and long saplings together under a slight overhang". Other comfortable beds were found in WA. Less luxurious nests have been found near Springbrook, Yellow Rock and in Royal National Park. One yowie was apparently living in a tunnel it had created in a lantana thicket. [Cases 115, 154, 199, 205 and 233]

Pat Ryan found a probable yowie nest between Valley of the Waters and Leura Cascades in the Blue Mountains: "In a rainforest gully, under a sandstone overhang, we found a bed made of grass that wasn't from the immediate area. It probably came from the top of the escarpment – 100 metres higher, at least. It had a faint animal smell and looked as if it had been slept on more than once. There was no sign of a fire or other human activity. The spot is close to several areas that have had a long history of Hairy Man encounters."

One "nest" was of particular interest because it was found by a party of scientists led by one of Australia's most knowledgeable bushmen, Major Les Hiddens, ABC-TV's "Bush Tucker Man". The discovery occurred while Major Hiddens was escorting the scientists to a very remote location near the Russell River, in the leech-infested, rain-sodden jungle between Millaa Millaa and Cairns. He described the incident in *The Townsville Bulletin*, December 8, 2004:

"In that country it takes a good hour to move 1 kilometre through the scrub, so it's fairly slow going ... the forward scout came upon this very strange

Major Les Hiddens. (Steve Strike)

construction ... that left us quite baffled ... Someone or something had constructed a rectangular sleeping mat on the ground made from fronds. It was perhaps a little over a metre long and a metre wide. This mat had been slept on that very night and the vegetation [calamus fronds] that made up the mat was extremely fresh. We all stood around ... this strange discovery ... None of us could come up with any sort of logical answer. We examined the ends of the fronds to see if they had been cut ... but ... they had been chewed off the main vine, not cut. Dr. John Campbell, our expedition archaeologist, said that 'If I were anywhere but here in Australia, I would have to say that was a primate nest'."

In July 1985 Corporal J. Webster of the Special Air Services Regiment sighted a yowie about 25 kilometres to the south. [Case 265]

Since the mid 1990s Paul Compton has found, photographed and analysed four such nests in the Glen Innes area. The beds, Paul told us, are all very similar and carefully constructed: "the grass is not just thrown together in a pile, it is actually placed in a pattern." The bedding material is generally about 15 centimetres [six inches] thick. "One bed had a pillow about 30 centimetres [12 ins] wide, made up of tussocks that had been pulled out, roots and all, and ... laid in such a way that the roots ... were at either end of the pillow. The yowie had also removed rocks from inside the bed and placed them in a pile ... they had only recently been put there, as the grass was still green beneath them." The area can be very cold, so it is interesting that all of the beds were located under north-leaning trees, as if to catch the winter sun. As we have seen, our data suggests that many yowies enjoy a nap between dawn and about 10 am.

Apart from searching them for hair, Paul left the beds intact and baited one particular site with food, hoping to keep the creature in the area. Initially, he left food for four weekends in succession: "While I was there I would place a large stick in the bed, just to see what reaction I would get ... on my

Paul Compton at one of the yowie beds. Note worn bark at base of tree. (Paul Compton)

return ... I would find all traces of food and the stick gone."

During those weeks, Paul was also busy preparing a camera trap (a video camera powered by a long-life battery, triggered by a motion detector) to place near the bed. As soon as he had it ready, however, Murphy's Law came into operation: the yowie had apparently, just that week, decided to move on.

"It would appear", Paul writes, "that they are nomadic ... because one of the beds that I keep an eye on is always neat and well maintained for a period of two to three months and then it deteriorates (decomposes) as the yowie moves on. He returns after a period and makes the bed again with fresh grass, tussocks and leaves [and after another two to three months] moves again. I imagine he would be looking for different foods as the seasons change."

Although the site is in rough country where it is very difficult to check and maintain the camera, Paul feels he has a reasonable chance of eventually obtaining a video clip of the creature or creatures.

Hair analysis

Occasionally, interesting looking hair has been discovered immediately after yowie sightings. Very little of it, however, seems to have been preserved, and we know of only two instances where samples have been sent away for expert analysis.

Shortly after he became involved in yowie research, Neil Frost found a strand of likely looking hair. It was long, reddish-brown, and snagged on a fence line, close to where a yowie had been seen. He duly posted it off to a well-known Australian scientist who is supposedly a hair analysis guru. The results, the "expert" told him, were clear: it was dog hair. Naively, perhaps, Neil accepted the verdict and didn't ask for the sample to be returned. Since then, a sceptical Mike Williams has sent hair of a circus lion to the same scientist – only to be notified that it, too, was canine.

After sifting through his yowie beds, Paul Compton sent four hair samples to an American scientist, Dr. Henner Fahrenbach, for analysis. Dr. Fahrenbach, now retired, was with the Oregon Primate Research Center for many years. He is an acknowledged expert in primate hair identification and also a leading authority on the sasquatch phenomenon.

Dr. Fahrenbach concluded that three of the samples were primate hair. Although he hastened to add that similar hair could be found on humans, he mentioned that two of the samples resembled supposed sasquatch hair that he had examined. One strand of fine, reddish brown to deep brown hair, he wrote, "as it stands, it would correspond to my general definition of sasquatch hair, though rather narrow". Three other hairs, "except for their generally grey cast rather than a reddish one, resemble North American sasquatch hair".

The evidence in a nutshell

On the negative side:

Rock-solid evidence is rare to the point of non-existence. No yowie bones, fossilised or otherwise, have found their way to a laboratory, no yowie vocalisations have been captured on audiotape and no photographs of yowies are clear enough to convince anyone who is inclined to be sceptical. Most alarmingly for yowie enthusiasts, apparent yowie tracks are found much less frequently than one would expect, and those that *are* found show (with a few notable exceptions) little consistency, not only in the number of toes, but also in their general shape.

On the positive side:

Testimonial evidence very strongly supports the notion that yowies really do exist. Eyewitness reports have been consistent from the colonial era to the present day, and Aboriginal yowie lore strongly corroborates those reports. The apparent range of the yowie, as indicated by the spread of sighting reports since 1847, makes sense in terms of Australia's ecology and geography.

The terror exhibited by livestock and wild animals during apparent yowie incidents must also be taken into account, as it indicates the hairy giants are not just figments of overactive human imaginations.

When one analyses the reports in which a time of day or season of the year is mentioned, unexpected but coherent patterns emerge. If the yowie phenomenon were simply the result of hallucinations, hoaxes and misidentification of ordinary animals, one would expect the great majority of reported sightings to occur at night. One would also expect reports to be distributed fairly evenly throughout the year. Such is not the case. Just as many yowies are reported during the day as during the night, very few are reported between 4 am and 10 am and considerably fewer are reported in summer than at other times of the year.

The eyewitness reports of yowies hunting and foraging, and Paul Compton's analysis of one enormous dropping strongly suggest that something much larger than man really is hunting, eating and defecating in the Australian woods.

Although he was properly cautious in his pronouncements, Henner Fahrenbach's analysis of the hair sent to him by Paul Compton is also compatible with the idea that the yowie is a hairy, man-like primate resembling the North American sasquatch.

Finally, Paul Compton's photographs and the expert testimony of other high-quality witnesses, like Major Les Hiddens, Dr. John Campbell and Pat Ryan, leave no doubt that large beds resembling primate nests really do occur in remote areas notorious for yowie activity.

Chapter 7

Who or What is the Yowie?

In this chapter we look at various theories that have been put forward to explain, or explain away, the yowie phenomenon.

The theories fall into four categories:

A. The creatures never existed.

B. They are real, flesh and blood animals.

C. They existed once, but no longer do.

X. They are psychic phenomena, akin to ghosts.

Category A: They never existed

There are several ways of explaining how people could come to believe in creatures that have never existed:

1. Folklore gone feral?

It has been suggested that today's yowie phenomenon is just an amalgam of European and Aboriginal folklore.

In the middle ages, when vast forests still existed there, Western Europe had its own "yowie" legend. Huge, hair-covered manlike monsters appeared frequently in medieval art and literature. They were often called "woodwoses" or "woodhouses", corruptions of the Anglo-Saxon term *wudewasa* meaning "wild man of the woods". Some medieval commentators, such as St. Augustine, thought they were a degraded race of humans, others saw them as a kind of Lord of the Animals and some believed they were the embodiment of evil forces. Some cryptozoologists now believe the legend grew from real, relatively recent encounters with Neanderthaloids. According to this theory, the Neanderthals died out much more slowly than is commonly believed, lingering on in the shrinking forests of central Europe well into historical times.

Be that as it may, apart from the fact that the *wudewasa* were usually depicted as

This drawing of a European wild man, by H. Burgkmair, dates from the early 1500s.

heavily bearded and armed with clubs, they weren't all that different from yowies.

It is sometimes suggested that before the arrival of the British, Aborigines thought of the yowie/yahoo primarily as an evil spirit rather than as a flesh-and-blood animal. Although many tribes said it looked something like a hairy man, opinions differed as to its size. Stories of the yaroma and mumuga – yowie-like, but not quite the same – suggest there was some confusion about even its basic shape.

As mentioned earlier, some colonial-era Aborigines believed that the yowie's feet were reversed on its ankles. Neil Frost suggests that some traditional Aborigines came to that bizarre conclusion out of sheer frustration. In the pre-colonial era the Hairy Man was probably as rarely seen and as devilishly difficult to track as it is today. It may have been the only creature traditional Aborigines couldn't catch – and perhaps didn't particularly want to.

While Neil's theory is quite attractive, it must be acknowledged that "reversed foot syndrome" is a classic folklore motif that crops up all over the world. It is, in fact, mentioned in several other ape-man legends. In Nepal, for instance, it is a standard feature of yeti lore, and Malaysian aborigines, the Orang Asli, believe the same thing about the hairy and super elusive *orang mawas*. Even on remote Andros Island, in the Bahamas, one of the authors (Tony) was told that in the event of finding the tracks of the fearsome *yay-ho*, he should "go the way the feet are pointing … his feet – they turned backwards". [26]

Way back in the Middle Ages, in fact, while writing about reports of "monstrous races of men", St Augustine mentioned that some supposedly had "feet turned backwards from the heel". [27]

Backward-pointing feet are frequently mentioned in folk tales other than those concerning ape-men.[28] Perhaps Carl Jung was right. He suggested many years ago that such folklore motifs emanate from a collective unconscious shared by the entire human race.

In any case, the suggestion is that prior to the colonial era, the yowie/yahoo was nothing more than a rather amorphous bogeyman – just one of a host of evil spirits that kept Aborigines close to their fires at night. When British settlers heard about it they supposedly ignored anything that smacked of the supernatural and focussed instead on its hairy, manlike aspects. Ancient tales of the *wudewasa* supposedly came to their minds, and British and Aboriginal folklore combined to create a hybrid legend that soon took on a life of its own.

The fact that Jonathon Swift had, in *Gulliver's Travels*, written about a fictional race of ape-men called, coincidentally, yahoos – living on an Australian island – probably affected the way settlers interpreted the Aboriginal stories. The discovery of large, man-like apes, orangutans, in nearby Sumatra must also have influenced their thinking.

Aboriginal society, meanwhile, was being shaken to its core. By the mid-1800s, many south-eastern tribes had been heavily influenced by the language, culture and folklore of the British interlopers. The Hairy Man beliefs of late colonial-era Aborigines, therefore, could have been as much a product of European superstition as they were of genuine Aboriginal tradition.

Nourished by rumours and hoaxes through the late 1800s and early 1900s, the new European/ Aboriginal legend was supposedly given a strong shot in the arm, in the modern era, by media coverage of the American bigfoot mystery.

It is a theory worth considering. The term "wild man of the woods" certainly occurred in several early yowie reports and the "Monstrouf Giant" handbill of 1790 suggests that at least some Britons were eager to believe something resembling the legendary European wild man might exist in the vast unknown of New South Wales. In one early illustration, the "Hairy Wild Man from Botany Bay" is, with his long beard and menacing club, virtually identical to some medieval depictions of the *wudewasa*.

'The Hairy Wild Man from Botany Bay.' Engraving, c. 1790. (Rex Nan Kivell Collection)

Although this theory gives us plenty to think about, it doesn't explain every aspect of the yowie phenomenon. Does the mere fact that it reminded British settlers of the *wudewasa* legend really mean the Australian Hairy Man didn't exist in any shape or form? Were these Britons – the same people who were busy conquering and ruling half the world – so weak-minded as to imagine hundreds of yowie encounters in every part of eastern Australia? What of the multiple-witness incidents and face-to-face encounters in broad daylight? And what about, in the modern era, the Greek and Middle-Eastern migrants who'd never so much as heard of the yowie, let alone the obscure *wudewasa* of medieval Europe, before experiencing their own sighting? [Cases 116 and 226]

Aborigines insist they have known of the Hairy Man since time immemorial and the cave painting of Turramulli seems to support that contention. Almost all colonial era Aboriginal references to the yahoo, even very early ones, describe it as a hairy, bipedal man-like creature.

In our experience, Aborigines take the yowie phenomenon very seriously; so seriously, in fact, that some absolutely refuse to discuss it with non-Aborigines. They think of it as being very much their "thing" – part of their culture. Some certainly attribute semi-magical powers to the Hairy Men, but that doesn't prove the creatures are mere myth. There are many magical stories about crows, too, but that doesn't mean crows aren't also real animals. Aborigines, in fact, believe everything – every single rock, plant and animal – has a spiritual dimension.

The suggestion that American bigfoot stories have influenced the yowie phenomenon to a large degree is equally problematic. As far as we know, only one such story appeared in an Australian paper in the 19th century. That was a reprint of a letter that was sent to the Antioch [California] *Ledger* in October 1870.

The correspondent, a resident of nearby Grayson, claimed he'd seen two five-foot-tall, gorilla-like, bipedal creatures a year earlier. One, which he surreptitiously observed for 15 minutes at a range of 20 yards, was "fully five feet high, and disproportionately broad and square at the fore shoulders, with arms of great length. The legs were very short and the body long. The head was small compared with the rest of the creature, and appeared to be set upon his shoulders without a neck. The whole was covered with dark brown and cinnamon colored hair, quite long on some parts, that on the head standing in a shock and growing close down to the eyes, like a Digger Indian's."

Apart from the "very short" legs, the Californian bigfoot, as described, matches the yowie to a "T". Even so, the notion that a single bigfoot story had a significant effect on the yowie legend seems rather far-fetched. As we have seen, the general description and behaviour of the Australian Hairy Man had been well known to Aborigines and settlers for many years before that story appeared.

Although there may have been some coverage that we failed to notice, the bigfoot mystery, as far as we know, wasn't mentioned again in the Australian press until the late 1960s – and then only briefly. It wasn't until the late 1970s, when the subject was covered in television shows such as "In Search Of ..." that most Australians became aware of it. Since then many Australians have become quite familiar with the legend, so we don't find it strange that a small number of modern-era yowie witnesses have compared the creatures to the bigfoot. Actually, as we have seen, a great many more have compared yowies to gorillas rather than to bigfoot. Just as many witnesses, in fact, have compared yowies to Chewbacca, from the movie *Star Wars*, as have used the term "bigfoot".

(Sceptics, incidentally, may even point to the subtitle of this book as evidence the yowie is a mere by-product of the bigfoot legend. In fact, our working subtitle was originally *In Search of Australia's Mysterious Apemen* but, in order to make the subject of the book more easily recognisable to non-Australian readers, we decided, in the end, to include the term "bigfoot".)

2. Feral humans?

In the *Australian and New Zealand Monthly Magazine* of 1842 it was suggested that the *yahoo* legend might have begun when Aborigines unexpectedly encountered dishevelled white castaways or runaway convicts. The theory has been dusted off several times over the years but we have never found it very convincing.

If there was ever a man who could have been mistaken for a hairy monster it was "the wild white man", William Buckley, a runaway convict who lived in the wilderness of southern Victoria from 1803 to 1835.

By the standards of his day, Buckley, almost six foot six inches [198 cm] tall and heavily built, was a giant of truly yowie-like proportions. When discovered by members of the Wathaurung tribe, he had been living rough for several months, was shaggy-haired, bearded, and almost completely

William Buckley, 'The Wild White Man'.

naked. They had found him by following, as his Aboriginal wife put it many years later, his "giant" tracks. The natives thought him "a strange-looking being", but, far from fleeing in terror, they talked to him, fed him, and welcomed him into their clan.[29]

Australian history books, in fact, contain several other accounts of ragged castaways and runaways being cared for by Aborigines, who often took them to be reincarnations of deceased relatives. Another thing that argues against the notion of a convict-on-the-lam being "the father of all yowies" is that, as we have seen, Aborigines have never said the creatures have beards.

In our previous book, *Out of the Shadows*, we included a story from the February 18, 1987, *Centralian Advocate*. It concerned an encounter between an Aboriginal family and a stark naked, behaviourally disturbed man at a remote location near Alice Springs. The man was very tall, obese and hairy – and the Aborigines supposedly took him for a yowie.

We have since been informed, by a local landowner, who was present when the man was taken into custody, that the story was a complete "beat-up": no one, at any time, thought the poor fellow was anything but human. [Case 174]

"Rambo"

Later the same year a similar "feral man"/yowie story appeared in the Sydney tabloids. This time it concerned seven young white men who told police they'd seen a hair-covered ape-man near Woronora Dam. Twenty-two-year old Jim Kikoudis said they'd been looking for the creature since hearing "werewolf-like growls" 18 months earlier. At about 10 o'clock on the night in question, they heard the usual growls, the sound of something crashing through the scrub and then, "there it was in front of us. It was about two metres tall but it wasn't standing straight. It walked across the road about 50 metres in front of us ... took only a few seconds to walk across ... and then it was gone".

The shaggy creature, which was fully illuminated by the men's spotlights, had a bullet-shaped head and was carrying a wallaby carcass. As it moved into the scrub it dropped the carcass, from which most of the flesh had been stripped.

Picking up the carcass, the men drove to Engadine Police Station. "[They] seemed pretty genuine", a constable said later, "you don't get eight blokes coming down to the station with a carcass unless they've got a good reason to". Two officers immediately drove to the scene and reported hearing noises in the scrub. They also examined what appeared to be fresh tracks. One newspaper referred to them as "giant tracks".

It sounded like a classic yowie report. Almost immediately, however, a Water Board employee attempted to ... er ... pour cold water on the story. He said the witnesses had probably glimpsed a hermit who was known to live in Heathcote National Park.

Other local residents agreed, saying that the hermit, 53-year-old Franjo Jurcevic, recently dubbed "Rambo", had lived in the bush for as long as anyone could remember. He emerged only once a fortnight. "He comes up on pension days and gets his supplies," said Heathcote Inn barman "Buck" Rogers. "I can see why people would think he's an ape-man. He has hair everywhere and wears a bandanna around his head. But he's not seven foot tall and he doesn't seem very ferocious." Publican Ces Partland agreed: "He may be a bit short with people but he's certainly no Rambo. He just takes his grog, along with fruit and veggies, and off he goes." [Case 175]

Although the journalists seemed happy with the "Rambo" = yowie theory, we have never found it very convincing. Although "Rambo's" presence in the general area can't be disregarded entirely, we doubt that he triggered the report in question. The witnesses noticed no clothes and didn't mention long head hair or a beard – they said only that the creature was hair-covered. Their reference to a "bullet-shaped" head doesn't suggest a great mop of hair held back by a bandanna, but it is reminiscent of the dome shape mentioned by some other yowie eyewitnesses. "Rambo" didn't loom two metres tall: he was apparently of average height.

There are other things to consider. The witnesses heard the "werewolf –like growls" (the same growls they'd been hearing in the area for 18 months) just before the creature appeared. The condition of the wallaby carcass, too, suggested predation by a wild animal. There was no indication that the animal had been killed by a bullet, arrow or spear, or that the missing flesh had been cut away with a knife.

Another thing that suggests this really was a yowie incident is that similar reports have emanated from the vicinity long before and long after 1987. Heathcote National Park is contiguous with Royal National Park. As detailed in chapter two, yahoo/ yowie stories have come from the area now covered by the parks as long ago as 1856 and as recently as 2000. In 1992, a good sighting occurred on the edge of Holsworthy Military Reserve, which joins the northern boundary of Heathcote National Park. [Case 189]

There is no denying that in the past 200 years many people, nearly always men, have retreated into the Australian bush to live as hermits. Although "Tarzan" Fomenko, who has lived in the jungles of Cape York since 1951, is our best-known "feral" man, at least one other individual spent a comparable time in the scrub. As documented by Max Jones in his marvellous book, *A Man Called Possum*, David James Jones, aka "Possum" (1901-1982), spent 54 years living off the land along the Murray River between Wentworth and Renmark.

There has never been any suggestion that unexpected encounters with either "Tarzan" or "Possum" ever prompted a yowie report.

During the 1800s, when Aboriginal society was almost destroyed by frontier violence and disease, many ragged survivors eked out an existence in rough country on the fringes of settlements. It has occasionally been suggested that unexpected encounters by British settlers with solitary Aborigines might have given rise to the yowie legend.

That theory, too, contains several serious flaws. In the colonial era most white pioneers were quite familiar with the sight of longhaired, naked or near-naked Aborigines. Although most Aboriginal men of that era were unshaven, yowie witnesses have almost never mentioned beards.

The Pardooks

One case that has a direct bearing on the "feral man" theory – and on a couple of other aspects of the yowie mystery – is the remarkable story of the capture of the *pardooks* on the West Australian frontier.

In 1878 Albert and Ethel Hassell moved to "Jarramungup", a sheep station on the edge of civilisation 140 kilometres north-east of Albany. A thoughtful, observant woman, Ethel became very friendly with the local Aborigines and, between 1878 and 1886, made copious notes on their customs, language and legends.

One day a mounted party led by Ethel's brother caught some very strange people in the desert to the north-east and brought them back to "Jarramungup". As they were led towards the homestead, Ethel's Aboriginal friends were "absolutely dancing with excitement". The captives, they said, were *pardooks*, people who lived in the heart of the desert. They were seldom if ever seen, but were known to exist because every now and then their very distinctive tracks were found on the edge of better country.

Ethel Hassell. (C.W. and W.A. Hassell)

There were three *pardooks*: an elderly man and two women. The man, said Ethel, was less than five feet [1.52 m] tall, and had a "low forehead, small eyes, flat nose, with very wide nostrils, so short an upper lip that it that it only seemed a small bridge below the nostril, but a large, heavy lower jaw, rather broad shoulders, but his body fell away at the hips and his legs were nearly spindles. His arms were long and he was hairy. It seemed a question whether he was a man or an animal."

Then Ethel noticed something even stranger: the *pardook* had five "perfectly formed" fingers plus a thumb on each hand. There were also six toes on each foot: "five very short toes and a long big toe which seemed to spread well away from the others. The women were just the same."

They had rough, reddish-black skin and extraordinarily thin legs: Ethel could encircle the calf of the younger woman with her middle finger and thumb. "I never thought", she wrote, "[that] any human being could have such thin legs." They wore nothing but animal skins tied around their waists.

Their eating habits were just as strange as their appearance: although they used fire to keep warm, they ate only raw meat. They had language but none of the station Aborigines could understand a word of it. The only one who knew anything about them was old Buckerup, the local "clever man", who had apparently seen their tracks on a few occasions. He said most *pardooks* had as many fingers and toes as the three captives, but some had extra digits on only one hand and one foot.

After a day or so the strange people were given bags full of meat and escorted back towards the desert, into which they vanished, never to be heard of again.

It is difficult to imagine more primitive, weird-looking folk than the *pardooks*. But while Mrs. Hassell said, "I did not think there could be human beings so nearly resembling animals", she seemed at no time to think of them as ape-men or monsters. It is difficult, therefore, to imagine that chance encounters with isolated, presumably inbred tribes like the *pardooks* could really have triggered the yowie phenomenon.

Finally, if the yowies were merely Aborigines disorientated by frontier wars and epidemics, why are they still seen today?

"We honestly thought it was some feral *guy*"

While most eyewitnesses, even those who have seen them at very close range in broad daylight, describe yowies as ape-like beings, there are a small number of apparently very sound witnesses who say they look much more man-like.

Our best report of this type comes from Gary Jones, of Glenmore Park, Sydney. Although he was only 16 or 17 years old when he encountered the creature in June 1989, he is confident his recollection of its appearance is accurate. The sighting occurred 15 kilometres south of Katoomba in the heart of the Blue Mountains wilderness.

"We [Gary and two life-long mates] often go trout fishing where the Kowmung River joins Coxs River just upstream from Lake Burragorang. It takes us six and a half hours to get there – half mountain bike riding and half bushwalking. The last 1,600 feet of the descent takes you about an hour because you have to tack down. It's very remote and there's a lot of wildlife down there: dingoes, wombats, 'roos, platypus.

"We got down to the river at about 2:15 pm, fished for about an hour, and started collecting firewood. So we were going along the riverbank and my mate said 'Check that out!' and, lo and behold, on the other side of the river about 30 metres upstream was a gentleman who was not wearing any clothes and who was basically as hairy as you could possibly imagine. At the beach I've seen blokes with really hairy backs ... well, if you can imagine the worst-case scenario – but all over. And I said, 'Oh, it's a bloke', and we had a short conversation and then walked up to get level with him."

The creature, which had its back to them, was standing in a most unusual pose: "He had one foot up on the bank and one in the water. He was doing the splits – but one leg was higher than the other. It's a very steep bank and the water there is about four feet deep. He was stooping down, drinking out of his right hand ... his left shoulder and that part of his back towards us. He didn't see us because we were behind him and on the other side of the river ... a rapid current moves the river stones and it's quite noisy.

"We walked up to directly opposite and we were only the width of the river, 14 metres maximum, from him when he saw us and stood up. To stand up from that angle you'd have to be fairly well conditioned. He showed no strain, no effort; he just stood up, very strong and straight.

"He was very tall, probably six foot six or seven inches and had long legs. I weigh 100-112 kilos [about 230 lbs]. He would have had 10-15 kilos on me, but he wasn't fat. This person was solid – really wide shoulders. Very muscly; a short neck."

Interestingly, Gary did not consider the creature's arms remarkably long: they seemed no longer, proportionately, than those of a large man.

"Its hair was a very dark brown, the same colour all over, until you got to his face ... Have you ever seen a black dog when it's been in the sun too long – it gets that reddy tinge to it? That's what his head and beard looked like. The beard and head hair were matted; it looked disgusting – you know – like some of those dreadlock-type things. The hair on his shoulders was about five to six centimetres long [two inches] ... it was thick cover.

"The only spots that were not covered with hair were the soles of his feet [which he glimpsed as the creature walked away]. And they were discoloured with dirt and mud, so there was really no way to tell what his skin colour was. Even his buttocks were covered in hair. I didn't see the inside of his hands, but I'd assume there would be no hair there.

"He had deep eye sockets. Dark eyes. Thickset lips; they were a dark colour. His hair seemed to extend all the way down to his eyes, and his beard all the way up to his eyes. I'd never seen a face like that before. It didn't look like a monkey or an ape or anything. Quite a long forehead, probably twice the height of my forehead. His head looked more pointed than, say, mine, but all the hair made it difficult to tell. He looked right at us. He had no expression – just blank. Didn't show any fear; made no sound. It was quite weird. And then he turned and took off. The bush was thick forest, but towards the edge there was prickly native scrub. A normal person would push past it or go around, but he didn't even stop – he ran straight through it.

"If you look at someone who's done weights. They have wide shoulders and, like, a 'V' shape – that's how it looked [from behind]. And solid buttocks – you know how the buttocks of sprinters get really big? That's what was like – but with hair all over.

"Our first instinct was to go across and check this bloke out, but when we rock-hopped across we found we had to struggle up onto the bank [that he had simply stepped up onto]. On the bank where he'd been standing there was a really, really, strong smell, like ammonia or urine. The closest thing I've smelled was a fox in season, or maybe if you go past the bat cage in a zoo.

"He had probably a 50 metre head start, but we could hear him running in the distance and we followed him up the ridge. That gradient is incredibly steep but he could move very, very quickly. I used to play football and we were all quite fit, but we were buggered after going only about 800 or 900 metres. We didn't have any chance of catching him. There were loose rocks underfoot and we could hear them rolling and crashing as it moved away. We were puffing and panting and we could hear it puffing and panting in the distance. I couldn't have run faster ... we were cut and scratched from the bushes and weren't making any ground on it at all.

"At this stage it would have been 3:15 or 3:30 and as you get up that side, because the sun sets to the west, you're on the inside of the valley, so it gets quite eerie. When we got up on the ridge we could still hear him in the distance. The urine smell was getting very strong. We got to a small clearing and it got worse. And that's when your mind starts reeling. You think, 'What is this thing? It doesn't smell too good. Are we going back to where it lives? What could happen?'

"It dawned on me that if we kept going it might be dark before we got back to camp and we didn't have torches. I said, 'Look, it's gonna get dark' and one of my mates said, 'Yeah, let's get out of here – we don't know whether there's others.' Walking down, we kept looking over our shoulders thinking, 'Is that thing following us?'

"On the way down, on the bushes that had cut us – there was his hair. It had stuck on the bushes as if a horse had gone through – big chunks of hair. Just normal hair; it was coarse, a dark brown colour, about five or six centimetres [two inches] long. I wish I'd been smarter and grabbed some of it, but we honestly thought it was some feral guy.

"When we got back to where he'd been standing, there was that really coarse river sand and there were definite footprints. His feet were probably size eleven, not absolutely huge. You could see five toes; there was a definite heel and an arch. It wasn't until we got back across the river and sat down and chatted, saying, 'What was it?' that one of my mates said, 'It's a yowie' and I said, 'No, I think it's just a wild man.'

"For the rest of the trip we were a bit worried – every stick that broke or any noise outside the tent, we were quite alarmed. We stayed three days. When we were fishing we could hear, on the ridgeline, rocks moving, and we all had the feeling we were being watched – felt eyes on us. It was extraordinary. We stayed fairly close together. We've been back there most years since and every time I'm half expecting to see this person, or hairy man – which we now think might be a yowie."

Although Gary finally uttered the "Y" word, he was clearly reluctant to accept that the creature he saw, though tall, wild, smelly and hairy, was anything less than fully human. Readers will have noticed that although he sometimes referred to the creature as "it" and "this thing", he more often used terms like "he", "him", "hairy person", and "wild man".

When pressed on the matter, he explained that he was, by nature, "probably the world's biggest sceptic. I'd heard about yowies [but] I thought this person had sort of left society and become hairy because he wasn't wearing any clothes ... had evolved to become hairy ... or it could have been some wild race of man that was undiscovered – because where we were was about as remote as you could get. It was human".

3. Hoaxes, lies and videotape

Over the years we have interviewed a few people who we suspect of fabricating their yowie reports for one reason or another. In some cases stories fell apart when "witnesses" inadvertently contradicted themselves. Another "witness" pretended to have had no prior knowledge of the yowie phenomenon when we knew from another source that he did.

Sighting reports by people who put themselves forward as experts, have marketed yowie-related material, or who have courted media attention, are, quite naturally, subjected to unusually close scrutiny. There's nothing unreasonable about that. If one of us ever claimed a sighting, we would expect our story to be subjected to the same degree of scrutiny. Unsupported reports by anyone with a discernable ulterior motive, particularly a financial one, will always remain, justly or not, under a cloud of suspicion.

At the risk of sounding outrageously gullible, however, we believe that the great majority of informants have told us the unvarnished truth. We have had to virtually wring stories out of some particularly shy people. Many have asked (and two have actually demanded, with not-so-veiled threats) that we keep their identities secret. We know of others who have seen yowies but who absolutely refuse to discuss the matter with anyone outside their families. By no stretch of the imagination could such people be seen as publicity-hungry.

Many informants, in fact, are quite transparently honest, and their excitement in describing what was one of the most exciting experiences of their lives is obvious. When someone looks you in the eye and tells you, in the presence of their spouse and children, that they've seen what could only have been a yowie, it is very difficult to doubt them. When David Holmdahl says, "I didn't imagine this ... I'll swear upon the Holy Bible, and I'm a Christian" or when Jerry O'Connor offers to swear to the truth of his story on his late daughter's grave, it is virtually impossible to disbelieve. [Cases 242 and 223]

But although we are certain David, Jerry and scores of other excellent witnesses are truthful, the possibility that *they* were hoaxed still exists. And it is true that hoaxes have played a small part in the yowie saga from the very beginning.

The first written reference to Australian hairy giants, the 1790 handbill depicting the "Monstrouf Giant" was, of course, clearly a hoax (although the place of its supposed capture, Botany Bay, is tantalisingly close to the site of several yowie encounters of the colonial and modern eras).

As mentioned earlier, it seems likely that Arthur Marrin's claim to have killed a six-to-seven-foot-tall hair-covered creature near Captains Flat in 1893 was also a hoax, although it may have been prompted by a series of genuine yowie reports in the area. [Case 32]

Another hoax really did fool a few people for a remarkably long time. It is interesting to note, however, that throughout the entire episode the terms "yowie" or "Hairy Man" were never used. Perhaps the perpetrators had never even heard of the yowie. In any case, the creature in question was always known as the "Byng Bunyip".

The story began on the night of September 3, 1959, when Max Spicer of Byng (near Bathurst, NSW), saw something standing in the middle of the road near the local cemetery. Stopping his car, he saw that the creature seemed to be about three and a half feet tall and covered with long, black hair. It had glaring red eyes and arms that reached to the ground. It ran into the cemetery.

After three Sydney men and others encountered the same thing on the Byng Road, armed parties began scouring the area and the "bunyip" received front-page coverage in the national press. One night a truckload of hunters spotted the creature, but when one of them accidentally doused the spotlight, it vanished into the shadows. It was supposedly encountered several more times but always escaped. After a while the sightings became fewer and the "Byng Bunyip" seemed to just fade away.

It surfaced again in 1987 when Denis Gregory, who had taken part in the "bunyip" hunts, sent an article to *The Picture* magazine. The article focussed mainly on another local, Freddy Nunn, who had supposedly been spending large amounts of his spare time since 1959 hunting for the "bunyip". Because *The Picture* proudly bills itself as "the magazine with more arse than class" and runs articles with sober headlines such as "These Tits Killed a Man!" we never knew quite what to make of the "Byng Bunyip". But because Byng isn't far from the sites of several other yowie reports, we thought that, jokey as it seemed, the story might possibly be genuine.

In 2002, however, our colleague Roger Frankenberg uncovered the awful truth. The "bunyip" had been a pint-sized teenage panel-beater, Graham "Darby" Offen. The hoax had been dreamed up by some of his mates, including Denis Gregory, and, as Darby said later, "they only picked me to be the bunyip because I was a little bugger."

Roger learned that Denis Gregory had, in fact, revealed most of the inside story in 1993 in his book *There's Some Bloody Funny People on the Road to Broken Hill.* He told Roger they'd dressed Darby up in a hairy hessian suit. His "big red eyes" were reflectors, as used on bicycles. He would wait in the middle of the road and when a car approached would jump around a little before scooting away. His mates had a car hidden in the bush with the boot up and the motor running. He'd leap in and they'd take off into the night.

One night the prank almost got Darby killed: "the small hole I looked through in the suit got stuck on my nose and I couldn't see a thing. I could feel the lights burning into me and I didn't know which way to run ... my mates were yelling to me to run but I didn't know where to and I thought, I'm a bloody goner this time. But I got off ... fell down a culvert and me mates grabbed me. That night was too close for comfort, so I decided to retire ... there was no way I was going to get my head blown off".

Freddy Nunn meets the 'Bunyip' face to face.
(Denis Gregory and Alf Manciagli)

Although he'd taken part in some of the "bunyip" hunts and actually sighted the hessian-clad horror, Freddy Nunn was supposedly not let in on the secret until about 1990. If that detail is true, then the "Byng Bunyip" is one yowie-type hoax that fooled at least one man for almost 40 years. [Case 78]

An interesting film was taken at 6 pm on August 28, 2000, in the Brindabella Mountains about 20 kilometres west of Canberra. Forty-four-year-old businessman Steve Piper claimed that, while recording footage to be used in a music video, he caught sight of what he first assumed was an unusually large kangaroo. "I zoomed in and couldn't believe my eyes – this massive creature was trampling through the underbrush." It was, he said, seven or eight feet tall and like nothing he'd ever seen. He filmed it for almost 27 seconds before it disappeared. A few days later he contacted Canberra-based researcher Tim the Yowie Man, who quickly became his agent and attempted to market the film worldwide.

Although the film is far from crystal-clear (it was taken during the late afternoon as the subject walked along a scrubby, shadowy gully), the subject certainly isn't any ordinary animal. It is uniformly dark from head to foot, bulky, long-legged and bipedal. It is difficult to tell whether its arms are any longer than those of a human. It can be one of only two things: either a yowie or a man in a gorilla suit.

Mr. Piper and Tim took us, along with Neil Frost and Ian Price [Case 160] and visiting American bigfoot researcher Dan Perez, to the film site on September 24. We found that the creature had been walking along an abandoned track that runs roughly parallel with the road on which Mr. Piper had been standing. On recreating the incident with human stand-ins, it soon became clear that the subject was not seven to eight feet tall as Mr. Piper estimated. It was, in fact, only the same height as Tim the Yowie Man and Ian Price – both about six feet tall.

Whatever the animal was, it moved very slowly, displaying a pronounced limp, almost dragging one foot. During 26.8 seconds it covered only about 24.5 metres [27 yards]. Because it was so close – only about 46 metres away – and since it moved so slowly, we asked Mr. Piper why he didn't pursue the creature. To this and all other questions, his response (he began to follow but became frightened) was brief but adequate. There was nothing about his demeanour that screamed "hoax" and he stood up well to an aggressive interrogation by Dan Perez, but didn't volunteer much without being asked.

The subject of the film *might* be a yowie. But since it is no taller than a man, no bulkier than a man in a fur suit and no faster than a man, it seems more likely it was a man in a gorilla suit. Perhaps a hoaxer knew Mr. Piper would be filming in the area on the day in question and decided to trick him. The hoaxer may have affected the very pronounced limp to disguise any obvious similarity to a human gait and to account for the creature's slow progress.

Even after being computer enhanced by Tim the Yowie Man, the best stills from the film are too unclear to be of any use in a book of this format. In any case, Mr. Piper has since disappeared into the teeming metropolis of Sydney, taking his murky, puzzling film with him.

While it is undeniable that hoaxes have occurred, we believe they have played only a very small part in the yowie saga. To create such a multifaceted, continent-wide phenomenon, an impossibly elaborate series of hoaxes, involving Aborigines and Europeans, stretching back for at least 200 years, if not for millennia, would have to have been perpetrated.

4. Delusions of dulagarls

It has been suggested that most yowie sightings are reported by people who are "fantasy-prone personalities" (FPPs), but our statistics suggest otherwise. While an extensive study conducted by American psychologists in 1983 found that women were several times more likely to fall into the FPP category than men, only 77 of our cases (28.5 per cent) involve female witnesses. Another consideration is that in 115 cases two or more witnesses were involved.[30]

That having been said, the human mind is, to put it mildly, a very strange thing, and we're pretty sure that a small number of "eyewitnesses" simply imagine their yowie encounters. Some of these fantasy-prone individuals seem totally convinced of the reality of their experience. Others may have some degree of awareness that their stories are false. In any case, some of them attract suspicion because, not content with just one sighting, they proceed to claim numerous subsequent encounters.

That is not to say we consider everyone who has claimed more than one sighting to be fantasy-prone. Far from it – some of our most credible witnesses have experienced multiple encounters. They are normal, reasonable people who just happen to live in noted yowie hot spots or who have spent a long time hunting the creatures. The individuals we consider fantasy-prone tend to display a range of odd personality traits. The most prominent of these are an absolute craving for attention coupled with an apparent inability to shut up. Their tales of frequent yowie encounters are often embellished with interminable stories about their accomplishments as hunters or fishermen, and about how little, compared to them, the "so-called experts" know.

Hallucinations?

It is often glibly suggested that all yowie sightings are simply hallucinations. The people who make this unthinking suggestion often add that the poor deluded witnesses should lay off the grog. Apart from the fact that alcohol isn't a hallucinogenic drug – unless you drink a truckload of it – there are several problems with the hallucination theory.

Driver fatigue can certainly produce hallucinations. Recent studies have shown that drivers go to sleep at the wheel with alarming frequency. Almost anyone who drives late at night, it seems, is liable to nod off for between one and two seconds without being aware of it. These sometimes-fatal "micro sleeps" are most likely to occur in the early hours of the morning on quiet country roads.

Researcher Paul Chambers thinks it no coincidence that many reports of UFOs, ghosts, monsters, phantom hitchhikers, time-slips and other strange phenomena come from people who were driving on such roads in the wee small hours. He quotes a laboratory study that found some people experience quite complicated dreams during micro sleeps. Although they often contain very bizarre elements, dreams experienced immediately after falling asleep tend to "take on a life-like nature in which the plot of the dream centres around the dreamers themselves and the activity they were doing when they fell asleep". [31]

Mr. Chambers' theory is well worth considering, but as he himself is quick to point out, "micro sleeps" cannot explain every reported encounter with strange animals.

Our files contain 46 cases of yowies being seen from moving vehicles, but in 24 of those cases more than one observer was involved. Eleven of the incidents, furthermore, occurred in broad daylight.

Mr. Chambers also points out that hypnopompic hallucinations (dreams that continue after the sleeper thinks he or she is fully awake) can explain many unusual experiences, such as apparent sightings of bedside ghosts and angels.

Again, it seems unlikely that many yowie reports are the product of such dreams. The vast majority of yowie sightings occur between noon and 3 am. Relatively few people report seeing the creatures between 4 am and 9 am.

One of the few stories that contain elements suggestive of sleep disorder or hypnopompic hallucination is that involving George Grey. His struggle with the small, silent, odourless, but very strong yowie is rather reminiscent of a type of terrifying, waking nightmare known as the "Night Hag". That having been said, several other people have reported seeing yowies in the same area. [Case 83]

Similarly, if Thelma Crewe was the only person to have reported a yowie near Woodenbong, it might be easy to suggest her moonlit sighting was actually a hypnopompic hallucination. Her experience, however, occurred only a couple of hundred metres from the Moloney residence, where a similar creature was seen at close quarters, a dog was mortally injured and several clear tracks were found. [Cases 105 and 111]

Another big problem with the hallucination theory is that in a great many instances animals, both domestic and wild, have become highly agitated when yowies are nearby. Do animals hallucinate?

If all eyewitnesses are hallucinating, how is it that their hallucinations are so remarkably similar? Kos Guines, as mentioned earlier, hadn't even heard of the yowie legend prior to his own sighting. How did his subconscious know what to conjure up? The same goes for Jason Cole and the immigrant family at Gatton.

What of our 115 multiple-witness cases? There is no denying that human beings are occasionally swept up in outbreaks of mass hysteria. Even so, we find it very difficult to believe that groups of Australians have, for no apparent reason, been spontaneously and simultaneously hallucinating sightings of big hairy ape-men for nigh on 200 years.

People of all different ages and ethnic backgrounds have reported seeing the creatures. Aborigines, of course, say their people have been bumping into them since time immemorial. If all those people imagined their yowie encounters, the country is clearly in the grip of a mass delusion comparable to the witch-burning crazes and dancing manias of medieval Europe. If that is the case, the term "hallucination" seems rather inadequate.

Haunted houses, fans and crooked corridors

Although it doesn't come close to explaining everything about the mystery, one natural phenomenon has been found to trigger reactions similar to those experienced by some yowie witnesses.

Since the mid-1980s, British engineer Vic Tandy has made interesting discoveries at two apparently haunted locations in Coventry. In both places people felt distinctly uneasy, hair rose on the backs of necks and some sensed "presences". Some fled and refused to return. Tandy found that parts of both buildings produced standing waves of 18.9hz infrasound. Anything below 20hz is inaudible to humans. In one building an extractor fan was responsible; in the other, a 14th century pub, it was air funnelling though a long, crooked corridor.

When considering whether infrasound might have anything to do with the yowie phenomenon, it is worth noting that sounds in the range of 18.9hz can cause, in addition to fear and anxiety, blurred vision, hyperventilation, headaches, imagined drops in air temperature, gagging sensations, nausea and post-exposure fatigue. As we have seen, extreme fear is a feature of many yowie reports. Gagging sensations, nausea, icy sensations in the spine, headaches and post-sighting fatigue have also been mentioned by a small number of witnesses. [e.g. Cases 123, 220, 223, 225, and 149]

But how could yowie encounters, which occur in the great outdoors, well away from electric fans and basement corridors, have anything to do with infrasound?

Mother Earth, mountain panic and the great god Pan

Vic Tandy points out that some elements of the natural environment – storms, tornadoes, and waves hitting the shoreline – are all fertile natural sources of infrasound. Earthquakes, volcanoes, shifting tectonic plates and certain types of wind [like the psychosis-inducing Mistral of the Rhone Valley] also create the ultra-deep, inaudible sounds. As mentioned earlier, although our statistics don't seem to support the notion, some Aboriginal groups say the Hairy Man is most often seen in wild, stormy weather. Some white pioneers had the same belief. The region of North America most associated with bigfoot reports is, of course, the earthquake-plagued and volcano-riddled western seaboard.

People have experienced panic attacks similar to apparently yowie-induced "nameless dread" in many parts of the world. In Europe, the panic strikes on lonely stretches of seashore, on moors and in woodland. Most often, however, it occurs in the mountains. Because of that the phenomenon is often referred to as "mountain panic".

"Mountain panic" survivors (some reportedly don't survive – having run headlong over precipices) speak of "an overpowering fear ... utterly cold in quality, and terrible because of its irrationality ..."; "suddenly I was seized with such terror that I ... in panic fled ... Something – utterly malign ... obscenely human, invisible ... was trying to reach me ... If it did I should die ..."; "Terror had seized me also, but I did not know what I dreaded ... we ran like demented bacchanals ...". Interestingly, one survivor stated that the "utterly overwhelming" terror caused birds and animals to flee alongside him and his companion.

There is something eerily appropriate in the notion that such panic attacks are caused by infrasound generated by the natural environment itself. Fortean researcher Patrick Harpur points out that the word "panic" derives from the name of the Greek god Pan: He who personifies the wilderness. In ancient times, Pan was said to occasionally emit a terrible shout, causing wayfarers to flee uncontrollably. [32]

So the mountain panic of Europe seems to be very much like the nameless dread experienced by some people in yowie hot spots. But many Australians who experience the fear also actually *see* the Hairy Man. If those sightings are hallucinations triggered by naturally occurring infrasound, shouldn't European victims of mountain panic also see hairy giants?

Well ...

The Big Grey Man of Ben Macdhui

Remarkably, at one of the places most notorious for mountain panic, Ben Macdhui in Scotland, the panic is sometimes said to be accompanied by sightings of an immense, rather yowie-like apparition. The Big Grey Man of Ben Macdhui has been described as 20 feet tall, hair-covered, with a large ape-like head. It supposedly walks with a heavy, lurching tread.

Fortean researcher and author Andy Roberts states: "digging deep in mountaineering literature, I discovered that this core experience [mountain panic] is relatively widespread in wild or mountainous areas, but has either been ignored or subsumed into the broader and more exciting area of ghost stories. This is a mistake because, whether paranormal or psychological in origin, there appears to be a very real phenomenon at work". [33]

On the face of it, the fact that naturally occurring infrasound can induce confusion, panic, nausea and chills seems to support the notion that the yowie is psychological in nature. There is, however, another way of looking at this interesting aspect of the mystery. We will revisit the subject later.

Although each of the above factors might account for a small number of yowie reports, it seems to us that none of them comes close to explaining the entire phenomenon. When combined, however, they become much more difficult to dismiss.

5. All of the above

Could all of the above factors – Aboriginal and European superstitions, hoaxes, misidentification of feral men and hallucinations – have combined to produce a kind of folkloric stew from which, in the colonial era, the first yowie emerged, steaming, stinking and blinking? In the modern era, tales of the American bigfoot may have further shaped our beliefs about the appearance and habits of the non-existent Hairy Man.

The suggestion that yowies never existed at all now begins to seem more plausible. It is undeniable that throughout history fear, superstition, hysteria, rumours and lies have often mixed together and fermented to create many strange brews. But is the theory strong enough to account for the remarkable similarities between the statements of witnesses of so many ethnic groups and ages, in so many different parts of the continent, in so many different eras?

The vast majority of people who report yowie encounters believe that what they saw were real, solid, flesh and blood animals. Some become quite annoyed when asked if they might have experienced a hallucination, or encountered a hoaxer or a hermit. Can, they ask, hallucinations leave gigantic footprints? Can hallucinations be seen, heard, smelt and felt by up to 20 people at a time? Are the terrified dogs, cats, cows and horses also hallucinating? Are Aussie animals, like their befuddled masters, susceptible to hoaxes and slaves to superstition? Is the average human hermit eight feet tall, three feet wide, and covered from head to foot in hair? Can hermits roar like lions and rip large bushes out of hard ground? Do their eyes reflect light? Would a hoaxer in a gorilla suit simply run away quietly after being blasted between the shoulder blades with a shotgun?

Category B: The creatures are real animals.

At this stage, given the absence of physical remains, we don't expect that many readers (apart from those who have actually seen one themselves) are willing to believe unreservedly in the yowies' flesh and blood reality. Anyone who has read the book to this point, however, must have been struck by at least a couple of cases in which the witnesses were clearly truthful and in which the creatures could not, by any stretch of the imagination, have been hallucinations, hoaxes or hermits. Any open-minded person, we believe, will find several more such cases in the Catalogue of Cases.

But if yowies are real flesh and blood creatures, exactly what kind of creature are they?

1: Orangutans, regular or super-size?

The nearest large apes to Australia are the orangutans of Sumatra, some 3,000 kilometres [1,864 miles] away. It is rarely suggested that yowie reports are triggered by sightings of orangutans that somehow found their way to this continent. Apart from the fact that they are hairy and rather man-like, the slow moving, knuckle-walking, herbivorous and tragically vulnerable pongids don't have a great deal in common with the predatory, bipedal, fast moving and highly elusive yowies.

In prehistoric times, giant orangutans existed in Asia, and it is possible such creatures were a little more yowie-like. There is, however, no evidence to suggest they got anywhere near Australia.

Only about three or four of our eyewitnesses compared yowies to orangutans. A great many more said they resemble long-legged gorillas.

2*: Gigantopithecus*?

A good candidate, particularly in terms of its size, is *Gigantopithecus*, a huge hominid that is believed to have become extinct about 500,000 years ago. It was very heavily built, had teeth that were more human-like than ape-like and may have been very hairy. If it was bipedal (the matter is still in doubt), it would have stood eight feet tall and may well have been very yowie-like. Unfortunately no evidence has been found to suggest that *Gigantopithecus* existed any further south than Vietnam, more than 3,000 kilometres from Australia.

3: An undiscovered marsupial?

Because of the marvellous phenomenon of convergence, nature produced, in Australia, many marsupials that closely resemble the mammals of other continents. The thylacine ("Tasmanian tiger"), for instance, was very wolf-like and thylacoleo was cat-like. Why should nature not have produced, in Australia, a marsupial that looked ape-like?

Although he is not wedded to the idea, Neil Frost has occasionally speculated that the Hairy Man, seemingly so perfectly adapted to the Australian environment, might be marsupial. That would, as he points out, remove the need to explain how the yowie's ancestors found their way to Australia. It might also explain some of the anomalies that have been noticed in supposed yowie tracks.

Graham Joyner has made the same suggestion from time to time. Graham, however, who is suspicious of all modern reports, believes that the Hairy Man, whether it was a marsupial or not, probably became extinct in the early 1900s.

Gary Opit points out that the mountain diprotodontid (*Hulitherium thomasttii*), which lived only 20,000 years ago in New Guinea and could have walked to Australia during the last ice age, might be seen as a possible yowie ancestor. The creature was bear-like, weighed up to 300 kilograms [660 lb], had highly mobile limbs, a short muzzle and a domed head. It would, however, have been much more slow moving than any yowie, narrow shouldered and almost exclusively quadrupedal.

One problem with the marsupial theory is that, as we will see later, yowies may be related to creatures that have been reported on continents not noted for marsupial life forms.

4: *Homo erectus/Homo floresiensis*: "Honey, I shrunk the kids"?

Homo erectus seems a very good candidate. The creatures were powerfully built, bipedal, had protruding brow ridges and were possibly quite hairy. It is now generally believed they reached the Indonesian island of Flores about 840,000 years ago. Flores is 700 kilometres [435 miles] from Australia, but during the ice ages sea levels were so reduced that some plants and animals were able to move between them.

There is one very appealing thing about the *H. erectus* = yowie theory: it could account for both the yowie and the junjudee legends. In 2004 Australian scientists Mike Morwood and Peter Brown discovered skeletons of tiny (three-foot-tall) hominids on Flores. Morwood and Brown believe that the little people – *Homo floresiensis* – were the descendants of *H. erectus*, which, over the course of its 840,000 years on the small island, slowly shrank in size. (Flores was also home to elephants, which, over a similar period of time, shrank to the size of cows.)

Some of the characteristics of the *H. floresiensis* skeletons are interesting, in light of their possible connection to the yowie and junjudee mystery: they had extraordinarily long arms, reaching almost to their knees; their faces projected forward and, although their skulls were very small, their teeth were large and prominent. One of the most exciting things about the "hobbits", as their discoverers dubbed them, is that the present inhabitants of Flores say the little creatures survived there until about 300 years ago. According to tribal tradition, the creatures, known locally as Ebu Gogo, had big eyes, flat foreheads and *hair all over their bodies.* [34]

Even during the ice ages, when sea levels were much lower than today, island-hopping from mainland Asia to Flores would have

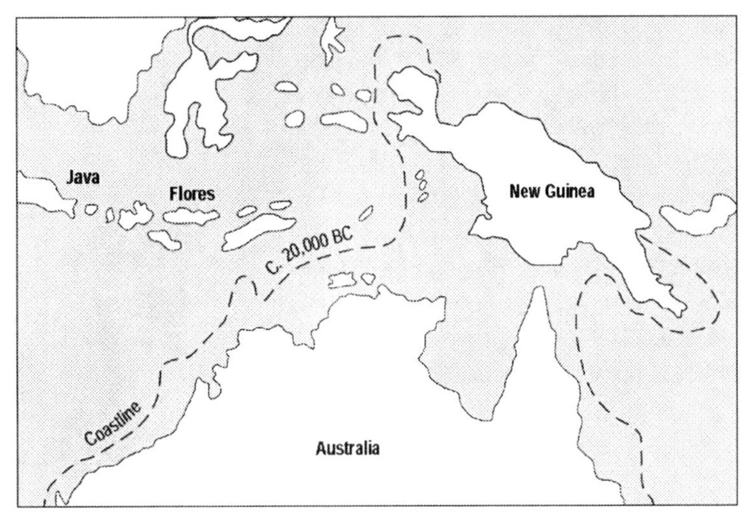

involved sea crossings, for *H. erectus*, of up to 24 kilometres [15 miles]. Some scientists (including Mike Morwood and Peter Brown, who discovered the "hobbit" skeletons) now think *H. erectus* may have been capable of building rafts, but such technology might not have been necessary. Each crossing could have happened by accident. After the tsunami disaster of 2004, several groups of Indonesians were swept tens of kilometres out to sea. To island-hop from Flores to Australia, *H. erectus* would have needed to make a crossing of 60 kilometres. That seems like a rather long jump – until we consider what happened to Rizal Shaputra, who drifted 160 kilometres [100 miles] on a tangle of tree branches after the recent tsunami. He told his rescuers that for the first few days of his ordeal there were "many" other people with him. During the last 840,000 years, hundreds of similar tsunamis must have occurred, as well as tens of thousands of cyclones and other cataclysmic weather events.

It is therefore tempting to speculate that while some *H. erectus* stayed on Flores, others blundered southward to this continent. While their stay-at-home cousins shrank on tiny Flores, they, finding themselves on a gigantic island teeming with lumbering megafauna, may have greatly increased in size (as another relation, *Homo heidelbergensis*, did on the Asian mainland). Hundreds of thousands of years later some *H. floresiensis* may have followed them. [35]

On their own arrival in Australia about 60,000 years ago, the ancestors of modern Aborigines would therefore have encountered both types of Hairy Man – the "big fellas" that they came to know by many names including *dulagarl*, and the "little fellas", the *junjudee*, aka *nimbin*, etc.

At first all three species may have coexisted quite happily. Over several millennia, however, the Aborigines colonised every part of Australia. Their fire sticks changed the flora, fauna and climate of the continent, converting hundreds of thousands of square kilometres of dense lowland forest to open plains and deserts. The introduction

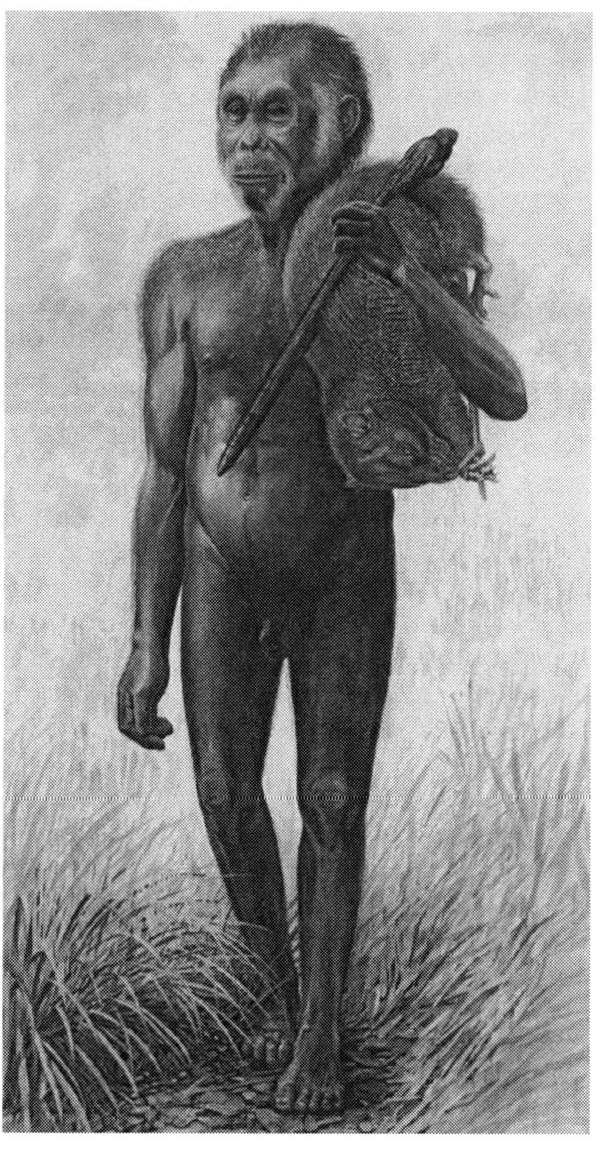

An artist's impression of Homo floresiensis. The creatures may actually have been covered, head to foot, in hair. (Courtesy of Mike Morwood)

of the dingo 3,000 – 4,000 years ago and their use of the animal as an aid in hunting may have given Aborigines a great advantage over *H. erectus* and *H. floresiensis*. Conflict would have been inevitable. After centuries of skirmishing with their technologically superior neighbours, the Hairy Men, greatly reduced in numbers, may have retreated to the places where it is easiest to hide: the deep forests and rugged mountains.

One item of Aboriginal lore, in fact, describes a very similar scenario. The Dharawal people, whose country extends from Botany Bay down to Jervis Bay and inland to Camden and Bowral, have a story about how humans, yowies and junjudees interacted in ancient times. The Dharawal refer to yowies as *dooligahs* and to junjudees as *kuritjah*. Both types were also referred to as *wattun goori* – "hairy men".

Originally, according to the legend, there was plenty of food for all, and the dooligahs, kuritjahs and humans got along quite well, even getting together sometimes for feasts and dancing. Eventually, however, a great drought dried up the rivers and waterholes. Many trees, plants and animals died. The little kuritjah survived by eating the nectar of flowers and humans subsisted on roots and tubers. Because of their great size, however, the dooligahs couldn't find enough food to sustain them and began preying on human children.

As the Dharawal warriors began donning their war paint and sharpening their spears to attack the dooligahs, the kuritjah came to their assistance: they tricked the dooligahs into entering the trunks of large Kurrajong trees (which in those days were all hollow) and then sealed up the trunks. Although it has recently died of old age and toppled over, one of the dooligah trees, which is estimated to have been more than 500 years old, can be seen at Mount Annan Botanic Garden on the south-western edge of Sydney.[36]

One problem with the notion that yowie and junjudee are descended from *H. erectus* is this: both *H. erectus* and *H. floresiensis* used stone tools and, apparently, fire. Could both the "big fellas" and the "little fellas" have lost or abandoned all their technology upon arrival in Australia?

Such a scenario isn't entirely implausible. In both cases the original immigrants are likely to have arrived clinging to storm-driven debris. Perhaps only isolated couples or very small groups – perhaps even groups consisting only of children – were swept ashore. A lot can happen in 500,000 years or so. As we have seen, the West Australian Aboriginal *Pardooks*, presumably a very inbred group descended from tribal outcasts, apparently forgot how to cook food within a relatively few generations.

The Tasmanian Aborigines, isolated for 12,000 years since the last ice age and numbering fewer than 8,000, also lost almost all of their technology. In *The Future Eaters*, Tim Flannery mentions that, when first encountered by Europeans, the Tasmanians had forgotten how to make fire. If a tribe's fire became extinguished, its members had no option but to eat raw meat until they managed to locate another group whose fire sticks were still burning.

Subsequent excavations of Tasmanian campsites revealed other strange things: while bone tools including awls and needles were in common use 7,000 years ago, their use slowly dwindled until, 3,500 years later, they had ceased to be used at all. "This suggests", Flannery wrote, "that stitched clothing [such as the possum skin robes worn by many mainland Aborigines] was lost from the material culture … at about this time". The knowledge of how to make hafted axes, boomerangs and spear throwers was also lost.[37]

It is interesting to think that, had they remained isolated for another 10,000 years or so, the Tasmanians may have also lost the knack of cooking and spear-making and been reduced to a rock-throwing, yowie-like lifestyle.

While there is no undisputed evidence that either *H. erectus* or *H. floresiensis* ever reached Australia, there have been some moments of excitement when it was thought such evidence *was* found.

The Kow Swamp people

In 1967, 40 large, very unusual skeletons were unearthed at Kow Swamp, in Victoria. They were thick-boned, large-toothed, and appeared radically different from those of modern Aborigines. The jaws were among the largest human jaws ever found and to Dr. Alan Thorne of the Australian National University the skulls, which displayed pronounced brow ridges and long, receding foreheads, appeared, in some respects, similar to those of *H. erectus.* As their skeletons were between 9,000 and 14,000 years old, the Kow Swamp people must have co-existed with the ancestors of today's Aborigines for at least 5,000 years.

Not everyone shared Dr. Thorne's opinion about the skeletons. Other scientists suggested that the shape of the skulls was the result of artificial skull manipulation (similar to the binding and shaping of infants' skulls that was practiced by some Native Americans). Anthropologist Gail Kennedy argued that, in the femur at least, the skeletons were perfectly modern with no *H. erectus* characteristics.

Whatever the Kow Swamp people were, they certainly weren't the yowies we have come to know and love: although broad and powerfully built, they didn't have arms down to their knees and weren't seven feet tall. Their remains, furthermore, were found in actual graves along with artefacts including a kangaroo-tooth headband.

Aussie elephants

Of course, the fact that *H. erectus* and *H. floresiensis* fossils haven't yet been found in Australia doesn't prove they never got here. The former director of the Australian Museum, Michael Archer, is aware of three elephant fossils found as far apart as Western Australia and New South Wales. Elephants have been known to swim across 45 kilometres of ocean. Island-hopping to Australia would have involved swims of only a further 15 kilometres.

It seems highly likely, therefore, that the massive creatures lived in Australia for some time in significant numbers until they were extinguished for some unknown reason – yet remains of only three have ever been found. [38]

Hairy Houdinis

Given Australia's vast size, the non-discovery, so far, of *H. erectus*/yowie/junjudee fossils may not be particularly remarkable. But if they have continued to exist through the colonial era to the present day, how is it that not a single Hairy Man, large or small, has been killed or captured and hauled into a museum?

Colonial era Aborigines were fleet of foot and very good bushmen, yet, tragically, more than 20,000 of them were shot dead between 1788 and the early 1900s. How is it that not a single, solitary yowie has been shot dead – if only by accident – in the last 220 years?

Sadly, about eight million large animals – and a couple of hundred human pedestrians – are killed each year on Australian roads. How is it that not a single yowie has ever been flattened by a truck? Despite occasional tantalising stories of ape-like skeletons found in caves or on the forest floor, no carcass or part of a carcass – not so much a finger bone – has ever found its way onto a laboratory bench.

Yowie enthusiasts have rationalised the complete absence of physical remains in various ways. The creatures have vast areas of forest in which to hide; they are solitary by nature; are largely nocturnal; can see in the dark; may be every bit as intelligent as *Homo sapiens*; are hard to find because they don't use fire or construct shelters and may carry away or even eat the bodies of their dead. Being as strong as, and much faster than, grizzly bears, they would be difficult to bring down; because they appear so man-like, people hesitate to shoot them.

Some researchers also cite cases where human beings have lived in the wilds for many years without being discovered. Scores of Japanese soldiers, for instance, hid in the jungle for decades after World War Two. Another relevant case involved the tattered remnants of the Yahi tribe, who, between 1874 and 1911, eked out a living, undetected, near Mt. Lassen in northern California. For the final few years they hid in two thorny gullies totalling only three square miles in area. [39]

Funny feet

After the absence of physical remains, the hardest thing for yowie enthusiasts to rationalise is the fact that there seems to be little uniformity in the shape of supposed yowie tracks.

Neil Frost, who has seen both three and five-toed tracks, suggests that since European settlement some groups of yowies may have become isolated and seriously inbred. If yowies are real, flesh and blood animals, we can think of no better explanation. The case of the six-toed *Pardooks* proves that the shape of human feet *can* be drastically affected by inbreeding.

The "ostrich-footed" Vadoma people of the Zambezi Valley provide another example. Centuries ago, a charismatic chieftain decreed that they should never marry outsiders. Consequently, quite a few Vadoma are born with feet that are split lengthways, forming only two large toes. The deformity, according to some observers, doesn't prevent them from "running like the wind", and actually helps them to climb trees.

Category C: They once existed, but no longer do.

This theory suggests that an unknown species of very primitive hominids – *Homo erectus* is often mentioned – once shared Australia with the ancestors of today's Aborigines but, instead of surviving to the present day, as per the preceding theory, the primitive species was wiped out by Aboriginal tribes in centuries of skirmishing.

The primitive hominids may have been terrifying opponents – strong, hairy, bestial and cannibalistic. Stories about them, passed down by word of mouth over millennia, may have formed the basis of the yowie legend. Aboriginal folk-memory can certainly stretch back a very long way. In the colonial era, South Australian Aborigines were still recounting stories about great fire and destruction coming out of Mt. Gambier, an extinct volcano that last erupted 3,000 years earlier.

Percy Trezise and Roland Robinson – skilled bushmen, prolific writers and great friends of the Aborigines – both subscribed to the theory and it is certainly worth considering. One very interesting story recorded by William Telfer (1841-1923), in fact, actually spells out a tale of intermittent warfare between Aborigines and a more primitive species. Those creatures, identified as *yahoos* by Telfer's informant, were believed to have been the original inhabitants of the country.

Telfer (a self-educated pioneer with no particular interest in punctuation) wrote: "then they [the Aborigines] have a tradition about the Yahoo they say he is a hairy man like a monkey plenty at one time not many now but the best opinion of the kind i heard from old Bungaree a Gunedah aboriginal he said at one time there were tribes of them and they were the original inhabitants of the Country before the present Race of aboriginals took possession of the Country he said they were the old Race of blacks he was of Darwins theory that the original race had a tail on them like a monkey he said the aboriginals would camp in one place and those people in a place of their own telling about how them and the blacks used to fight and the blacks always beat them but the yahoo always made away from the blacks being a faster runner mostly Escaped the blacks were frightened of them a lot of those were together the blacks would not go near them as the Yahoo would make a great noise and frighten them with sticks he said very strong fellow very stupid the blacks were more Cunning getting behind trees spearing any chance one that Came near them this was his story about those people ..." [40]

As detailed in chapter one, an Aboriginal story connected to the Mullumbimby stone circles also tells of large-scale warfare between humans and yowies.

Anthropologists have long noted that one of the most widespread themes in Aboriginal folklore is the contest between eagle-hawk and crow. It has been suggested that that these battles, in which the eagle-hawk is almost always triumphant, recall a centuries-long contest between the ancestors of the Aborigines and an earlier wave of more primitive people. Perhaps the "crows" were *H. erectus*.

One of the strengths of the "yowie is dead" theory is that it accounts for the ubiquity of Aboriginal yowie lore while at the same time explaining why no one can catch the critters. A glaring weakness is that it fails to take into account the hundreds of excellent modern-era eyewitness reports.

By this stage, most readers will probably have opted for either Theory A or Theory B. Some, impressed by the quality and quantity of the eyewitness testimony, will have decided that, despite the lack of physical evidence, the Hairy Man must be a real, albeit extremely elusive, creature. Others will have accepted the proposition that Aboriginal myth, European legends, misidentification of feral men, hoaxes, wishful thinking, and hallucinations have combined to create a lurching, looming, folkloric figure, which, though totally unreal, is so appealing to our subconscious minds that we "see" it everywhere.

In our opinion, neither theory is entirely satisfactory. While each accounts for some aspects of the yowie legend, neither fully explains – or explains away – the entire mystery. To us, the yowie phenomenon is a true conundrum.

The Hairy Men apparently can't be killed, captured, clearly photographed or otherwise proven to exist, but they have, nevertheless, been *seen*, *heard*, *smelt* and *felt* by hundreds of credible people of all races and ages for the better part of 200 years. When you step back and look at the phenomenon dispassionately, it is obvious the creatures *can't possibly* exist. When you step forward and examine it closely, however, it is equally obvious they *must* exist.

If Captain Kirk and Mr. Spock of "Star Trek," instead of us, were presented with this problem, Mr. Spock would simply furrow his noble brow for a few moments and say, "A fascinating riddle, Jim, but I think I see the answer ..." We mere earthlings, however, are pretty well flummoxed.

There is, in fact, a way of explaining this conundrum ... sort of ... but it is an explanation that raises almost as many questions as it answers.

This theory suggests that the elusive hairy giants are not quite what they seem, but are actually just one part of a much deeper mystery. It suggests that yowies, though real enough to be seen, leave tracks and kill animals, aren't entirely "real" in the usual sense of the word. According to this theory, the Hairy Man is some kind of psychic phenomenon.

At this point some readers will start rolling their eyes skywards. They will see the "paranormal theory" as a clumsy attempt to solve one mystery by invoking another. Most yowie researchers and yowie eyewitnesses, in fact, also find the idea impossible to accept.

As mentioned in chapters one and six, however, while Aborigines say the Hairy Men are real, they also occasionally mention aspects of their behaviour that, to the Western mind, suggest the supernatural. This is particularly noticeable in regard to the little junjudee. Readers may have noticed that odd details in some non-Aboriginal eyewitness testimony could also be seen as hinting at the paranormal.

At this point, we should say that while we have collaborated for 25 years and seen eye-to-eye on most things, we have never seen this particular aspect of the mystery in quite the same way. Tony is much more inclined than Paul to entertain the possibility that the phenomenon is in some way paranormal. While he is not particularly enamoured with the theory, Paul nevertheless agrees that it shouldn't be dismissed out of hand. He is quite happy to have it spelt out here, along with all the other theories, for the consideration of readers.

Category X: They are paranormal creatures

"It is an old maxim of mine that when you have excluded the impossible, whatever remains, however improbable, must be the truth." – Sherlock Holmes

"The dulagal lives in another realm." – Percy Mumbulla

The yowie – international man of mystery

If giant ape-men were reported only in the mountains of eastern Australia and in one or two other rugged areas, such as north-western North America and the Himalayas, it might be reasonable to argue that they are real, albeit amazingly elusive, creatures.

The fact is that apparently uncatchable ape-men have been reported in every state and territory of Australia (including the island of Tasmania) and in virtually every state and province of the USA and Canada. They supposedly frequent not only the mountains of Nepal, India and Tibet, but also parts of many other countries, including Russia, Mongolia, China, Vietnam, Thailand, Malaysia, Indonesia, New Guinea, Guatemala, Panama, Colombia, Brazil, Argentina, Kenya and Congo.

For centuries, thousands of armed men – hunters, trappers, explorers, prospectors, soldiers – roamed most of the areas concerned, wiping out whole species of animals and races of people, and looting everything that wasn't nailed down. Thousands of less predatory people – farmers, foresters, naturalists, rangers, tourists – have followed in their footsteps. Yet in no museum anywhere is there a yowie/yeti/bigfoot carcass to examine – not so much as solitary finger bone.

It is very difficult to imagine how the hairy ape-men could have avoided being captured, killed or clearly photographed in *all* of those areas – unless they have an avoidance technique that is way beyond our ken.

Yowie and sasquatch – the terrible twins

Despite the fame of Nepal's colourfully named Abominable Snowman, North America's bigfoot/sasquatch is by far the best documented of the world's legendary ape-men. Having read many books about the creatures, spent many months searching for them and interviewed dozens of American eyewitnesses, we have come to an odd, but inescapable, conclusion: in virtually all aspects of their appearance and behaviour, sasquatches are identical to yowies. The reactions they trigger in other animals, in indigenous people, contemporary eyewitnesses, believers and sceptics, are also remarkably similar to those created by their Australian cousins.

Anyone can verify the truth of our assertion by comparing the data in the previous chapter with the profile of the "average" sasquatch in any good book about the North American mystery. Size, colouration, stench, vocalisations – almost every detail is, if not identical, then remarkably similar. Sasquatch genitalia, for instance, like those of yowies, are almost never observed, but American eyewitnesses, like their Australian cousins, almost always assume the creatures they encounter are male. In *Bigfoot! The True Story of Apes in America,* veteran bigfoot researcher Loren Coleman devoted an entire, very interesting, chapter to the matter of bigfoot's rarely seen appendage and other aspects of the creature's sexuality. [41]

Even one of the ways sasquatches mark their territory is identical to yowie behaviour: in *Raincoast Sasquatch*, Robert Alley writes, "Sasquatches ... have long been credited with twisting the tops [of saplings] to mark their territory".

Remarkably, like Australian Aborigines, some Native Americans also believe in two types of hairy ape-men – the huge, yowie-like sasquatch and a similar but much smaller junjudee-like creature. The Tlingit of Admiralty Island, Alaska, call their local little Hairy Men the "Dwarves of Pybus Bay". Loren Coleman has collected several reports of the smaller type and suggests they might be chimpanzee-like *dryopithecines*. Non-native as well as Native Americans have encountered both types of creature. One white man who encountered the smaller type was a prospector known as "Cowboy" Watson. On West Hecata Island, Alaska, in 1948, the creatures constantly harassed him: "Those little, black, hairy devils drove me out of there. They're small, maybe not much bigger than chimpanzees, three and a half, maybe four feet high – but man, are they tough!"[42]

As with yowie witnesses, the great majority of people who encounter sasquatches are convinced they have seen real flesh and blood animals. Over the years, however, many American and Canadian researchers have noticed aspects of the bigfoot mystery that contain hints of the paranormal. Interestingly, several of those "high strangeness" elements are remarkably similar to some of the weirder aspects of the yowie phenomenon.

One characteristic the sasquatch and the yowie have in common is their ability to scare the living daylights out of humans, dogs and other animals.

The frighteners

No doubt an unexpected encounter with a shambling, eight-foot-tall ape-man would give anyone a bit of a jolt, but the level of fear exhibited by some yowie witnesses seems out of all proportion to the situation. Such extreme fear reactions could be seen as an indication that the creature is some kind of psychic phenomenon.

We have already mentioned that feelings of dread can be triggered by naturally occurring infrasound. Putting that aside for the moment, it is worth reminding ourselves of the sheer intensity of the fear experienced by some yowie witnesses.

Aborigines were often said to be mortally afraid of yowies. Peter Cunningham, for instance, noted that his Aboriginal friends never ventured far from their fires at night for fear of them. Similar fears were recorded in South Australia in the 1840s, and in 1871 a group of Aborigines abandoned their camp near Kempsey after sighting a yowie. [Case 16] Although most non-Aboriginal eyewitnesses don't exhibit anything beyond the high level of excitement one might expect, a significant number seem to be almost frightened out of their wits. Two young men admitted to *crying* with fear when approached by the hairy giants. [Cases 181 and 250]

"R", who encountered a huge, red-eyed creature on yowie-haunted Mt. Kembla, gave a particularly graphic description of the terror he experienced: "My immediate reaction was to flee. I had this feeling I was being hunted – an immense feeling of dread ... like someone had drained all the life or blood out of you ... an awful, awe-inspiring feeling ... all your senses come alive ... it is very hard to explain. I remember thinking, 'this thing's going to come over here and rip your head off!'" [Case 156]

When Dean Harrison was approached by a yowie he felt no particular fear until he suddenly experienced "this almost indescribable chill which ran from my head to my toes. This hugely terrifying sensation just overtook my entire body ... I felt *really* unnerved ... I had to physically force myself out of this paralysed state". "Sue" experienced a similar overwhelming terror and deathly chill. It was, she said, as if something sinister had physically clamped onto her spine with an ice-cold grip. [Cases 218 and 225]

"J", a very experienced hunter, said that on seeing a yowie he became "a ball of jelly, unable to move". For nearly 10 years afterwards he could hardly bear to think about the incident. "Now", he says, "I NEVER, NEVER, EVER go into the deep bush by myself unarmed". [Case 183]

Several other people have found it impossible to shake the fear. After his encounter, Patrick Maher sold the family farm and moved to another area. "Sue" now avoids the forest altogether. Twenty years after his sighting, Senator Bill O'Chee experienced cold chills just thinking about it. After 13 years, George Fairweather still gets "a funny, wobbly feeling" in the legs. Adam Bennett and "L", who encountered yowies in 1985 and 1991 respectively, still can't discuss the events without feeling very distressed. [Cases 40, 115, 168 and 181]

In several instances people who encountered yowies at close quarters became utterly paralysed by shock or fear. [e.g. Cases 123, 172, 183 and 253]

If only one or two witnesses reported experiencing such extreme reactions, it might be possible to dismiss them as hysterical, deluded individuals. As we have seen, however, during the past 150 years a great many people – sometimes couples and groups of people – have reported similar feelings of abject terror.

A hair-raising incident that occurred at the Abercrombie River in 1876 involved multiple witnesses and suggestions of the paranormal. When the first witness was approached by the hairy monster, she "became almost spellbound, screamed and screeched, but [was] unable to run". When it approached again, "a fearful commotion amongst the females and a kind of supernatural terror amongst the men took place". Local Aborigines, it was said, "will never camp within miles of this death-like chasm". [Case 22]

Many sasquatch/bigfoot eyewitnesses, their dogs and livestock, have experienced similar feelings of extreme terror.

"Nameless dread"

It is easy enough to understand people becoming frightened when face-to-face with huge ape-men. Less easy to imagine is why some people are struck with the same feelings of terror *without actually seeing the creatures.*

This strange phenomenon has been reported in good faith by numerous Australian and North American witnesses and cannot simply be ignored. Some years ago, borrowing a phrase from horror writer H.P. Lovecraft, we began referring to the apparently causeless fear as "nameless dread". Sometimes the horrible feeling strikes people completely out of the blue; sometimes a strong feeling of being watched precedes it. Sometimes, indeed, the feeling of being watched doesn't escalate into full-blown nameless dread.

The bizarre and unsettling phenomenon is mentioned in Aboriginal lore as well as in the testimony of non-Aborigines. Percy Mumbulla and another Aboriginal elder, Guboo Ted Thomas, said that the "Doolagarl ... frightens people by hypnotising them and putting them under his spell. You sometimes feel him watching you." [43]

While hunting near Batemans Bay in about 1960, Laurie Allard had a close encounter with a yowie that was carrying a dead wallaby over its shoulder. [Case 77] Significantly, about six months earlier, in the same area, he had experienced a classic case of nameless dread.

"I was cutting timber by myself when all of a sudden the hair went up on the back of my neck and I felt so uncomfortable. The feeling was so bad I packed up and left. Then, when I got down to the main road, I thought, 'Why did I do that?' I think now that the creature was there: it was the same feeling [as he experienced when he saw a yowie six months later], the same presence.

"About three weeks later I was working in the same spot. My future wife came up and she took a shotgun and a couple of dogs up a gully to see if she could shoot a wallaby for dog food. And she got the same feeling, sensed the presence of this thing and became very frightened and came back. The dogs were so frightened they were almost tripping us up. I believe [the yowie] can frighten a dog. I think maybe that's its defence. And that's why a lot of people don't see it – the fear it can put in you. I don't know how it would do it; I think it's more likely to be psychic, because I didn't smell anything."

Many other people have reported similar experiences. Some feel just a slight uneasiness, such as when the hair bristled on the back of Bradley Stratford's neck just prior to his sighting. Others, like "Sue", experience unbearable, icy terror.

Some sasquatch researchers theorise that the fear is triggered when a person registers, without realising it, an imperceptible scent released by a sasquatch. Robert Alley explains: "The human reaction to [nameless dread]", he writes, escalates from a sensation of being watched "to the point of wishing strongly to leave the area, often with all of the associated sympathetic nervous system reactions, such as piloerection (hair standing on end), perspiration, etc. The automatic nervous system is capable of doing this at the brainstem level, without one even knowing what one may have seen, smelled or heard at a subliminal level to produce this protective avoidance behavior ... The number of scent molecules is strong enough to trigger an increase, through the lower brain, in adrenalin, but not enough to register on the higher centers in the brain." [This theory is] "popularly referred to by sasquatch field researchers as 'the pheromone theory'". [44]

It is a good theory, but those pheromones must home in like cruise missiles. As early as 1965, Lee Trippett of Oregon, stated, "He [the bigfoot] can terrorise you from the far side of a mountain." [45]

Interestingly, one bigfoot researcher, Duane Hibner, told us that nameless dread could be resisted. His experience suggests that the feeling is induced psychically rather than by pheromones. One day, shortly after he and his wife Ramona had experienced bigfoot sightings, he heard something large moving directly towards him through the trees. As his dog ran in panic-stricken circles and hurtled away, Duane, a big, broad shouldered woodsman, was suddenly overcome by a feeling of terrible dread: "like something was saying 'Run! Get out of here!'"

As he half-turned to run, he thought, "Hey – what's this? I've got no reason to run. I'm not afraid of the woods!" As soon as that thought struck him – and he turned to literally face his fear – the massive creature abruptly changed course and retreated. [46]

Feeling Abominable

It is well worth noting that the elusive Himalayan yeti, too, seems able to terrify people from a distance.

Bill Grant. (Tony Healy)

A Scottish scientist, Bill Grant, told us that as he was approaching a tiny lake on the Nepal/Tibet border in 1983 he was suddenly immobilised by overwhelming fear. A voice-that-was-not-a-voice commanded him to go no further. Much as he struggled to do so, the veteran expeditioner simply couldn't take another step forward. He retreated and, hours later, returned to find the psychic barrier had been lifted. Proceeding cautiously, he discovered a line of huge five-toed tracks along the muddy shore. [47]

Infrasound revisited: Tigers

It has recently been discovered that the roar of an attacking tiger sometimes contains, in addition to its audible components, frequencies of 18hz (cycles per second) – just below the human threshold of 20hz. As we already know, sounds in the area of 18hz – be they generated by faulty fans, winds, storms or earth tremors – can induce feelings of confusion and terror very similar to the nameless dread experienced in yowie hot spots.

The infrasound component of the tiger's roar is physically stunning. It "rattles and shakes people", says Dr. Elizabeth von Muggenthaler of the North Carolina Fauna Communication Research Institute, "because it happens so fast, just a split second ... you never have any thoughts about running away because you're so glued to the moment". [48]

Is it possible that yowies, like tigers, emit ultra-low sounds, inaudible to humans, to disorientate or paralyse their prey? Might they use the same sounds to frighten away humans and their dogs?

Some yowie researchers have, in fact, reported hearing extremely low, rumbling growls in yowie hot spots. Researcher Tim Power described the noise as resembling that of several large boulders rolling very slowly over each other.

In light of all the above, an August 5, 2000, sasquatch incident near Jarosa Mesa, Colorado, becomes doubly interesting. It began when a woman who was camping out alone "felt like I was being watched". Her dogs became extremely nervous. Then an eight-foot-tall bigfoot walked to within 12 feet of the camp. As she and her dogs froze, terrified, it emitted "a low rumbling sound".

The woman, a musician, said the sound "was so much on the low range. I got the feeling that there were frequencies I couldn't hear". A few minutes after the creature turned and loped away all the local bird and insects, which had been totally silent, resumed their vocalisations. Several yowie witnesses have also noted the total cessation of bird and insect noise. [e.g. Cases 183 and 62]

Significantly, perhaps, the woman also said that just prior to the incident "there had been a lot of electrical storm activity ... lightning". [49]

Hairy hypnotists

As mentioned earlier, Aborigines believe that yowies can "frighten people by hypnotising them". Some Aborigines also believe the creatures can use hypnosis to induce sleep or confusion.

According to Percy Mumbulla, "If you're walking up the mountain and you start to feel a bit sleepy, that's his presence. He'll try to put you to sleep but if you sit down and have a bit of a snooze, he'll get at you ... he has this mesmeric power that makes you drowsy and a bit confused about where you are and what you're doing".[50] As we now know, inaudible sounds in the range of 18.9hz can cause people to feel confused and tired.

When Laurie Allard found himself face-to-face with a yowie near Batemans Bay, it was standing out in the open, in broad daylight, only 20 metres away. It had a dead wallaby over its shoulder and was clearly not human. Despite that, Laurie simply couldn't accept what his eyes were seeing. Although he doesn't use the term "hypnosis", he acknowledges that throughout the minute-long encounter and for some time afterwards he was very confused: "Funny thing: it gave me a really hairy, scary feeling ... I didn't really absorb it." Despite the evidence of his eyes, "I thought it was my mate ... and I spoke to it: 'have you got a wallaby, Bill?'" [Case 77]

Although several witnesses have been thrown into some degree of confusion by their yowie encounters, it is hard to find cases that support the Aboriginal belief that yowies can induce sleep. Only two spring to mind.

Adam Marion's mother told us that, immediately after blurting out the story of his yowie encounter, the 12-year-old went straight to bed and slept soundly for two hours – something he'd never been known to do in daylight. When the huge, grimacing ape-man approached their camp, one of surveyor Charles Harper's assistants collapsed in a faint. While it is not unknown for people to faint from shock, it seems decidedly strange that, as Harper reported, the man "remained unconscious for several hours". [Cases 149 and 35]

Interestingly, in the early 1970s, during a dual-witness sasquatch encounter near Easterville, Manitoba, another man fainted dead away. He later suffered recurring nightmares and couldn't sleep without a loaded gun beside his bed.[51] His reaction would not have surprised the Tlingit and Haida people of southern Alaska: like some Australian Aborigines, they believe the hairy giants can cause unconsciousness.[52]

Any reader who cares to fossick through our Catalogue of Cases might find other odd details that could be construed as smacking of hypnosis. The only non-Aboriginal eyewitness to specifically allude to hypnotism, however, was Thelma Crewe who watched two yowies at very close range in 1976. She said she was "mesmerised" by the creatures as they stared back at her. [Case 105]

Causes illness or death

As mentioned in chapter one, Pickett Hill, near Coffs Harbour, is the most significant cultural and spiritual site of the Gumbaynggir people. They believe a "Barga hairy man" protects the area. "A lot of people who go up there get sick."

As mentioned in chapter five, many Aborigines believe junjudees – the rather magical little Hairy Men – keep a close, protective eye on sick children. If, however, they are ridiculed or talked about too loosely, they are liable to punish the talkative person or his family with illness.

Interestingly, some non-Aborigines have also linked strange illnesses to the proximity of the larger variety of Hairy Man.

In 1971, Richard Gilson was operating tin mines on the Carrai Plateau, where George Grey was attacked by a yowie three years earlier. Shortly after he and his employees built their camp, they began to hear strange noises at night. "It was sort of a deep howl ... a short, deep-pitched noise. I've travelled the world and mined in jungles and I've never heard a noise like it."

Large footprints also appeared: "like a webbed-foot creature ... a big duck. There were four toes which seemed to have a sort of claw". Although they were 20 strong, the miners became decidedly nervous. No one ever stayed in the camp alone and doors were bolted at night, which was just as well: one night something tried to get into one of the huts, leaving claw marks halfway up the door. Then several men were struck by a mysterious illness: "It was like some sort of brain trouble. We thought it might have been the water, but it was found to be okay." [Case 88]

When a yowie that smelt "like a burnt mattress" was lurking outside their house, "Sue" and her teenaged son weren't only scared out of their wits – they also experienced headaches and stomach pains. Like Percy Window in 1978, the three brothers who were attacked by a yowie near Byron Bay in 1905 were said to be "ill after their fright". [Cases 225 and 40]

Although the 1981 Dunoon incident seemed quite straightforward at the time, it too had a rather strange aftermath. The mother of the oldest witness told us that her son, who was the only one to actually make eye contact with the yowies, suffered health problems for months afterwards. He experienced recurring nightmares and his eyes were constantly sore and inflamed. After conventional treatments failed, the woman asked a visiting faith healer to look at him. To her great relief, the "laying on of hands" was immediately effective and he recovered completely. [53]

We know of no Native American lore that suggests sasquatches induce illness. However, one story Tony heard while searching for the yowie-like *orang mawas* of the Malay Peninsula described such an effect. In 1961, after a huge, hair-covered man-like monster was seen near Simpang Renggam, a heavily armed party comprised of some of Malaya's best trackers, hunters and soldiers followed its enormous tracks through a swampy region known as Ayer Hitam. After four days hot on the animal's trail, every member of the party came down with an illness unlike anything they'd ever experienced. [54]

Mercifully, we know of no verifiable case where yowies have (as per the Aboriginal belief) caused people to die. The only witness to hint at such a dire occurrence was Jean Maloney of Woodenbong. She remarked that shortly after yowies were seen within the town's residential section, several local Aborigines were killed in a terrible road accident. [Case 111]

Tracks

One way in which the bigfoot phenomenon differs from that of the yowie is that in North America tracks are discovered much more frequently than they are here. Although the much greater incidence of snow, higher rainfall, and softer ground in North America goes some way to explaining the great disparity, the extreme rarity of yowie footprints still seems very odd.

Tracks attributed to yowies vary wildly from three to four to five toes. Descriptions given by people who have actually observed the feet are also inconsistent. This lack of uniformity in foot shape is the most difficult thing for yowie enthusiasts to rationalise. Significantly, perhaps, the sasquatch, too, has something of a foot problem.

The majority of bigfoot tracks are pretty consistent in shape: like huge, flat, five-toed human feet. But a sizable minority – sometimes discovered immediately after apparently genuine sightings – consist of weirdly shaped three and four-toed footprints. Some of the three and four-toed American footprints look very much like some Australian tracks.

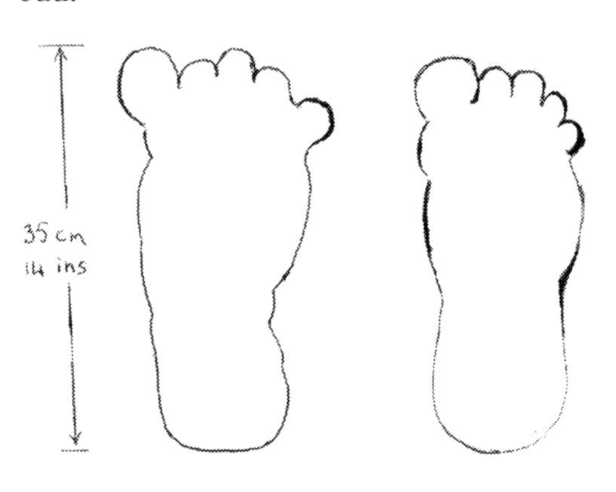

The 1998 Springbrook yowie track (left) compared to a 'normal' five-toed bigfoot track found at Bluff Creek, California, in 1968. (American sketch courtesy of Chris Murphy)

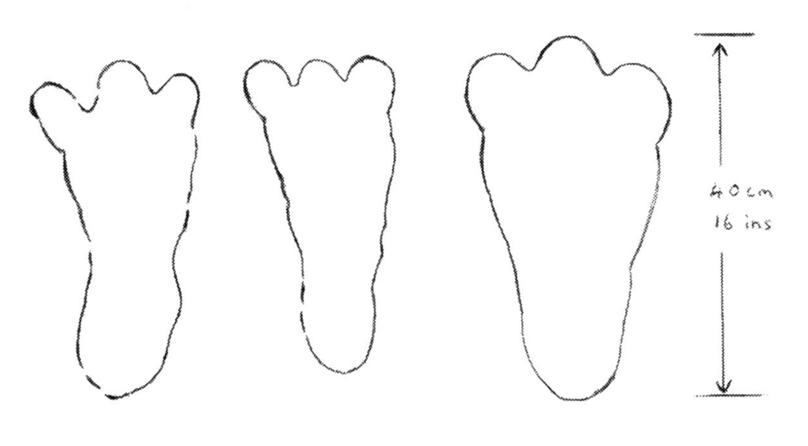

Left Track - Nerang, QLD 1975 (Case 125).

Middle - Blue Mountains, NSW, 2001 (Case 223).

Right - Tylertown, MS, USA, 1977. (American sketch courtesy of Peter Guttilla and Lucy West)

Most bigfoot hunters are rather nonplussed by tracks other than the "normal" five-toed variety, and some dismiss them as hoaxes. Others think they indicate that the sasquatch is no ordinary earthly animal.

Jeepers creepers

As mentioned earlier, several witnesses say the eyes of the yowie reflect the light of fires and torches. Strangely, sometimes they *don't* reflect. Worse, in a small proportion of both yowie [e.g. Case 156] and sasquatch reports, the creatures' eyes are said to glow in the dark *as if lit from within.* [55]

The yowie/sasquatch/black panther connection

Another aspect of the yowie and sasquatch mysteries hints strongly at the paranormal: in both Australia and North America, localities that produce ape-man reports also frequently produce reports of large, uncatchable "black panthers".

In 1978/79, during 11 months on the trail of the sasquatch, Tony heard 12 reports of black panthers – all of them in noted bigfoot hot spots. In two cases black panthers were seen in the same small fields that sasquatches had previously strolled across. Some witnesses assumed they'd seen black-pigmented pumas but that couldn't have been the case. Melanistic pumas do not exist: no black puma has ever been killed, captured or photographed.

It seems simply beyond reason that two totally different types of large, apparently uncatchable animals should crop up time and again in the same limited areas.

In our 1994 book *Out of the Shadows*, we devoted a lengthy chapter to the "alien big cat" phenomenon in Australia. Since at least the 1880s, thousands of people have reported seeing what they assume to be exotic big cats in various parts of the country. Most are

Logo of the New England Bigfoot and Black Panther Research Alliance, Brandon, Vermont, USA.

described as black panthers. They have killed thousands of sheep – and even horses – and have left many large, feline footprints. It has been impossible not to notice that, as in North America, these panthers are often seen in areas that produce a lot of hairy ape-man reports. All of the yowie hot spots covered in chapter four, for instance, are noted for panther sightings. In several instances black panthers have been reported within a couple of hundred metres of the site of a yowie encounter. Despite the thousands of reported sightings, no alien big cat has ever been killed, captured or clearly photographed anywhere in Australia.

Stop Press: An associated mystery has recently been solved. Over the years, many bushmen have reported seeing or shooting feral domestic cats that have supposedly grown to great sizes. Proof was always elusive. Recently, however, our colleague Michael Williams found a hunter who killed and took convincing photographs of a cat that measured 1.8 metres [6 ft] from nose to tip of tail. DNA tests arranged by Michael proved that the daunting animal was indeed the descendant of wayward domestic moggies. (*Australian Shooter*, November 2005.)

Hairy wallbangers: a polt connection?

As if suggestions of a yowie/black panther connection were not weird enough, the possibility exists that our furry friends might also have something in common with poltergeists.

The fact that noxious odours, stone throwing and wall banging – hallmarks of poltergeist activity – also occur frequently in yowie events may or not be significant. In one remarkable case, however, the polt/yowie connection is totally unambiguous.

In 1946, when George Nott and his family moved into a long-abandoned property near Wilcannia, they heard thumping sounds in the ceiling. Doors swung open, objects flew, and so many pebbles fell on the roof that they "sounded like a heavy shower of rain". Immediately thereafter, huge five-toed tracks appeared in the yard and a "big hairy monster ... a bloody big gorilla" twice invaded the house. [Case 70]

The Wilcannia report is the only one in our files that specifically links polts and yowies. While investigating the Northern Territory's greatest-ever yowie outbreak near Acacia Hills [Case 220], however, we couldn't help noticing that one of Australia's most dramatic poltergeist episodes was occurring simultaneously only eight kilometres away at Humpty Doo. Eight kilometres, in Northern Territory terms, is just around the corner.[56]

"Beam them down, Scotty"

In North America a link between bigfoot and UFOs has been suggested on many occasions. The theory, in fact, has led to a kind of schism between bigfoot hunters who treat the mystery as a purely zoological phenomenon, and those who think it is something a darn sight weirder.

Although he resisted the notion, Tony finally had to acknowledge that six widely separated bigfoot hot spots he visited in the USA and Canada in 1978-79 also produced many UFO reports. Four of the locations, in fact, were said to have experienced quite intense episodes of UFO activity.[57]

Between June 1973 and February 1974 a phenomenal outbreak of weirdness in western Pennsylvania yielded 118 bigfoot and no less than 600 UFO reports. On a dozen occasions low-flying UFOs were reported immediately before or after bigfoot sightings.[58]

Although a similar connection between the Australian Hairy Man and UFOs is rarely suggested, a couple of poorly documented cases lend some support to the idea.

In January 1975, people near Goolma, NSW, supposedly observed a UFO descending. During the next two weeks, there was an intense wave of yowie reports, which suddenly ceased after a second UFO was seen.[59] Five months later at Tailem Bend, South Australia, two groups of people allegedly saw a giant, hairy ape-man with what looked like a lantern in his hand. A third group saw a huge illuminated dome-shape in the same field. [Case 100]

Short of photographing half a dozen yowies emerging from a flying saucer carrying their lunch boxes, we can see no way of ever establishing with certainty a UFO/yowie link. Elements in two or three of our best cases, however, do make us wonder. The O'Connor property, for instance, which has experienced repeated yowie "yard invasions", has been the site of another strange visitation. In August 2002, neighbours observed a large (12 feet in diameter) dazzling light hovering noiselessly just above the O'Connor residence. Similarly, the wife and son of "Jim", who encountered yowies repeatedly in south-east NSW, told us that a huge, humming saucer-shaped craft had descended on one of their previous residences, shaking it to its foundations. Jim fired at it with a .303. Two days after his wife saw a yowie from their back window, Dan "M" and researcher Gary Opit observed an apparent UFO from the front step of the same house. [Cases 223, 187 and 272]

Dimensions

American researchers Peter Guttilla and Dr. Alan Vaughan have come up with a very interesting theory to explain why UFOs, bigfoot and other strange phenomena are often seen in the same areas. The theory also accounts for the phenomenon of UFO "flaps" and bigfoot "outbreaks" (large numbers of sightings occurring in a particular locality at a particular time, followed by long periods where nothing is reported).

With the help of physicist James Spottiswoode, they analysed 224 UFO and bigfoot reports and established that such incidents don't occur at random. Sightings, they found, increase markedly during times of decreased activity of the earth's geomagnetic field. They hypothesise that the field normally forms "a natural barrier to a hidden universe in other dimensions … When the barrier of geomagnetic activity is low…the human mind [more easily expands] consciousness into another dimension to obtain psychic information".

For the benefit of lay people, Guttilla and Vaughan explain that Earth's magnetic field "is thought to result from spinning of the molten liquid outer core. Strongly influenced by the Sun's magnetic field, the geomagnetic field continually fluctuates. Solar storms generate an enormous increase in geomagnetic activity. This fluctuating activity, which is measured in three-hour periods by observatories around the world, is averaged into a global geomagnetic activity index and published regularly in Boulder, Colorado". Their preliminary findings, they claim, "give overall odds against chance of 3,000 to 1 for our extradimensional hypothesis".

Although many will find their theory mind-boggling, Guttilla and Vaughan point out that many mainstream scientists are not averse to speculating about, and running calculations on, other dimensions. They mention that in February 1998, *Scientific American* published "a new theory that in addition to 'our' 10-dimensional universe there is a 10-dimensional parallel universe".[60]

Guttilla and Vaughan, furthermore, insist their theory is scientifically testable. They are now collecting more reports (which need to contain the exact date and time of the event) to process through their computers. They predict future calculations will show that sightings of other mysterious creatures (such as black panthers) and a whole range of psychic phenomena also peak during times of low geomagnetic activity.

Dr. Michael A. Persinger, of the Department of Psychology, Laurentian University, Sudbury, Ontario, has engaged in studies similar to Guttilla and Vaughan's. He agrees that multiple reports of weird phenomena, often clustered in "window areas", occur when Earth's geomagnetic field is weak. He, however, stops well short of suggesting the "sightings" are necessarily the result of glimpses into or intrusions from other dimensions, and appears to favour the idea that they may be hallucinations triggered by the geomagnetic fluctuations. That theory, of course, fails to account for all physical traces, such as bigfoot, yowie and yeti tracks.[61]

Several veteran bigfoot researchers share Guttilla and Vaughan's belief that the bigfoot is "a visitor from elsewhere". For many years, in the face of constant ridicule, the always controversial Jon Beckjord has been saying that the sasquatch seems to be "a different type of life form, operating on a different chemical, physical and biological order than ordinary animals. They're a type of earthly alien."[62]

Jack Kewaunee Lapseritis has also been pilloried by many "mainstream" bigfooters. In *The Psychic Sasquatch* he wrote at length about the interdimensional nature of the creatures and of their connection with the UFO phenomenon. "There is an interdimensional aspect to bigfoot", he insists. "It's the answer." Mr. Lapseritis, who has many Native American informants, claims to have actually communicated (both in person and psychically) with the hairy giants. Although his claims are extraordinary – he says, among other things, that the creatures are capable of shape shifting – his many years of field research in both the Himalayas and North America, and his obvious conviction, demand respect.

After more than 40 years of research, John Keel, of *The Mothman Prophecies* fame, believes the hairy giants are "paranormal or paraphysical apparitions". Their footprints, he points out, "sometimes number three and four toes, a physical impossibility for a primate ... They chase cars and carry off cattle ... scream like banshees and, despite their great height, girth and weight ... melt into nothingness when the posses turn out."[63]

In the past 30 years or so thousands of sightings of mysterious, seemingly uncatchable big cats – mostly black panthers – have occurred in Great Britain. Even more weirdly, since the mid-1990s a small number of sightings of bigfoot-like creatures have also been reported there. British bigfoot witnesses, like their benighted colonial cousins, have reported "intense feelings of fear", strange booming noises and unusual electromagnetic effects.

Interestingly, cryptozoologist Jonathan Downes who, with five companions, actually saw one of the shaggy ape-men at Borlam Lake, Northumberland in 2003, doesn't believe it was "real" in the usual sense of the word. "I do not know what I saw", Jon said later, "but logic tells me it was not flesh and blood. This forest could not sustain a primate, and there is no known higher primate that can survive the cold weather. I believe it is something of a parapsychological nature ... something which is defined by laws of physics that we don't understand. A population of a hitherto unknown species of higher primate could not under any circumstances be living in a country park only 30 miles from Newcastle city centre." Jon refers to such apparently real, but uncatchable and utterly implausible creatures as "zooform" phenomena.[64]

Not being scientists, we struggle to understand how the Earth's geomagnetic field operates at all, let alone how it could affect the psychic abilities of humans or barriers between dimensions. That having been said, we know of several yowie and bigfoot reports in which apparent electrical effects have been mentioned.

Electromagnetic radiation can cause conjunctivitis. As previously mentioned, a boy who made eye contact with a yowie at Dunoon in 1981 suffered from inflamed eyes for months afterwards. American researcher Steve Mizrach has written, "As in many UFO cases, there have been Bigfoot sightings that have left witnesses with ... eye burn."[65]

As detailed in chapter four, since 1997 Jerry and Sue O'Connor have experienced repeated "yard invasions" at their Blue Mountains property. Their case is one of the strangest in our files. A few years ago we might not have given it much credence. We might have assumed Jerry and Sue were overimaginative, the victims of a clever hoax, or hoaxers themselves. In the past five years, however, we have come to realise they are strong-minded, courageous people who are very unlikely to have fallen for a hoax. Their sincerity is obvious. Jerry swears, "on my daughter's grave", to the truth of what they say, and both he and Sue are willing to undergo lie detector tests at any time.

Because the O'Connors' story is so mind-boggling, it is important to keep in mind that everything they have reported, even the very weirdest of details, has been reported elsewhere in Australia or overseas.

They have seen huge, hair-covered ape-men close to their house by day and by night. The creatures leave both three and five-toed tracks, thump walls, and can detect infrared light. They avoid camera traps so easily that it seems, as Jerry puts it, "the buggers can read your mind". Both Jerry and Sue experience what they describe as weird, rather sickening, "electrical" sensations in their backs when the creatures are nearby. The feeling is often centred on the region of the kidneys. Sometimes it creates a burning sensation, more often it feels icy cold, like "liquid nitrogen". Sometimes it induces a degree of paralysis, leaving them barely able to talk. Sometimes it leaves them feeling very thirsty. The "electro" feeling sometime wakes them from a deep sleep. They are certain yowies somehow "project" it.

It is interesting to note that the wall behind the head of the O'Connors' bed – the one that the yowie has apparently thumped so frequently – contains an external electricity meter and fuse box. On at least one occasion the yowie actually hit the box, knocking the lid open. Interestingly, bigfoot researcher Stan Gordon has noted that the American ape-men, also, seem to be attracted to electricity meters. [66]

As mentioned previously, some Aborigines say the Hairy Man is most often seen in stormy weather. In light of that, the following may well be of some significance: since 1997 the O'Connor's house has been struck by lightning on no fewer than four occasions. The fuse box, three video recorders and several other appliances have been destroyed. On one occasion, as their television set exploded, Jerry saw a ball of plasma emerge from the screen, fly across the room and land on a couch. As mentioned earlier, neighbours once saw a large, dazzling, round light hovering noiselessly just above the O'Connors' roof.

The property is situated on a razorback ridge. As four nearby trees and the houses of their two closest neighbours have also been struck, it seems logical to assume that minerals in the ridgeline attract the lightening. There is, however, an extra degree of strangeness: the O'Connors' previous residence, in Sydney, was also struck by lightening. (It is worth mentioning that at a certain Australian farm that has for several years suffered the depredations of uncatchable black panthers, the homestead has been struck by lightning on a dozen occasions). [67]

The O'Connors noticed an odd pattern to the yowie visitations: they occurred most frequently just before Sue's monthly periods. Interestingly, the same behaviour has been noted in North America. Stan Gordon wrote that during a wave of bigfoot activity near Nowata, Oklahoma, in 1974, the creature showed "a certain peculiar interest in women", especially menstruating women. [68]

Like sasquatch researcher Kewaunee Lapseritis, Sue O'Connor claims to have established a telepathic link with one of the creatures. The individual she contacted seemed to be their most regular visitor, and was a female. It seemed to say that it resided in the "Black Dimension" but was, nevertheless, a benign "being of light" which was drawn to her and her garden. It was immensely old, essentially immortal.

The suggestion that the hairy giants are extra-dimensional, unpalatable as it may be to most researchers – and indeed to most eyewitnesses – could certainly explain some of the odder aspects of the yowie/bigfoot/yeti phenomenon. It might account for their apparent invulnerability to bullets, their ability to induce debilitating fear in men and dogs, their ability to produce a choking stench like burnt electrical wiring, their sometimes glowing, sometimes not glowing eyes, their frequent proximity to uncatchable black panthers, their occasional proximity to poltergeists, their apparently very malleable foot shape, their sheer elusiveness and many other baffling characteristics.

The inter-dimensional theory could also explain some of the odder things Aborigines say about yowies – such as their occasional assertion that the creatures' feet are reversed on their ankles, and that although real, they can never be caught. Percy Mumbulla once remarked, without elaborating, "the dulagal lives in another realm".[69]

Some Native Americans refer quite unambiguously to the bigfoot's inter-dimensional origin. At a Miccosukee Indian village in the Florida Everglades, Patrolman Don Osceola and tribal leader Bobby Tiger told Tony that saquatches are "solid enough when they're here ... they make tracks, kill animals ... but they can disappear totally".

The Miccosukee term for the creatures is *yati wasagi*: "*Yati* means 'man', *wasagi* means 'lost', 'separated', 'different' – something like that. What [our] people are saying, really, is that these things are from another dimension ... they may be things we are not meant to find out about." Significantly, perhaps, Bobby Tiger linked the hairy giant phenomenon to that of UFOs. "All these things", he believes, "are part real and part not." Some Klamath Indians near bigfoot-haunted Bluff Creek, thousands of miles west of the Everglades, also strongly suspect a bigfoot/UFO link.[70]

As far back as the 1930s a pioneer of Australian cryptozoology, R.W. McKay, suggested the Hairy Man might be paranormal. Noticing the apparent yowie/panther connection, he referred to them as "something supernatural". "Whatever these animals are", he wrote, "they have something protecting them ..."[71]

Fred Beck, last survivor of the 1924 "Ape Canyon Incident", in which a group of miners was besieged by several sasquatches at Mt. Saint Helens, stated: "We genuinely fought [them] and were quite fearful ... but I was always conscious that we were dealing with supernatural beings ... the other [men] felt the same." Something else Fred Beck said is interesting in light of Guttilla and Vaughan's theory about the earth's geomagnetic field and in view of what was conveyed to Sue O'Connor about the "Black Dimension": sasquatches, he said, "are from a lower plane. When the condition and vibration is at a certain frequency, they can easily, for a time, appear in a very solid body."[72]

But if the hairy giants are inter-dimensional apparitions, why do 90 percent of yowie incidents contain absolutely no hint of the paranormal? Why are the vast majority of eyewitnesses left with the firm conviction that the creatures are real, solid animals? Even Neil and Sandy Frost, who have experienced 20 years of yowie visitations in the same area as the O'Connors, see no reason to suspect the creatures are anything other than flesh and blood.

And if yowies are gentle, intelligent, ageless, saucer-riding or vortex-surfing shape-shifters, why do they bother rampaging through the bush in all weathers, screaming like banshees, tearing kangaroos apart, and wolfing down grubs, termites, garbage and maggot-riddled dog carcasses?

Is the paranormal theory simply too strange to be true? Well ... it is no stranger than what most of the world's religions ask their adherents to believe. No stranger, either, than what scientists tell us about quantum physics: that some particles, for instance, can be in two places at the same time, that the mere act of observing particles can change them, that time can slow down, and that solid matter doesn't actually exist – it's all just empty space and energy.

The theory certainly accounts for a lot of the stranger aspects of the yowie mystery. Like all the other theories, however, it doesn't seem to explain *everything* about the baffling phenomenon. It is also somewhat creepy. The worst thing about it, of course, is that unless some clever psychic persuades a yowie to materialise on the front steps of the Sydney Opera House, it seems impossible to prove. Because of that, most researchers are reluctant to whole-heartedly embrace the notion.

American researcher Jerry Clark advises caution. He says that while it is foolish of sceptics to try to explain away "high strangeness" cases by pulling patently implausible "logical explanations" out of a hat, it is just as unwise for believers to "explain" them by resorting to "scientifically meaningless or overtly supernatural 'theories' based on a host of unverifiable assumptions about the nature of reality".

Clark's attitude is purely fortean. Charles Fort (1874-1932) was a pioneer investigator of the unexplained whose books are as amusing as they are thought provoking. His spirit lives on in *Fortean Times*, a monthly journal of strange phenomena founded by Londoner Bob Rickard in 1973.

American writer Terry Colvin defines forteanism in this way: "[it] is not a 'belief' in any pet hypothesis, nor commitment to any one particular explanation of anomalous phenomena ... [it is] an open-minded scepticism about reported phenomena ... and a willingness to consider a wide range of possible explanations ... True forteans ... see the generally accepted scientific world picture neither as perfectly complete, accurate and infallible nor as completely worthless and misguided." [73]

We like to think of ourselves as forteans. The various theories relating to the existence or non-existence of the Hairy Man all have some merit. In the past 28 years we have had a lot of fun bouncing those ideas around, but in the tradition of Fort have always refrained from engraving any one theory in stone.

Having examined apparent yowie tracks, interviewed more than 120 mostly very credible eyewitnesses and compared their stories to those of Aborigines and white pioneers, we find it almost impossible to believe the creatures don't have some kind of objective reality. At the same time, we acknowledge that hoaxes, folklore-gone-feral, shoddy reporting, wishful thinking and superstition have also had a role in shaping the phenomenon. The inability, so far, of anyone to clearly photograph, capture or kill the creatures, and the many possibly paranormal aspects of the mystery also give us pause.

After nearly 30 years of research all we can say is this: the creatures *seem* to be very real indeed ... most of the time. And we believe they *are* real – but perhaps not necessarily "real" in the full "Western" sense of the word. Perhaps the elusive Hairy Man is, as Dr. McCoy of "Star Trek" might say, "life – but not life as we know it".

Obviously, we would love to see it proven that yowies (and their little cousins the junjudees) are living, breathing, flesh and blood creatures and that they really *are* out there somewhere, lurking in the bush.

We live in hope. But meanwhile, until all the facts are in, our yowie file remains open.

Endnotes

1. "A Squatter", *Reminiscences Of A Sojourn in South Australia*, Kent and Richards, London, 1849, pp. 135-136.

2. Christine Nicholls (ed.), *The Pangkarlangu and the Lost Child, A Dreaming narrative belonging to Molly Tasman Napurrurla*, Working Title Press, Kingswood, SA, 2002.

3. *The Sun*, 24 November 1912, p. 13, cited by Graham Joyner, *The Hairy Man of South Eastern Australia*, Union Offset, Canberra, 1977, p. 21.

4. Several 19th century writers mention this odd belief. Alexander Harris, for instance, heard it from Hunter River Aborigines in the 1830s. P.J. Gresser, who recorded Aboriginal lore in the Blue Mountains and around Bathurst in the 1940s and '50s, wrote that the yahoo or yowie "was said to be an animal of large proportions whose body was covered with masses of long hair and whose feet were reversed, the toes being where the heel should be. The belief ... is well authenticated".

Alexander Harris, *An Emigrant Mechanic, Settlers and Convicts, or Recollections of Sixteen Years Labour in the Australian Backwoods*, London 1847; P.J. Gresser, *Manuscripts Relating Principally to the Aborigines of the Bathurst District, Bathurst, 1964, pp.167-71.*

5. Mrs. Mamie Mason, interviewed by Patricia Riggs, *Macleay Argus*, 2 Oct 1976; letter to *The Sun,* 24 Nov 1912, from A. B. Walton, cited by Joyner (1977), p. 21. When he was a boy, in the Braidwood district, Aborigines told Mr. Walton the "yahoo" was only rarely seen.

6. R. H. Mathews, "Ethnological Notes on the Aboriginal Tribes of New South Wales and Victoria", *Journal of the Royal Society of New South Wales*, 1904, pp. 345 and 361-3, cited in *Illawarra & South Coast Aborigines, 1770 – 1850*, compiled by Michael Organ, 1990.

7. Lee Chittick and Terry Fox, *Travelling with Percy, A South Coast Journey*, Aboriginal Studies Press for the Australian Institute of Aboriginal and Torres Strait Islander Studies, Canberra, 1997, p. 82.

8. *The Advocate*, Coffs Harbour, 15 June 2000.

9. Chittick and Fox, p. 82.

10. Graham Joyner, "The Meaning of *yahoo* and *dulugal*. European and Aboriginal perspectives of the so-called 'Australian Gorilla'", *Canberra Historical Journal*, No. 33, Mar 1994. (Graham points out that during the 19th century, the term "yahoo" was occasionally used in connection with other hairy man-like creatures. In 1814, for instance, when a large ape, possibly an orangutan, was exhibited in England, it was billed as "the Great Yahoo or Wild Man of the Woods".)

11. Gresser, pp.167-71.

12. Luise A. Hercus, *The Languages of Victoria: A Late Survey*, Part 2, pp. 392, 444 and 474, Australian Institute of Aboriginal and Torres Strait Islander Studies, Canberra; also *The Queanbeyan Observer*, 7 Aug 1903, both cited by Joyner (1977) pp. 13 and 25.

13. Ron Heron, "The Dreamtime to the Present, Aboriginal Perspectives", MS, College of Indigenous Australian Peoples, Southern Cross University, Lismore, NSW, 1991, pp. 42-43.

14. Cunningham, Peter, *Two Years in New South Wales; A Series of Letters, comprising Sketches of the Actual State of Society in that Colony; of its Peculiar Advantages to Emigrants; of its Topography, Natural History, &c. &c,* Henry Colburn, New Burlington Street, London, 1827.

15. Harris, p. 97.

16. "Superstitions of the Australian Aborigines: The Yahoo", *Australian and New Zealand Monthly Magazine*, 1842, pp., 92-96, cited by Joyner (1977), pp. 4-5.

17. Ernest Favenc, *The History of Australian Exploration from 1788 to 1888.* Sydney, Turner and Henderson, 1888, pp.188 and 202.

18. Rex Gilroy, *Mysterious Australia*, Nexus Publishing, Mapleton, QLD, 1995, pp. 21-27 and 253-255; Rex Gilroy, *Australian UFOs Through The Window Of Time*, URU Publishing, Katoomba, NSW, 2004, pp. 257-258.

19. Roland Robinson, *Black-Feller White-Feller*, Angus and Robertson, Sydney, 1958, p. 121; *The Australian*, 26 May 1973.

20. "Aboriginal Dialects: Cooma Sub-District", *Science of Man*, 23 Aug 1904, Vol. 7, No. 7, p.104, cited by Joyner (1977), p.4.

21. M. Feld, "Myths of Burragorang Tribe', *Science of Man* 3(6), 1900, cited by J. Smith, *Aboriginal Legends of the Blue Mountains*, p. 89-90; Bernard O'Reilly, *Cullenbenbong*, W. R. Smith and Paterson, Brisbane, 1961, Ch. 9.

22. Gerry Bostock interview with Tony Healy, 27 Jan 2005.

23. Heron, pp 42-43

24. *Fraser Coast Chronicle*, 28 Jan; 4,5,10 and 30 Feb 2000; letter from Brett Green to *The Gayndah Gazette*, 9 Feb 2000.

25. *Courier Mail*, 29 Jan 1994.

26. Tony Healy, "Monster Safari", MS, Canberra, 1983, pp. 317 and 641.

27. *The Works of Aurelius Augustine*, Vol. 2: *The City of God*, Vol. 2, pp. 116-117, cited by D. Jeffrey in *Manlike Monsters On Trial*, M. Halpin and M. Ames, (eds.) p. 48.

28. In *Dolpo, The World Behind The Himalayas*, for instance, Karna Sakya relates the tale of the *kichi-kinny*, a man-eating ghost that takes the form of a beautiful girl. She seduces men, leads them to a secluded spot and tickles them to death. Eager young men can avoid that not-so-gruesome fate if they think to glance at the lady's feet, which "have the heels in front and toes behind".

29. John Morgan, *The Life and Adventures of William Buckley*, xxvi.

30. Psychologists Theodore Barber and Sheryl Wilson conducted the study. It was cited by Jenny Randles in "F For Fantasy Prone Personality", *Fortean Times*, No. 196, May 2005.

31. *Conscious Cognition*, Vol. 7 (1) pp. 67-84, 1998, cited by Paul Chambers, "Lighter Sleepers", in *Fortean Times*, No. 192, Jan 2005.

32. Patrick Harpur, "The Landscape of Panic", in *Fortean Times*, No. 141, Dec 2000.

33. *Fortean Studies*, Vol. 5, 1998, p. 152.

34. *Nature*, Vol. 431, pp. 1055 and 1087; *New Scientist*, 30 Oct 2004; Richard Freeman, "For fear of little men", *Animals & Men*, No. 35, pp. 19-20; *The Sydney Morning Herald*, 6 Dec 2004, pp. 1-2; *National Geographic*, Nov 2005.

35. Although not quite yowie-size, one *Homo erectus* descendant, *Homo heidelbergensis*, which flourished for millennia in Europe and Asia, stood six feet tall on average and was more muscular than modern humans. *H. heidelbergensis* is sometimes referred to as "Goliath".

36. *Register of Historic Places and Objects*, The Professional Historians Association of New South Wales Heritage Register, SHI Report No. 4671012, 26 Aug 2001.

37. Tim Flannery, *The Future Eaters*, pp. 264-270.

38. James Woodford, *The Wollemi Pine*, Text Publishing, Melbourne, 2000, p. 119.

39. Theodora Kroeber told the remarkable story of Ishi, the last surviving member of the Yahi or Mill Creek tribe, in *Ishi In Two Worlds*, University of California Press, 1961.

40. William Telfer, "The Early History of the Northern Districts of New South Wales", also known as The Wallabadah Manuscript, c. 1900, pp. 32-34.

41. Loren Coleman, *Bigfoot: The True Story of Apes in America*, Paraview Pocket Press, 2003, Ch. 13, "Sex and the Single Sasquatch".

42. Robert Alley, *Raincoast Sasquatch*, Hancock House, Surrey, BC, pp. 123-124 and 252; Loren Coleman, *Mysterious America*, Paraview Press, New York, p. 289.

43. Guboo Ted Thomas and Percy Mumbulla, quoted in *Umbarra, Introduction to the oral traditions of the Yuin people of the Wallaga Lake region*, Wallaga Lake Cultural Centre, NSW, 1999.

44. Alley, p. 213;

45. *San Francisco Chronicle*, 8 Dec 1965.

46. Healy (1983), pp. 349-350.

47. Interview with Tony Healy, June 1999.

48. "Infrasounds Amazing", *Fortean Times*, No. 145, Apr 2001, p. 21.

49. Eyewitness interview with BFRO investigator T.E. Stein, c. 2001.

50. Chittick and Fox, p. 82.

51. Healy (1983), p. 229.

52. Alley, p.162.

53. Tony Healy and Paul Cropper, *Out of the Shadows, Mystery Animals of Australia*, Ironbark/ Pan Macmillan, 1994, p. 188.

54. *Straits Times*, 10, 11 and 12 Feb 1961; witness interviews with Tony Healy, Mar 1980.

55. In email exchanges with Michael Williams in 2001, veteran bigfoot researcher Peter Guttilla stated, "it's a common assumption … that 'glowing eyes' always means reflective … Not so. In many instances the eyes are intrinsically (self) glowing in a way similar to bioluminescence".

56. Tony Healy and Paul Cropper, "Stone Me!" *Fortean Times*, No. 116, Nov 1998, pp. 34-39.

57. Healy (1983), pp. 419-422. The locations were in the vicinity of Bluff Creek, northern California, Lake Okanagan, British Columbia, Whitehall, New York, and Brooksville, Florida.

58. Stan Gordon, "UFOs in Relation to Creature Sightings in Pennsylvania", a paper presented to the Mutual UFO Network Symposium, 1974.

59. Frank Anderson, "The Yowie Mystery", in *Bigfoot, Tales of Unexplained Creatures UFO and Psychic Connections*, Page Research. (No further bibliographical information available).

60. Alan Vaughan and Peter Guttilla, "A Testable Theory of UFOs, ESP, Aliens and Bigfoot", *Mutual UFO Network Journal*, Nov 1998, pp. 8-10.

61. Michael A. Persinger and Gyslaine F. Lafreniere, *Space-Time Transients and Unusual Events*, Nelson-Hall, Chicago, 1977.

62. *The Sun*, Middlesex, Massachusetts, 12 Oct 1980.

63. John Keel, *The Complete Guide to Mysterious Beings*, Tor Books, 2002.

64. "The Hunt For The Borlam Beast", *Animals & Men*, No. 29, pp. 35-44; *Fortean Times*, No. 169, Apr 2003, pp. 24-25; The Centre for Fortean Zoology Director's Report, Dec 2003.

65. Steve Mizrach, "Superspectrum Blues, The trans-electromagnetic nature of elements in the Fortean continuum". (Via email – no further bibliographic data available).

66. Gordon (1974) p. 140.

67. Confidential information provided by researcher Michael Williams.

68. Gordon (1974) p.140.

69. Chittick and Fox, p. 82.

70. Healy (1983), pp. 183-184 and 272.

71. Letters from R. W. McKay to Rod Estoppey, 13 Nov 1934 and 22 Apr 1940.

72. Fred Beck, *I Fought The Apemen of Mt St Helens*, Washington, 1967, p. 10.

73. The Clark and Colvin quotes are from a Terry Colvin article in *Cryptolist Digest*, No. 838.

Acknowledgements

Much of the information in this book came our way courtesy of other researchers, some of whom began collecting yowie stories more than a hundred years ago.

Hunter Valley pioneer Peter Cunningham, bushman Alexander Harris and ethnographer Horatio Hale documented fragments of Aboriginal yowie lore in the 1820s, '30s and '40s. William Telfer and R. H. Mathews did the same in the late 1800s. John Gale, founder of *The Queanbeyan Age*, interviewed eyewitnesses and published their stories in the early 1900s.

Credit for being Australia's first modern-style yowie hunter/researcher should probably go to the cattleman and poet Sydney Wheeler Jephcott (1864-1951) whose classic investigation of George Summerell's October 1912 sighting is detailed in the third chapter of this book. Jephcott collected at least a dozen other eyewitness reports, some dating from the early 1880s. Sadly, all but his account of the Summerell case appears to have been lost.

During the 1930s and early '40s a little-known pioneer of Australian cryptozoology, R. W. Mackay, developed a strong interest in the yowie phenomenon and some of his musings have survived. Unfortunately, a manuscript in which he discussed the yowie and other Australian zoological mysteries in detail has, like Jephcott's yowie file, disappeared.

From the 1940s to the '70s the bushman and writer Roland Robinson collected Aboriginal yowie lore in south-east New South Wales and north-east Victoria. He documented some of it in *Black-Feller White-Feller* (1958) and *Aboriginal Myths and Legends* (1966).

P.J. Gresser, another knowledgeable friend of the Aborigines, collected stories of hairy giants in the Blue Mountains during the 1940s and '50s. He was, as far as we know, the first researcher to link the term *yowie* to the mysterious ape-men. During approximately the same era, historian Errol Lea-Scarlett collected stories of the Hairy Man in the vicinity of Queanbeyan and Captains Flat. Credit must also go to Rex Gilroy who, almost single-handedly and in the face of considerable ridicule, reintroduced the yowie legend to the general public in the mid-1970s.

Graham Joyner, of Canberra, has made a great contribution. In his book *The Hairy Man of South Eastern Australia* (1977), he recorded 29 early references to the mysterious creatures in Aboriginal lore and colonial-era journals. In so doing, he proved that Aborigines have believed in the hairy giants for a very long time and that many white pioneers also encountered them.

We would like to acknowledge our debt to all of those people and to many other yowie hunters, colleagues and friends who have helped us over the years. These include:

Laurie Allard, John Appleton, Bigfoot Field Research Organisation, Janet and Colin Bord of Fortean Picture Library, Gerry Bostock, Steve Carter, Mike Cassidy and Eve Grace, Frank Cauchi, Bill Chalker, Rose Chapple and the Australian Rare Fauna Research Association, Lee Chittick, Andre Clayden, Loren Coleman, Paul Compton, Michael De Vere; Ray Doherty, Jon Downes, Graham Inglis and Richard Freeman of the Centre for Fortean Zoology, Netta Ellis, Kevin Fielder, Brian and Pru Foskett, Bryan Fox, Roger and Judy Frankenburg, Gulf Coast Bigfoot Research Organisation, Lyall Gillespie, Dave Glen and family, Brett Green, John Green, Peter Guttilla, Tony Harris and Annie Donahoo, Cleve Hassell, Geoff Healy, David Hearder, Major Les Hiddens, Liz Hurley, Dave Jackson and Barbara Gray, Peter and Gaye Kable, Kewaunee Lapseritis, Helmut and Siggy Loofs-Wissowa, Dave McBean,

Bernie Mace, Gary Maguire, Rob Millar, Gerry McMahon and Jim Montgomery of Guru Communications, Peter McMahon, Geoff and Vicki Nelson, Jerry and Sue O'Connor, Gary and Carmel Opit, Burris Ormsby, Dan Perez of *Bigfoot Times*, Tim Power, Dave Reneke, Patricia Riggs, Steve Rushton, Pat Ryan, Rino Seselja, Kyle Slabb, Malcolm Smith, Lisa Stack, Martin Stannard, Edwin Stratford, Cecil Thompson, Tim the Yowie Man, Meryl Tobin, Percy and Steven Trezise, Roger Virtue and Jackie Hickman, Graham Walsh, Bronwyn Watson, Merrilee Webb, Vern Weitzell, Russ Wenholz, David Window.

In particular we would like to thank Graham Joyner, Neil Frost and Dean Harrison.

Had it not been for Graham's pioneering research, we may never have realised the yowie phenomenon was more than just a legend. Without the many colonial-era reports and items of Aboriginal yowie lore that he uncovered and put on the public record, the early chapters of this book would be much less interesting. Similarly, chapter four and the Catalogue of Cases would have been a lot slimmer without the contributions of Neil and Dean.

Neil, who lives in the heart of the yowie-infested Blue Mountains of NSW, has uncovered and shared with us a veritable goldmine of data. Dean, who set up Australia's first yowie-related web site in 1998, and who has led well-equipped expeditions to dozens of yowie "hot spots", has also shared a massive amount of information.

We owe a special debt of gratitude, also, to ace investigator of the unexplained Rebecca Lang, who edited the manuscript, and to her partner, cryptozoologist Michael Williams, whose energy and enthusiasm never fails to inspire.

Appendix A
A Catalogue of Cases
1789 to 2006

Although this catalogue consists of 282 cases, some deal with two or more events that allegedly took place in the same area at around the same time. It therefore contains details of more than 300 yowie sightings. We personally interviewed 120 of the eyewitnesses.

While many of the entries are brief, others are quite lengthy. The more detailed ones generally relate to "classic" yowie encounters of the colonial or early modern eras, to multiple-witness cases or to those we have personally investigated. Some cases have been allowed greater space because they contain unusual or amusing details, some because they occurred in interesting locations, and others simply because we know, like, and trust the people involved. Seven cases involve the discovery of apparent yowie footprints rather than actual sightings of the creatures. To set the record straight, we have also included (and, we hope, discredited) half a dozen reports that have received a degree of publicity but which are almost certainly hoaxes.

Because it is impossible to know whether the small hairy creatures reported by a minority of witnesses are adult junjudees or juvenile yowies, we have, for the purposes of this exercise, lumped together all Hairy Man reports, both big and small.

All cases from number 263 onwards are out of sequence, chronologically. Some of those reports were conveyed to us shortly before the book went to press; others were rediscovered at the last minute after having been lost for years in the bowels of our filing systems. In any case, they were appended to the catalogue after our system of footnotes was completed and too late to be used in statistical analysis.

Case 1. 1789. Botany Bay, NSW.

A handbill circulated in England in about 1790 contained a fanciful story about a nine-foot-tall hairy wild man supposedly captured at Botany Bay and taken to Britain. [See Ch. 2] Although the story is obviously a hoax, it is interesting to note that Botany Bay is only eight kilometres north of Port Hacking, where several encounters with giant, hairy "yahoos" were reported in the 1850s and 1860s. The area has also produced two very credible yowie reports in the modern era. [Cases 5, 6, 9, 10, 159 and 233]

A defcription of a wonderful large WILD MAN or monftrouf GIANT, BROUGHT FROM BOTANY BAY, handbill circa 1790, reproduced in the *Sydney Morning Herald*, Sept 1, 1987, p. 17.

Case 2. Circa 1847. About 20 km west of Yass, NSW.

In 1903 a Ngunnawal/Kamberri elder, "Black Harry" Williams (1837-1921) told his friend George Webb that that when he was about 10 years old he saw a large group of Aborigines kill a Hairy Man near the junction of the Yass and Murrumbidgee Rivers. It was like a black man, but covered all over with grey hair. Two warriors dragged it downhill by its legs. [See Ch. 1] It is worth noting that, 140 years later, James Basham of Cootamundra encountered a gigantic yowie at Taemas Bridge, about 12 kilometres from the site of Mr. Williams' experience. [Case 169]

Letter from George Webb to *The Queanbeyan Observer*, Aug 7, 1903, cited in Graham Joyner, *The Hairy Man of South Eastern Australia*, Canberra, 1977 and in Ann Jackson-Nakano, *The Kamberri*, Canberra, 2001. Credit: Graham Joyner.

Case 3. Circa 1848. Near Cudgegong, NSW.

A shepherd in W. Sutton's employ claimed he had seen a Hairy Man in the scrub north of Cunningham's Creek. His dogs, which hunted everything else, ran from the creature with their tails between their legs. Significantly, perhaps, a feature just west of Cunninghams Creek has been known since at least the 1860s as Monkey Hill.

Lismore Northern Star, May 17, 1878.

Case 4. 1849. Phillip Island, VIC. Day.

A party led by a Mr. Hovenden allegedly fired at a large creature that was "in appearance half man, half baboon". [See Ch. 2]

The Argus, Oct 25, 1849; *The Melbourne Morning Herald*, Oct 29, 1849.

Case 5. 1856. Port Hacking, NSW. Day.

When Captain William Collin and a man named Massey were camped at Port Hacking, Charles Gogerly, who lived on the southern shore, told them he had seen a 12-foot-tall "yahoo" or "wild man of the woods" near their tent. [See Ch. 2]

Captain William Collin, *Life and Adventures (of an Essexman)*, Brisbane, H.J. Diddams & Co., 1914; Frank Cridland, *The Story of the Port Hacking, Cronulla and Sutherland Shire.* Sydney, Angus and Robertson, 1924; "The Port Hacking Wild Yahoo", *Sutherland Shire Historical Society Quarterly Bulletin*, No. 42, November 1982. Credit: Steve Rushton and David Hearder.

Case 6. 1860s. Heathcote, NSW.

A man claimed he saw a 12-foot-tall yahoo in the abandoned village of Bottle Forest. Many other local residents claimed to have seen the creature. They supposedly believed it could travel by land or water thanks to its webbed feet and suctioned under soles, and that it could belch forth fire.

Bottle Forest was only eight kilometres from Mr. Gogerly's yahoo-haunted property. Another hairy giant was reported about eight kilometres to the south-east in 2000. [Case 233]

A letter to the *St. George Call*, Jun 8, 1907. Credit: Steve Rushton and David Hearder.

Case 7. 1860s. Near Braidwood, NSW. Day.

Miss Derrincourt told of encountering "something in the shape of a very tall man, seemingly covered with a coat of hair ... what the people here call a Yahoo or some such name". [See Ch. 2]

William Derrincourt, *Old Convict Days,* T. Fisher Unwin, 1899, p. 316. Credit: Netta Ellis.

Case 8. 1865 or 1866. Cordeaux River, NSW.

In an article about an ape-like animal seen near Avondale in 1871 [see Case 14], a journalist mentioned that Mr. B. Rixon had seen a similar animal at the Cordeaux River five or six years earlier. Several reports have come from the same area in recent years. [e.g. Case 249]

The Illawarra Mercury, Apr 14, 1871, cited in Joyner, pp. 1-2. Credit: Graham Joyner.

Case 9. Late 1860s. Heathcote, NSW. Night.

An unnamed Spaniard encountered a "hideous yahoo" near the abandoned village of Bottle Forest. His usually savage dogs ran to him in terror. [See Ch. 2]

Patrick Kennedy, *From Bottle Forest to Heathcote – Sutherland Shire's First Settlement.* Credit: Steve Rushton and David Hearder.

Case 10. Late 1860s. Between Sutherland and Waterfall, NSW.

Patrick Kennedy also records that from the late 1860s, residents of the area between Sutherland and Waterfall began hearing strange and fearsome noises at night from "The Thing", also known as the "wild man of the bush". Several people who claimed to have seen it said "The Thing" resembled a tall, hairy creature that was neither man nor beast. At times its cry sounded "like someone screaming in pain".

According to *The Sutherland Shire Historical Society Bulletin* of October 1973 and July 1975, "The Thing" made occasional appearances until about 1915. [See Ch. 2]

Credit: Steve Rushton and David Hearder.

Case 11. Circa 1869. Avondale, NSW.

Two children observed a monkey-like creature the size of a 13 or 14-year-old boy.

The Illawarra Mercury, Apr 14, 1871 and *The Empire* (Sydney), Apr 17, 1871, cited in Joyner, pp.1-2. Credit: Graham Joyner.

Case 12. 1860s (?). Northern NSW. Dusk.

"Geo. Long and his mate had yarded their sheep at sundown for fear of the blacks, and were sitting down to their supper, when the dogs came running in to the hut in great alarm; the shepherds drove them out but presently they returned, evidently frightened by something outside. Long rose and peeping cautiously out of the door he saw under a tree at a distance, an object like a blackfellow only considerably larger. Whispering to his companion to bring him the gun, he raised it to his shoulder, and aiming at the breast of the blackfellow he fired. At once the object began to move, and with huge strides fled away across the river, making a clattering noise on the shingle, as if with the hoof of a horse. Unable to satisfy themselves what it was, they next morning obtained the assistance of a blackfellow, and proceeded to examine the track. They discovered several footprints, and as soon as the black saw the first of them he exhibited every symptom of terror, and muttered *dibbil dibbil.* They said, 'Nonsense, *dibbil dibbil* does not care for white men.' 'No,' he said, 'but *dibbil dibbil* has come to look for blackfellows.' The marks were like those of an emu's feet, but there was one long claw which penetrated several inches into the ground."

Naseby, C., *The Aborigines of Australia; stories about the Kamilaroi tribe.* Communicated by C. Naseby to John Fraser. Maitland: Mercury Office, 1882, pp. 9-10. Credit: Graham Joyner.

Case 13. Early 1870s. Ettrema Gorge, near Nowra, NSW. Midnight.

A *Sydney Mail* correspondent wrote that settlers in the vicinity of Ettrema Canyon (now part of Morton National Park) had long believed the gorge to be the abode of Hairy Men. He quotes explorer John Chaffey as saying:

"It may be 40 years back that a party of prospectors from the Shoalhaven ventured a few miles up the mouth of the Ettrema, and the first night they camped received a shock that hastened their movements. About midnight their dogs, with savage snarls and terrified looks, rushed into the tent and crouched beside their masters, then, making a bolt for it, cleared out: and the startled prospectors, looking out, saw what in the dim half moonlight they believed to be hairy men creeping around their tents. They were probably the large badgers or wombats which abound there."

Sydney Mail, Oct 9, 1912. Credit: Graham Joyner.

Case 14. April 1871. Avondale, NSW. Dusk.

George Osborne, of Dapto, watched an ape-like creature climb down a tree. Two children had seen a similar animal two years earlier. [See Ch. 2]

The Empire (Sydney), Apr 17, 1871 and *The Illawarra Mercury*, Apr 28, 1871, cited in Joyner, pp. 1-2. Credit: Graham Joyner.

Case 15. 1871. Bulli Mountain, Near Wollongong, NSW.

"A person who has resided on the Bulli Mountain for several years positively asserts that an animal similar to that seen by Mr. Osborne, but considerably larger, has been seen in the bush in that locality more than once, and by different persons, and that no dogs can be found to face it."

The Illawarra Mercury, Apr 28, 1871, cited in Joyner, p.2. Credit: Graham Joyner.

Case 16. July 1871. Belgrave, near Kempsey, NSW.

A group of Aborigines abandoned their camp and hastened to Warneton after being terrorised by a gorilla-like animal near Belgrave. Two young white men then encountered the creature. [See Ch. 2]

The Empire, Sydney, Jul 20, 1871; *Sydney Morning Herald*, Jul 21, 1871; *Queanbeyan Age*, Jul 27, 1871.

Case 17. December 1871. The Jingeras, a section of the Gourock Range below Captains Flat, NSW.

"From the fastnesses of the Jingeras, adjacent to or in the district of Manaro, comes the startling intelligence that a 'wild man' has been seen in that place. A little girl, the granddaughter of Mr. Joseph Ward, senior, of Mittagong, asserts that she has met an old man, whose back is bent, and body covered with a thick coat of hair, in height (to use the girl's words), about the same as her grandfather. The strange being in question had nails of a tremendous length on his hands, and he seemed desirous of shunning the girl. The main points of the assertion are given with remarkable earnestness by Mr. Ward's granddaughter; nothing can shake the simple outlines of her story. Confirmatory of the above incident, is the statement made by Mr. Kelly, of the Jingeras, who says that he has himself seen the 'wild man'. Anent the above, there is a tradition among the settlers of this place that the mysterious monster, the 'yahoo', is a denizen of the mountain country where the 'wild man' has been discovered, and that it is only observable in stormy weather, or on the approach of bad seasons."

The Manaro Mercury (Cooma) Dec 9, 1871 and J.A. Perkins, "Monaro District Items", MS 936, National Library of Australia, p. 1121, cited in Joyner, pp. 3-4. Credit: Graham Joyner.

Case 18. August 1872. Lake Cowal, south-west of Forbes, NSW.

While boating on Lake Cowal, a party of surveyors saw a strange animal about 150 yards [135 m] away. It looked "like an old man blackfellow with long, dark-coloured hair". As it swam it rose out of the water so that they could see its shoulders, and occasionally submerged as if in pursuit of fish.

Sydney Morning Herald, Aug 24, 1872.

Case 19. September 1872. Braidwood district, NSW. Evening.

During the course of eight or nine days in early September, several people reported seeing a large chattering animal at a place known as Giant's Cave. It was variously described as resembling a monstrous wombat, a bear or a gorilla. [See Ch. 2]

Sydney Morning Herald, Sept 16, 1872; *Manaro Mercury and Cooma and Bombala Advertiser*, Sept 21, 1872.

Case 20. Mid-1870s. Pyramul, NSW.

Young Pat Wring witnessed a battle between his dogs and a hairy, bipedal creature with arms as large as a man's thighs. [See Ch. 2] Shortly thereafter, a girl from a neighbouring farm encountered what was thought to be the same animal:

"A settler's daughter having gone for the cows, an older sister, thinking she was long away, went out to assist her. On turning the corner of a bush fence, about a quarter of a mile from the hut, she suddenly stood face to face with the stranger. No doubt both were frightened, as they stood watching each other, until the sister called out that she had all the cows, when the hairy creature turned about and walked leisurely away."

Lismore Northern Star, Apr 13, 1878; *Freemans Journal*, May 17, 1878.

Case 21. 1876. Walla Walla Scrub, 40 km west of Crookwell, NSW.

"The Milbury Creek correspondent of the *Bathurst Free Press* says: A resident of this place returned from the Fish River some forty miles from here, a few days ago, and told me that he had been informed by a respectable settler in that quarter that a party of sawyers, working in the Walla Walla scrub, came upon the dead body of an unearthly looking animal, human or inhuman they could not tell. It stood about 9 feet in height, with head, face and hands, similar to a man's; one of its feet resembled the hoof of a horse and the other was club-shaped; the body was covered with hair or bristles like a pig. For many years past it had been believed by the settlers of that wild part of the country that the Walla Walla scrub was inhabited by a monster commonly called "the hairy man of the wood," or what all the blacks stand so much in dread of – the Yahoo. Horses and cattle are said never to have been known to enter or remain in the scrub."

Australian Town and Country Journal, Nov 4, 1876, cited in Graham Joyner, *More Historical Evidence for the Yahoo, Hairy Man, Wild Man or Australian "Gorilla"*, MS, Canberra 1980. Credit: Graham Joyner.

Case 22. November 1876. Rocky Bridge Waterholes (at the present site of Wyangala Dam) NSW. Day.

"Fourteen days ago, and not more than ten miles from here, towards the head of the Lachlan River, on Coolamba station (Hammond's), in one of the most secluded and melancholy spots imaginable, imperceptibly a terror of awe creeps over every one that has to pass through this far and wide known gorge or death-chasm of the river. While a lad of the name of Porter ... was shepherding a flock of his father's sheep ... an inhuman, unearthly-looking being was seen ... coming towards him from the high, rugged and precipitous rocks. The dogs ... would not attack, became timid, and crouched around the lad's legs, who became horror-struck with fear; left the sheep to their fate, and ran, together with his collies, for home.

"On Saturday last ... a fishing party of young men and young women went to the Rocky Bridge waterholes for a night's sport. These waterholes are famed, far and near, for quality and quantity of fish. Two hours before sun-down the young men and some of the women went to set their lines, leaving one of their young friends to boil the billy and prepare supper. While engaged, the young woman was suddenly startled by observing a man, as she naturally imagined at first sight ... coming towards the fire, but on walking closer, discovered the appearance to be unsightly and inhuman, bearing in every way the shape of a man with a big red face, hands and legs covered over with long, shaggy hair - – from fright became almost spellbound, screamed and screeched, but unable to run. The men, on hearing such unearthly cries, left their fishing lines and ran with all speed towards their comrade. On reaching the fire, the monster of alarm was only distant some fifty yards ... it stood for a minute or two and turned away and made for the rocks.

"Two of the men armed themselves with a tomahawk and cudgel and followed this extraordinary phenomenon of nature for a short distance up the rocky and rugged mountain; when suddenly it turned round ... They also halted, being about sixty yards from the object of terror, commanding a full view of his whole shape and make, resembling that of a big slovenly man. The head was covered with dark grissly hair, the face with shaggy darkish hair, the back and belly and down the legs covered with hair of a lighter colour. This devil-devil – or whatever it may be called – doubled round, and hurriedly made back towards the fire ... On seeing him coming, a fearful commotion amongst the females and a kind of supernatural terror amongst the men took place. In the meantime, before reaching the camp, it sidled away towards the inaccessible rocky mount.

"The names of two of the men who witnessed and took part in the scene are Porter and Dunn, well-known settlers on the Abercrombie and Lachlan Rivers. Mr. Laner, another settler from the Lachlan, has informed me the other day that the neighbours all round have organised a party to go in search of the human monster, and hunt him down, dead or alive.

"It is well known to the old settlers for the last 30 odd years, that the blacks will never camp within miles of this death-like chasm of the Lachlan, though they come long distances to fish on adjoining waterholes, but leave before sundown to camp miles away."

In 2003 a party of shooters encountered a similar creature in the same area. [Case 260]

Australian Town and Country Journal, Nov 18, 1876, cited in Joyner (1980). Credit: Graham Joyner.

Case 23. October 1877. Sutton Forest, NSW.

Patrick Jones and Patrick Doyle reported seeing a hairy, seven-foot-tall, bipedal creature between Cable's Siding and Jordan's Crossing. It made a tremendous noise and left huge tracks spaced three feet apart. [See Ch. 2]

Sydney Morning Herald, Oct 12, 1877.

Case 24. May 1881. The Jingeras. South of Captains Flat, NSW.

"The Cooma Express relates that the Jingera hairy man has again turned up. It was seen on Saturday last by Mr. Peter Thurbon and one or two others. This is its first appearance for some considerable time past. The animal, if such it be, has the appearance of a huge monkey or baboon, and is somewhat larger than a man."

The Goulburn Herald, May 24, 1881; Errol J. Lea-Scarlett, *Queanbeyan and the Country of Murray (including the Australian Capital Territory): Notes on History & People*, Vol. 12, 1965, MS 1461, National Library of Australia, p. 353, cited in Joyner (1977) p.6. Credit: Graham Joyner.

Case 25. December 1882. Between Batemans Bay and Ulladulla, NSW. Day.

H.J. McCooey gave a detailed description of a five-foot-tall "Australian ape" that he observed from a distance of only 20 metres. [See Ch. 2]

Letter from H.J. McCooey to "The Naturalist" column, *The Australian Town and Country Journal*, Dec 9, 1882. Credit: Graham Joyner.

Case 26. Early 1883. Between Mount Keira and Mount Kembla, NSW.

"It is stated by several persons residing in the locality that a gorilla has again been seen in the mountain ranges ... between Mount Keira and the Mount Kembla coal tunnel. Different residents there aver that they have caught sight of the strange animal on separate occasions and in various places. They describe it as resembling a man, but covered with long hair, and having long sharp claws."

The Illawarra Mercury, Jan 19, 1883, cited in Joyner (1977) p.3. Credit: Graham Joyner.

Case 27. 1883. Between Keera and Cobbadah, north of Barraba, NSW. Night.

William Telfer encountered a creature "something like the gorilla in the Sydney museum". (See Ch. 2)

William Telfer, *The Early History of the Northern Districts of New South Wales*, (Also known as "The Wallabadah Manuscript") [c. 1900], University of New England Archives A147/V213, pp. 32-34. Credit: Graham Joyner.

Case 28. Circa 1885. Flea Creek, Brindabella Mountains, NSW. Night.

William and Joseph Webb, of Urayarra (now Uriarra), told John Gale, the founder of *The Queanbeyan Age*, that they shot at a gorilla-like animal while mustering cattle. [See Ch. 2]

John Gale, An Alpine Excursion, Queanbeyan, 1903, pp. 85-89, cited in Joyner (1977) pp. 10-11; handwritten account given by Eric McDonald to Lyall Gillespie, Mar 10, 1979. Credit: Graham Joyner and Lyall Gillespie.

Case 29. 1885. Parkers Gap, south-east of Captains Flat, NSW.

According to historian Netta Ellis, an 1885 edition of *The Town and Country Journal* reported that a hairy man or orangutan had been seen at Parker's Gap in the Gourock Range.

Netta Ellis, *Braidwood, Dear Braidwood*, p. 161.

Case 30. 1886. Faulconbridge, Blue Mountains, NSW.

While collecting firewood, the wife of the caretaker of Sir Henry Parkes' property heard "a commotion amongst the fowls ... on looking up, before her stood a Thing about seven feet high. The black hair growing on its head trailed weirdly to the ground, and its eyeballs were surrounded with a yellow rim. It was – the hairy man!" Before she could call her husband, the creature departed, taking with it several fowls and leaving "a track three inches deep".

Although the licensee of the Royal Hotel at Springbrook offered a fifty-pound reward and search parties went out, the monster got clean away.

The Illustrated Sydney News, Oct 3, 1889.

Case 31. Late 1880s. Einasleigh River, QLD. Night

An old settler on the Einasleigh River (a tributary of the Gilbert) told adventurer Arthur Bicknell that he'd seen a strange animal prowling around. It walked on its hind legs, was as tall as a man, had long arms and huge hands, and made a strange moaning noise. He referred to it as the "wood devil" and suspected it was responsible for the disappearance of several of his dogs.

Arthur Bicknell.

'The wood devil.' An illustration by J. B. Clark based on sketches by Arthur Bicknell.

That evening Bicknell, armed with a revolver, climbed a tree overlooking the place the animal had been seen: "I had not long been in the tree when I heard the peculiar moaning noise ... and also the crashing and breaking of the brushwood ... it was nearly dark and the moon only just rising. I could see nothing. The noise had an unearthly ring about it. I am not easily frightened, but ... slipping down from my perch ... I waited a moment to fire two or three shots ... in the direction the sounds came from, and then turned and bolted for the house. If the devil himself was after me I could not have made better time.

"When the old man appeared [the next morning] we got him to go with us to the place ... There sure enough he lay, as dead as any stone, shot through the heart ... He was nothing but a big monkey, one of the largest I have ever seen, with long arms and big hands, as the old man had described. These huge monkeys or apes are common in Nicaragua, but this one was certainly the largest I ever came across."

Strangely, Bicknell didn't seem to think it remarkable that an unusually large – a "huge" – Central American monkey, presumably a Spider Monkey, should have found its way to the wilds of northern Australia. Spider Monkeys grow no taller than 3 feet 7 inches, don't kill dogs, are arboreal, aren't noted for making moaning noises, and are almost always quadrupedal.

In J. B. Clark's illustration, based on a sketch by Bicknell, the "wood devil" is man-sized and has a short tail. Bicknell's failure to preserve even the skull of such a strange animal seems highly suspicious.

Arthur C. Bicknell, *Travel and Adventure in Northern Queensland*, Longmans, Green and Co., London and New York, 1895, pp. 173-178.

Case 32. Late October 1893. Near Captain's Flat, NSW. Day.

Arthur Charles Marrin claimed to have killed a six to seven-foot-tall, hairy, bipedal creature to the south-east of Captains Flat. He supposedly carted the body back to his factory at Braidwood, where local journalists examined it. [See Ch. 6]

Goulburn Evening Penny Post, Oct 28, 1893, cited in Joyner (1977) pp. 7-9. Credit: Graham Joyner.

Case 33. November 1894. Between Snowball and Jinden, NSW. 3 pm.

Johnnie McWilliams reported seeing a six-foot-tall, heavily built, hair-covered "wild man" that ran away slowly and "bellowed like a bullock". [See Ch. 2]

The Queanbeyan Observer, Nov 30, 1894; *The Cooma Express*, Nov 30, 1894, cited in Joyner (1977) pp. 9-10. Credit: Graham Joyner.

Case 34. Late 1800s. Near Stanthorpe, QLD. Day.

Cecil Thompson recalled that his aunt Emma, at the age of about seven, "was sent to look for the cows, and found 'young monkeys' up the she-oak trees, throwing nuts. So grandfather had to go and get the cows with her".

Several other people encountered Hairy Men on or near the same farm in later years. [See Cases 36, 59, 61 and 66]

Paul Cropper interview with Cecil Thompson, 1997.

Case 35. Late 1800s or early 1900s. Currickbilly Range, south-east NSW. Night.

Surveyor Charles Harper and his companions encountered a huge, hair-covered, man-like animal that approached their campsite. [See Ch. 3]

The Sun (Sydney), Nov 10, 1912, cited in Joyner (1977) pp. 18-20. Credit: Graham Joyner.

Case 36. Circa 1900. 10 km south-east of Eukey, QLD. Night.

Cecil Thompson recalled that his great uncle Tom often spoke of a large creature he encountered in his shed. "It had been eating carrots and the dogs put it up. It rushed past him into the night and made a grab at him as it passed. Granddad always maintained it was a Hairy Man."

Several other people have encountered Hairy Men in the same area.

Paul Cropper interview with Cecil Thomson, 1997.

Case 37. 1901. Brindabella Ranges, NSW. Afternoon.

In 1903 a close friend of John Gale told him of an incident that occurred two years earlier.

"Mr. Cox was ... camped alone ... on a shooting expedition. He was, he said, enjoying his billy of tea in the afternoon, when his attention was drawn to an enraged cry, between a howl and a yell, in the thick scrub of a gully close by. He instantly seized his rifle and ... saw a huge animal in an erect posture tearing through the undergrowth ... it was out of sight before he could bring his rifle to his shoulder. He distinctly heard the crashing of the undergrowth in its flight, and he followed after it. Its speed was greater than that of its pursuer; but as it fled its howling and yelling continued. That it was no creation of an excited imagination – (and from what I know of Mr. Cox, he is not a likely subject of wild hallucinations; but, on the contrary, a remarkably cool, intrepid fellow, too well enlightened and educated to magnify a simple fact into a chimera) – is confirmed by this, that in his pursuit he met several wallabies tearing up the gully in such alarm that, though passing close by, they took not the least notice of him. These were followed presently by a herd of cattle similarly scared."

John Gale, *An Alpine Excursion*, Queanbeyan, 1903, pp. 85 - 89, cited in Joyner (1977) p.11. Credit: Graham Joyner.

Case 38. Early 1900s. Burnt Bridge, near Kempsey, NSW. Dusk.

Several elderly people told *Macleay Argus* editor Patricia Riggs that a yowie once attempted to abduct two-year-old Chris Davis. His father managed to rescue him. [See Ch. 3]

Macleay Argus (Kempsey), Oct 2, 1976. Credit: Patricia Riggs.

Case 39. 1901 or 1902. Currambene Creek, near Jervis Bay, NSW. Day.

When he was about 10 years old, Henry Methven came face-to-face with a two to three foot tall Hairy Man. Henry and his adult companions followed the creature's five-toed tracks. [See Ch. 5] In 1985, the Maron family saw a large yowie in the same area. [Case 161]

Macleay Argus (Kempsey), 6 Jan 1977. Credit: Patricia Riggs.

Case 40. 1904. Byron Bay, NSW. 10:30 pm.

Mrs. Sarah Ratcliff said that her father, Patrick Maher, encountered a yowie while riding home to Tyagarah from Byron Bay. After crossing Belongil Creek, the horse became nervous and when Mr. Maher looked back he saw something "like a big hairy man with no neck" standing on a heap of stones. It was seven or eight feet tall. The horse bolted and the creature gave chase, but Mr. Maher lost sight of it when he was thrown from the saddle. He ran into a cane plantation, stayed there all night, and was still badly shaken when he reached home in the morning. The property was sold as quickly as possible and the family moved to Tumbulgum.

About a year later, while riding home from a dance at Byron Bay, three brothers were pursued by a beast that leapt on to the rump of one of their ponies. It hung onto the saddle until the party galloped off the road onto the home track. The rider's coat was torn, the horse so badly injured it had to be destroyed. The injured boy was hospitalised, and according to the report, all the brothers "were ... ill after their fright."

Daily News (Tweed Heads), Jun 20, 1982; Jim Brokenshire, *The Brunswick, Another River and its People*, Brunswick Valley Historical Society, 1988, p.176.

Case 41. June 1909. Mudgee, NSW. 5 pm.

"During the past few weeks the residents of the 'Bar' have been disturbed from their slumbers by noises, resembling at times a person choking, and at others a women screaming and then crying. These strange cries remained a mystery till Thursday last, when at about 5pm they were again heard, and shortly afterwards several persons residing in the locality were astonished to see a peculiar animal, five feet high, standing on his two legs, and at the same time brushing away with his claw-like hands the long unkempt-looking hair from his eyes. The animal is covered with long white hair and when seen was uttering the cries which have been disturbing the peace of the neighbourhood. The hairy man, or whatever he is, was only seen for a minute, and disappeared as suddenly as he came in sight."

A few days after the above story appeared, the *Mudgee Guardian* reported that police believed the sightings were the work of a prankster dressed in a goatskin. The supposed prankster and his costume were never tracked down.

Mudgee Guardian, Jun 10, 22 and 24, 1909; *Robertson Advocate*, undated clipping (probably June) 1909.

Case 42. October 12, 1912. Between Bombala and Bemboka, NSW. Noon.

George Summerell rode up close to a seven-foot-tall hair-covered ape-man that was drinking from a creek. A neighbour, Sydney Wheeler Jephcott, made plaster casts of large hand and footprints. [See Ch. 3]

In 1997 ranger Chris McKechnie saw a similar creature in the same area. [Case 216]

Sydney Morning Herald, Oct 23 and 24, 1912, cited in Joyner (1977) pp. 14-17. Credit: Graham Joyner.

Case 43. Circa 1914. Near Suggan Buggan, VIC.

Somewhere between Suggan Buggan and Gelantipy, an Aboriginal couple, Big Charley and his wife, were attacked by a *dulagar*. [See Ch. 1]

Aldo Massola, *Bunjil's Cave: Myths, Legends and Superstitions of the Aborigines of South-East Australia*, Lansdowne Press, Melbourne 1968.

Case 44. 1923. Nulla Creek, about 50 km north-west of Kempsey, NSW.

After encountering a Hairy Man, timber-feller Jack Brewer abandoned his job and left the area. [See Ch. 3]

Macleay Argus, Sept 28, 1976. Credit: Patricia Riggs.

Case 45. 1923. Watsons Creek, NSW.

While working in rugged bushland, Tamworth grazier Henry O'Dell noticed large footprints in the sand and then spied a strange creature hanging by one arm from a tree 30 metres away. It seemed about six feet tall. Although it was extremely hairy, its eyes, mouth and nose were visible. His workmate Keith Blanch also saw the creature before they both quickly retreated.

O'Dell also said that a friend, Tom Chapman had once shot a female "ape-creature" at Wild Cattle Creek before fleeing in fear. He left the carcass where it fell. [See Ch. 6]

Psychic Australian, Aug 1977. Vol 2, No 8, summarising a story from a Tamworth newspaper, probably the *Northern Daily Leader*.

Case 46. Circa 1924. Southern Highlands, NSW. Night.

Val Whalan, of Huskisson, told how her grandfather and four of his children, including Val's mother, encountered a yowie as they were checking their rabbit traps. After hearing a crashing sound in the bush they hid and "smelled a terrible, foul odour, and saw a huge, hairy beast like an ape; his smell was blowing to them so he didn't know they were there. He came across the creek within yards of them. They were terrified, not moving until the beast was well away ... My Granddad would not go around the traps at night after that."

Telegraph (Sydney) undated clipping, probably Aug 7, 1987.

Case 47. 1925. Near Murray Bridge, SA.

Mrs. P. Lindsay said that when she was 11 years old, she saw a creature that is known to Aborigines of the lower Murray River area as *mooluwonk*. It was more then 10 feet tall, had long black hair, dark red eyes, large teeth and webbed feet and hands.

Under the direction of Mrs. Lindsay and seven other Aboriginal elders, artist Brian Vercoe produced an excellent sketch of the creature.

The Advertiser (Adelaide), Jul 5, 1973.

Brian Vercoe and his sketch of the mooluwonk.

Case 48. 1928. Near Palen Creek, QLD. 10 am.

Bob Mitchell, of Redcliffe and two friends saw two yowies while riding through the Border Ranges. [See Ch. 3]

Brisbane Sunday Mail, Nov 9, 1980. Credit: Graham Joyner.

Case 49. 1920s or early 1930s. "The Three Ways", south-east NSW.

Fred Howell often told his grandson, Billy Southwell, about seeing a yowie on an old bullock track that led from the south coast up to the Monaro.

On reaching a fork in the road known as "The Three Ways", Fred and his mate hobbled their team and took buckets to a waterhole. On the way back their blue cattle dog charged ahead and began circling the dray, barking furiously. On top of the dray was a very angry looking Hairy Man – Fred referred to it as a "yourie" – throwing bags of grain around. It jumped down and ran into the scrub. Fred saw what might have been the same creature at the same spot on one or two other occasions.

Tony Healy interviews with Billy Southwell, 1986 and 1998.

Case 50. Circa 1930s. Burnt Bridge, near Kempsey, NSW. Dusk.

Ten-year-old Mamie Moseley and her cousins Zelma Moseley and Tom Campbell, had a close encounter with a foul-smelling yowie near the site of the Chris Davis abduction. [See Ch. 3]

Macleay Argus, Oct 2, 1976. Credit: Patricia Riggs.

Case 51. 1930s. Epping, NSW. Day.

A local man said that, as a boy, he'd seen a five to six foot tall, bear-like creature sitting on a rock in the bush at Epping. (Epping, now in Sydney's inner western suburbs, was then near the extreme western edge of the city.)

Witness interview with Paul Cropper, 1982.

Case 52. 1930s. Megalong Valley, Blue Mountains, NSW.

Mrs. Lola Irish, of Sydney, wrote that while she and her brother were holidaying in Katoomba in the 1930s, their landlady told them about her sighting of a "giant hairy ape-man". While returning to a campsite near the "Ruined Castle" rock formation in the Megalong Valley, she'd seen the creature disappearing into the bush. It carried off some stores from the campsite.

Letter from Lola Irish to the *Sydney Morning Herald*, Sept 14, 1978.

Case 53. 1930. Maria River, about 10km south of Kempsey, NSW. Day.

While collecting wildflowers, 12-year-old Melba Cullen encountered a seven-foot-tall yowie. It was very broad-shouldered and covered in long, tan-coloured hair. [See Ch. 3]

Macleay Argus, Sept 25, 1976. Witness interview with Paul Cropper, Jul 15, 2001.

Case 54. 1930. Dunmore, just south of Wollongong, NSW.

"Says the Kiama *'Independent'*: A considerable scare is felt on some farms at Dunmore in the vicinity of Connolly's range, as some strange animal prowls in the vicinity, and has at a distance been seen above Terrangong Swamp. While fox shooting, Mr. N. Hambly got the best glimpse of it and thinks it is a wild bear, that perhaps has escaped from some menagerie or circus."

Because the region in which it was sighted has produced many reports of hairy ape-men, it seems possible, if not quite likely, that the people who glimpsed the Dunmore "bear" actually saw a yowie.

Moss Vale Post, Apr 25, 1930.

Case 55. 1931. North of Moore River, WA. Afternoon.

In August 1931, three young Aboriginal girls, Molly, Gracie and Daisy, escaped from the Moore River Native Settlement and in an epic journey walked 1600 km [994 miles] north to Jigalong, on the edge of the Little Sandy Desert.

One afternoon, as a storm was brewing, they were startled by the sound of heavy footfalls coming their way. As they lay hidden in a thicket the footsteps came so close that they could feel the ground vibrating. Suddenly, a huge man-like creature emerged from the Banksia scrub and ran past. He was of seemingly gigantic stature and his massive footfalls were audible well after he disappeared from sight.

"He was a big one alright," said Daisy later, "had a funny head and long hair". "It was a marbu alright," agreed Molly. "A proper marbu."

Doris Pilkington / Nugi Garimara, *Rabbit Proof Fence*, University of Queensland Press, 1996, pp. 84-85.

Case 56. 1932. Scartwater, northern QLD.

Prospector Bert Bagnall told journalist Tony Lambert that he saw a "bunyip" while hunting crocodiles in a lagoon on Scartwater Station. He had just shot a big saltwater crocodile when he looked up and saw a hair-covered, ape-like animal standing on the opposite bank. It took off into the bush at great speed.

People Magazine, Jul 28, 1992.

Case 57. Circa 1932. Lake Condah, VIC. Day.

When an Aboriginal man, Andrew Arden, shot a rabbit in an area known as the Stony Rises, a small, hairy man-like creature seized the carcass and ran away. [See Ch. 5]

Massola, p.150.

Case 58. February 1932. Eurobin, VIC. Night.

STRANGE ANIMAL AT LARGE. MAN ATTACKED IN PADDOCK.

"MYRTLEFORD, Friday. William Nuttall of Myrtleford, accompanied by his sister and a companion, was returning from Bright last night, and near the Eurobin railway station he alighted from his horse and entered a paddock ... the others riding on a short distance. A strange animal, which snarled, attacked Nuttall and tore his shirt to ribbons. The horse broke away, and when Nuttall ran the wild animal followed him to the railway line, where a wire fence apparently stopped it. Nuttall described the animal as being about 7ft. in height, with [a] round, hairy head and four tusks. It stood on two legs. It is believed to be an animal which escaped from a travelling circus when it was at Yackandandah some time ago. The animal has been seen by residents in different parts of the district, its tracks being plainly discernable. It is said to resemble an ape. Parties are out searching for it."

The Argus, Feb 27, 1932.

Case 59. February or March 1934. Near Stanthorpe, QLD. Dusk.

While working on the family farm 12-year-old Cecil Thompson and his brother Ernie encountered a yowie as it was inspecting a bag of freshly harvested peas. [See Ch. 3]

Mr. Thomson has interviewed several other people who have encountered yowies in the area. [Cases 34, 36, 61 and 66]

The Stanthorpe Border Post, Nov 1 (probably 1992); Cecil Thompson interview with Paul Cropper, 1997.

Case 60. 1935. Lismore, NSW. 9 pm.

A local man told reporter Gary Buchanan that, when he was 10 years old, he and his grandparents watched a yowie as it walked within 25 yards of their farmhouse. His grandfather had seen the same creature a few years earlier. [See Ch. 3]

Lismore Northern Star, Jul 7, 1977.

Case 61. Circa 1935. Near Stanthorpe, QLD. Afternoon.

In 1997, Cecil Thomson told of his cousin Walter Beddoe's encounter with a yowie:

"[He] was coming to town with his horse one afternoon. It was a bit misty and he struck a Hairy Man on the Stanthorpe to Mt. Tully road, near what we now call Hairy Man's Bend. It was only about three feet away and made a grab at the bridle and then strode over the fence. It was apparently tall enough to stride over the top wire."

Paul Cropper interview with Cecil Thomson, 1997.

Case 62. 1930s and 1940s. Deua River, NSW.

Twice in 20 years, Rodney Knowles found large, five-toed, non-human footprints in a place known locally as Yahoo Valley, about 15 miles [24 km] west of Moruya. [See Ch. 4]

The Queanbeyan Age, [undated clipping] Sept 1976. Witness interview with Tony Healy, Sept 1976.

Case 63. 1938 or 1939. Between Nanango and Maidenwell, QLD. Afternoon.

While hunting wallabies in thick vine scrub, teenagers Clyde Shepherdson and Clarrie Parsons came face-to-face with a fearsome ape-man. It was about six feet [183 cm] tall and covered in rusty-yellow hair. [See Ch. 3]

Witness communication with Dean Harrison. Witness interviews with Paul Cropper and Tony Healy, Apr and Jul 2001. Credit Dean Harrison.

Case 64. Circa 1930s or 1940s. Dungay Creek, about 25 km west of Kempsey, NSW. Day.

When she was about 10 or 11 years old, Mrs. Joan Delaforce (nee Clarke) was walking with her sister and brother near the family farm when one of their dogs began acting strangely.

"[It] ... ran up into the bush and started barking and growling as if it were scared. When we got a bit closer, we got a good look at this thing. It was standing about six feet tall and was dark brown in colour with long hair about the neck. We didn't get a good look at the face. We didn't stay around long enough."

On another occasion, Joan and her siblings saw something moving about in the tractor shed. When they approached, the creature they'd seen earlier came out and walked away.

Another incident occurred as the children were walking from their homestead to the school bus stop: "I suppose the distance was about half a mile ... on the next hill. What we saw were two big creatures and three small ones. That was the last time I saw them, but my brother and sister have seen them since. When you have seen it, you will never forget about it ... I am telling you that there *is* a Hairy Man ... in the bush."

Macleay Argus, Oct 2, 1976. Credit: Patricia Riggs.

Case 65. Circa 1930s or 1940s. Sebastopol, 40 km north-west of Kempsey, NSW.

Tom Carroll, of Gladstone, told the *Macleay Argus* that his late uncle, Joe Carroll, saw a Hairy Man while shooting in the scrub near Sebastopol:

"Joe was ... almost within 50 ft when he saw this creature about 5 ft tall with long hair and heavily built with long arms. When the creature saw him he got up from a sitting position and walked away with a steady gait."

On the following day Joe took some men from Moparrabah to the site and they followed the creature's tracks for some distance. A degree of corroboration came from Leslie McMaugh, a local landowner, who said that his horses sometimes snorted and galloped wildly around his paddock at night, as if frightened by something. It is worth noting, also, that Sebastopol is only 16 kilometres north of Kookaburra, where George Grey was attacked by an ape-man in 1968. [Case 83]

Macleay Argus, Oct 14, 1976. Credit: Patricia Riggs.

Case 66. Circa 1940. Near Eukey, QLD.

Cecil Thompson recalled how his older cousin, a timber-getter named Teddy Collie, saw what was apparently a yowie:

"[He] came home sweating, his horse was sweating. As he put it, he'd seen a 'big baboon' on the road from Ballandean to Eukey. The spot is still known as 'Baboon Gully'."

Paul Cropper interview with Cecil Thompson, 1997.

Case 67. 1940 or 1941. Near Eukey, QLD. Afternoon.

Cecil Thompson also told of his younger sister Leila's encounter:

"In about 1940 or '41, when she was about seven or eight years old ... she had a cubby house of sticks and leaves up on the hill on my dad's farm. One Sunday afternoon ... she came home white-faced [and] ran and told her mummy, 'There's a man up there in my cubby house!' Later she saw a picture of a baboon in a magazine and said, 'That is just like the man in the cubby – sitting down, he was.'"

The Stanthorpe Border Post, Nov 1, (probably 1992); Paul Cropper interview with Cecil Thompson, 1997.

Case 68. 1940s. Kempsey, NSW. Day.

Kempsey resident Kevin Davis said that as a high school student he'd come within five or six paces of a 1.2 metre [3 foot 8 inch] ape-man near the present site of Kempsey airport. [See Ch. 3]

Witness interview with Paul Cropper, 1995.

Case 69. 1940s. Petroy Plateau (now in New England National Park) NSW.

Albert Mowle told family members that, while he was prospecting for gold, hairy, long-armed gorilla-like creatures had come to his campsite to forage for food scraps. One, which appeared to be a female, kicked a kerosene tin around. Michael Mowle, who passed the story on to journalist Patricia Riggs, said that although his great uncle Albert was fond of telling tall tales, testimony of other old bushmen lent corroboration to the "gorilla" story.

"Something", he wrote, "terrified the horses belonging to several chaps who were mustering cattle nearby ... their normally docile animals bolted, maddened by fear and were found ... days later at Georges Creek. One highly experienced bushman said that the only other time he'd seen horses act in this manner was when they had caught the smell of wild animals from a circus."

Macleay Argus, Oct 21, 1976. Credit: Patricia Riggs.

Case 70. 1946. Near Wilcannia, western NSW. Night and day.

George Nott. (Martin McAdoo)

As soon as George Nott and his family moved into a long-abandoned homestead, strange things began to happen. Huge five-toed footprints appeared, horses became seriously spooked and then a "bloody big gorilla or somethin'... about six foot tall, broad, and sort of brownish fur all over him" repeatedly entered the house. On one occasion it grabbed Mrs. Nott by the neck and seemed to try and drag her outside. The family retreated to an out-station but the hairy horror followed them: Mrs. Nott woke one night to find it standing over the bed. After Mr. Nott chased it out of the house it stamped around in the darkness, "bellerin' like a bull". Later, one of the daughters saw it in broad daylight – 3 pm – on the verandah of the homestead.

Wilcannia is a long way away from what is considered "normal" yowie country, and the case contains other strange elements. In the main homestead the Notts heard noises "like a man walkin' about in the ceiling"; objects flew across rooms and so many pebbles fell on the roof that they "sounded like a heavy shower of rain". [See Ch. 7]

Martin McAdoo, *"If Only I'd Listened To Grandpa" – Recollections Of The Old Days In The Australian Bush*, Lansdowne Press, Sydney, 1980. Credit: Louise Cassidy.

Case 71. 1949. Between Ravenshoe and Millaa Millaa, QLD. Evening.

Vera Hepple told researcher Meryl Tobin that her two brothers came home one evening "sheet white" and claimed they'd seen a huge hairy man. The seven-foot creature had been watching them as they played on the edge of Purcell Brook.

Email from Meryl Tobin to Paul Cropper, Nov 1, 2002. Credit: Meryl Tobin.

Case 72. Early 1950s. North Aramara, QLD.

Young Michael Meech encountered a strange animal while searching for cows with Errol and Bevan Johnson. Unusually, 40 milkers had "gone bush" on the highest part of the Johnson property. On sighting the boys, the herd stampeded right past them, tails held straight out.

The boys' dogs, one of which had the reputation of being a fearless fighter, also suddenly turned and ran, hackles up and tails between their legs.

"Then we saw it. I can only describe it as a wooden wine barrel size object covered in jet-black hair with indistinguishable facial features. It was moving … very quickly towards us with a peculiar upright lumbering gait and we could hear … guttural grunting sounds. I'm not sure who was first to reach the paddock some two miles below … the cows or the dogs, but … I am sure we boys would not have been far behind.

"No one to my knowledge had encountered or heard much of the yowie at that time."

Witness communication with Strange Nation website http://www.strangenation.com/au/Casebook/realyowies.htm

Witness interview with Paul Cropper, 2000. Credit: Rebecca Lang.

Case 73. 1953. South-west of Kempsey, NSW.

Neil Bowen, of Kempsey, wrote that his brother-in-law, Colin Fuller, then 17 years old, and his mate Joe Wright had watched a hair-covered bipedal animal moving through a cleared area adjoining thick rainforest. This was in the "Molly Milligan" (Marlo Merrikan Creek) area.

Letter from Neil Bowen to Paul Cropper and Tony Healy, 1995.

Case 74. May 1954. Mt Hope Station, Booie, QLD.

Three youths were chased by an upright, six-foot-tall, hair-covered creature that emerged from a cave. They said it had a long tail and what appeared to be an "apron" draped around its waist. It was suggested a cranky old wallaroo had frightened the boys. [See Ch. 6]

The Herald (Melbourne), Jun 7, 1954.

Case 75. October 1957. South of Gympie, QLD. Night.

A Cooroy resident, J.D. McDonald, his wife and his father-in-law encountered a huge man-like creature eight miles south of Gympie on the old highway. It was more than eight feet tall, covered in whitish hair "like a polar bear" and seemed to have large, flat feet. It emerged from scrubland and ran with astounding speed across the road in front of their car.

Witness interview with Steve Rushton, Feb 17, 1994. Credit: Steve Rushton.

Case 76. 1958 or 1959. Letitia Peninsula, near Tweed Heads, NSW. Early afternoon.

Kyle Slabb said that, when they were boys, his father and his uncle disturbed a Hairy Man beside Letitia Road. The irate creature ran at them brandishing a stick. In 1991 Kyle had a close encounter with a similar creature in virtually the same spot. [Case 185]

Tony Healy interview with Kyle Slabb, Apr 30, 2001.

Case 77. Circa 1958. Near Batemans Bay, NSW. Day.

While hunting with his friend Bill Taylor, Laurie Allard came face-to-face with a yowie. It was covered with grey hair, stood about five foot six inches [168 cm] and was carrying a dead wallaby over its shoulder. [See Ch. 3]

Witness interviews with Tony Healy, 1998 and 2003.

Case 78. September 1959. Byng, 16 km south-east of Orange, NSW.

Dressed in a hairy hessian suit complete with red reflectors for eyes, pint-sized Darby Offen startled a few motorists near Byng, prompting groups of men to go hunting for the "Byng Bunyip". [See Ch. 7]

Unreferenced Sydney tabloids, Dec 29 and 30, 1959; *The Picture*, Jul 25, 1989; Denis Gregory and Alf Manciagli, *There's Some Bloody Funny People on the Road to Broken Hill*, 1993. Credit: Roger Frankenberg.

Case 79. Summer 1962. Near Mount Buller, VIC. Day.

A Knoxfield (Melbourne) man wrote to yowie witness Maria Speer [Case 200] to say he once glimpsed a large, hair-covered creature in Howqua Valley. It was moving away through dense bush. Although he didn't see its legs, he was sure it was "too thick above shoulders (sic) to be a kangaroo."

Letter from witness to Maria Speer, early 2000. Credit Maria Speer.

Case 80. 1963. 14 kilometres south of Tenterfield, NSW. Dusk.

While driving along an old bullock track, Elsie Mitchell, wife of Robert Mitchell [Case 48] saw three or four large creatures:

"It was dusk and there were lots of trees along the side of the track. At one point I noticed what I thought were three or four trees cut off about six feet [183 cm] from the ground. I remember thinking it was a funny way to lop trees – then they moved. There was also a strong smell like rotting flesh and a low rumbling noise. I was a bit scared because I didn't know what it was. Some years later I read a report about a yowie sighting and it mentioned that distinctive smell. That convinced me I had seen yowies."

Sunday Mail (Brisbane), Nov 9, 1980.

Case 81. 1964 or 1965. Wyandra, 95 km south of Charleville, QLD.

Soon after seeing a strange, hairy, man-like creature crossing a road near Wyandra, a postman supposedly quit his job and left the area. Other locals, including Frank Colgin and Mrs. Summerfield, reported finding several sets of very weird-looking footprints. According to a newspaper report, the tracks were "almost eight inches by six inches [20 x 15 cm] ... they had three front pads – two with three toes and one with two toes. There was also a rear pad, or it might have been a heel ..." Despite the incomprehensible nature of the footprints, it was theorised that a large ape-like creature was on the loose.

The Sun (Sydney), Aug 1, 1965.

Case 82. Between 1965 and 1968. Drummer Mountain, VIC. Night.

The incident occurred as an Aboriginal family was driving home to Bermagui from Victoria on the Pacific Highway. The youngest family member, Colin Andy, was then about 10 or 12 years old.

Late at night they reached the top of Drummer Mountain, about 20 kilometres east of Cann River, and stopped for a toilet break. As the adults walked into the bush, Col strolled down the road behind the car. There wasn't a house light or campfire to be seen, but the road was lit up brilliantly by a full moon.

As he stood there, he saw a strange creature walk out of the scrub about 80 metres away and cross the road. Though bipedal, it was not human; its arms were so long that its hands almost touched the ground, and it "walked hunched over – like an ape." Because he was born into the Yuin tribe, Col was familiar with the *dulagar* legend. For some reason that he is still at a loss to explain, however, he didn't feel like telling anyone what he had seen until about the mid 1980s. When he finally confided the story to his mother, she agreed he had almost certainly seen a *dulagar*.

Witness interview with Tony Healy, Mar 27, 1998.

Case 83. September 1968. Kookaburra, 50 km west of Kempsey, NSW. Between midnight and 1 am.

While working at Kookaburra, an isolated saw-milling settlement on the Carrai Plateau, George Gray slept in a small hut surrounded by dense scrub.

One dark night he was woken by the sensation of something pressing down on his chest, and realised he was being attacked by a hair-covered ape-man. Although only about four feet [1.2 m] tall, it was extremely broad, powerful, and apparently intent on dragging him outside. It had hair "like a Phyllis Diller wig"; a hairless copper-coloured face; a big, flat nose and round, human-like eyes. "The thing was looking straight into my face. The funny thing was it didn't seem to be angry. The hair was a dirty grey colour [but] seemed to be clean. There was no smell. No smell at all. That's a funny thing".

There were other odd details. The creature's big "stubby" hands were seemingly five-fingered, but its arms, though thick, seemed very short. Even stranger, "the skin ... was sort of loose ... like it had no muscles ... like trying to hold something slippery. I could feel the bones. I couldn't feel the flesh at all." It didn't make a sound, "didn't seem to be breathing". But it could still shake him like a dog. As they wrestled on the floor, Mr. Gray could see, in the bright moonlight, that the hairy horror seemed to have webbed toes. After ten desperate minutes it abruptly ran out the door. His two young sons, in a nearby room, had heard the commotion but were too frightened to investigate. [See Ch. 3]

Macleay Argus, Sept 4 and Sept 18, 1976; *Sydney Sun Herald*, Sept 12, 1976. Credit: Patricia Riggs.

Case 84. Early 1970s. Mt. Clunie, 12 km north-west of Woodenbong, NSW.

In response to reports of sightings at Woodenbong, the following brief letter appeared in Lismore's *Northern Star:*

"Sir - I read the story about the two yowies sighted by Mrs. Crewe, of Woodenbong. I lived in Woodenbong for twelve years and have heard of previous sightings. The creatures were seen on Mt. Cluney [sic] and were described as 'hairy men' and as having an ape's body with the head of a man."

Louise Mackney, Lismore

Lismore *Northern Star*, Jul 11, 1977.

Case 85. Early 1970s. Cullendulla, NSW. Dawn.

Truck driver George Birch observed two yowies, one of which appeared to be female, standing beside the Princes Highway. [See Ch. 3]

Witness interview with Paul Cropper, Jun 29, 1993.

Case 86. August 7, 1970. Blue Mountains, NSW. 3:30pm.

In the early 1970s, researcher Rex Gilroy announced that he'd seen a yowie a couple of years earlier, on the western slope of the Ruined Castle, four kilometres south of Katoomba. Since then, several differing versions of the story, most of them containing statements attributed to him, have appeared in magazines and newspapers. He accounts for the contradictions by saying he has been extensively misquoted. Frustratingly, however, it isn't only journalists who have got the story wrong.

In June 1977, Rex wrote that the hair covered "ape-like" creature appeared to be four to five feet tall. It moved swiftly and disappeared "within a few seconds". In 1980, he wrote that it stood five to six feet tall. Again, it "disappeared within seconds". He added that it was 40 yards away, and that, as he only saw it from behind, he "never caught sight of the creature's face". In 2001, he remembered it being 15 metres away, and added that it was male, that it had "big eyebrows" and long, trailing head hair. He watched for "four to five minutes" as, oblivious to his presence, it moved slowly across a slope, carrying what looked like a digging stick, apparently fossicking for roots.

Despite having previously written that he never caught sight of the creature's face, Rex presented, in *Giants From The Dreamtime*, a very detailed close-up sketch of it, complete with nostrils, teeth, ears, and eyes. The sketch bears a remarkable resemblance to an artist's impression of a Neanderthal's face that had previously appeared in a Time/Life book on the prehistory of mankind.

Rex Gilroy, "Gorilla Giants at Katoomba", *Psychic Australian*, June 1977; Rex Gilroy, "Why Yowies are Fair Dinkum", *Australasian Post*, August 7, 1980; Rex Gilroy, *Giants From The Dreamtime*, pp. 173 and 189.

Case 87. September 1970. Lake Wells, WA.

In about 1950, Peter Muir, a well-educated 20-year-old white man, went to live with nomadic desert Aborigines in the vicinity of Lake Wells. He underwent several tribal rituals, had his nose pierced with a bone and acquired deep initiation scars.

In 1964, he began working for the Agriculture Protection Board as a dingo trapper and was soon regarded as one of the best bushmen in Western Australia. Senior APB officer John Kerr said "he is probably the best we have ever had ... He knows all a white man can know of the desert and most of what the Aborigines pass down through their tribal system." He was responsible for controlling dingoes over five million acres of desert. No one, it seems, could have been better qualified to identify footprints in desert sand. Tracks that he found and photographed in September 1970, however, were like nothing he'd seen before. There was, apparently, a long line of fresh, bipedal tracks. They were fifteen inches [38 cm] long, about seven inches [18 cm] wide and the weirdest shape imaginable: a lumpy, almost square "sole" with two long, ugly toes. At least one toe showed signs of having a large claw. When he asked desert Aborigines about the tracks they said they belonged to the *tjangara* or Spinifex Man. The 10-foot-tall creature was believed to carry a heavy club and to be a man-eater.

Basil Marlow, of the Australian Museum, admitted the tracks were like nothing he'd ever seen. Alex Jones, of the Manjimup Native Game Sanctuary, however, said they were probably those of a deformed camel. They bear no resemblance to any other supposed yowie tracks that we know of. However, because they were found by one of Australia's best trackers, and because they were linked with the *tjangara* by desert Aborigines, they can't be casually ignored.
Sydney

Sunday Mirror, Sept 6, 1970; Sydney *Sun Herald*, Sept 20, 1970.

Case 88. Circa 1971. Carrai Plateau, about 50 km west of Kempsey, NSW.

While tin mining on the Carrai Plateau (where George Gray was attacked by an ape-man three years earlier) engineer Richard Gilson and his employees heard nerve-rending howls and discovered large four-toed tracks around their camp. Something large tried to break into a hut, leaving claw marks on the door. [See Case 83]

Macleay Argus, Oct 2, 1976. Credit: Patricia Riggs.

Case 89. 1971. Blackheath, Blue Mountains, NSW. Day.

While playing in the bush near the old Blackheath aerodrome, nine-year-old Victoria Trimble and several other children became extremely uneasy and felt they were being watched. Running to a nearby road, Victoria glanced back and saw a brown, man-like creature running just inside the tree line. It appeared to be hunched-over and moved very fast.

Witness interview with Paul Cropper, May 1, 1994.

Case 90. Mid-1971. The Jingeras, south of Captains Flat, NSW. Night.

While spotlight shooting, 22-year-old Jim Banks and his mate Stan Hunt saw a heavily built, seven-foot-tall, hair covered, bipedal creature running away across a paddock. Jim fired two bullets into its back. It threw up its arms, let out "an unearthly sort of squeal" and kept running. [See Ch. 3]

Jim and Stan had no idea that the site of their encounter was in the heart of an area that was, in colonial days, notorious as a haunt of the Hairy Man [Cases 17, 24, 29 and 32].

Witness interview with Tony Healy, 1993.

Case 91. Circa 1973. Near Bexhill, 6 km north of Lismore, NSW. Dusk.

While camping beside a creek, teenager Mark Pope and two friends were visited by – and shot at – a huge, hairy, man-like creature.

"The tent was open at the front, so we had a clear view out. That's when we saw it ... just standing there ... 20, 30 metres from the tent, in a clearing. There was light behind it ... we could see him silhouetted. It was facing us, so you couldn't see any facial features. The face was dark ... matted hair ... it was probably close to eight feet high. We could see its hands quite clearly. [Its arms] were almost to its knees. The head seemed to be just sitting on the shoulders. It seemed to be a bit pointed; it wasn't rounded.

"We were just dumbfounded; silent and frozen. It was probably less than a minute, then it just slowly moved off ... but at a tangent, so if anything it got a bit closer. As it went up the side of the tent it was out of view. It never did anything to threaten us, but we were kids and it was pretty big. Well, one of the guys picked up a shotgun and fired it through the side of the tent, which was more or less where it would have been ... but we never heard noise out of it, but you could hear it moving off quicker than it had been. So I don't think it got hit but, certainly, I think it was startled."

Twenty-four years later Mark encountered what may have been a much smaller yowie near Woodenbong NSW. [Case 221]

Witness interview with Paul Cropper, Jan 30, 2000. Credit: Dean Harrison.

Case 92. 1973. West of Ashford, NSW. Night.

While camped with friends above the Ashford Gorge on the Macintyre River, a Brisbane man was woken in the middle of the night by a very large, human-shaped figure attempting to open his tent. When he called out to his friends, it moved back into the bush.

Letter to Paul Cropper, Feb 12, 1995.

Case 93. August 1973. Burleigh Heads, QLD. 11 pm.

Four teenagers reported being chased by a bipedal, man-sized, hair-covered creature on the western edge of West Burleigh Heads. [See Ch. 4]

Gold Coast Bulletin, undated clipping; Brisbane *Sunday Mail*, Aug 5, 1973; Sydney *Sunday Telegraph*, Aug 5, 1973.

Case 94. Winter 1974 or 1975. Killawarra, NSW. 3:30 pm.

One sunny afternoon Alwyn Richards and his sister rode their horses up a hillside near Alwyn's property. Ahead was a narrow strip of forest, then a cleared firebreak and beyond that a large area of rough scrub.

The horses normally loved to charge up the slope but on this occasion they shied violently away from the wooded area. Alwyn eventually dismounted and led his reluctant steed through the trees with his sister following. On the other side they were amazed to find a huge, shaggy creature standing in the firebreak. It was nine to 10 feet [about 3 metres] tall; broad chested, very muscular, well proportioned and covered in long, untidy looking hair. Although its arms were longer than those of a man, it appeared more human than ape-like. Parts of its face not obscured by hair appeared to be black.

It stood staring at them for what seemed like several minutes and Alwyn approached to within about 30 metres. A "terrible burning smell" was evident. The animal finally turned and walked away. It moved surprisingly quickly considering its size, and stepped right over a four-foot [1.2 metre] tall fence without breaking stride.

Alwyn walked straight over to where it had been standing. No clear tracks were evident, but plants were squashed and the ground was distinctly warm "like where a cow has been lying down".

Alwyn Richards. The yowie stepped over this fence without breaking stride. (Tony Healy)

Witness interviews with Geoff Nelson, Paul Cropper and Tony Healy, Jun 1993. Credit: Geoff Nelson.

Case 95. Circa 1974. Near Montville, QLD. Night.

As Mark McDonald of Cooroy and a mate drove slowly along the foggy Montville to Maleny road, a huge hair-covered ape-man jumped from the embankment and ran across in front of them. The incident bothered Mark so much that, even 20 years later, he could hardly bear to talk about it. Interestingly, Mark's parents and grandfather had a similar experience near Gympie in 1957. [Case 75]

Witness interview with Steve Rushton, 1994. Credit: Steve Rushton.

Case 96. Circa 1974. TAS.

A resident of Cooee, Tasmania, wrote to yowie witness Maria Speer [Case 200] to report two apparent yowie sightings from the island state.

"My daughter, who was about 5 yrs old, saw in dense scrub, something she described as a big ugly hairy man, and was very frightened ... We never saw it ourselves but after she calmed down she said it had been standing on a log watching us, but by the time she drew our attention to it, all we saw was the thick bushes moving and the sound of something large & heavy going through the scrub.

"Nothing was said about this to anyone ... But about 2 mths later two men who were logging in that same area, told us they saw, well, to use their very words, ... the weirdest thing they had ever saw in the bush in their years of logging. They were loading logs, it was raining quite heavily and windy, they almost had the truck loaded when the other chap said 'Look! What the hell is that,' and on the muddy road not far from them stood this ape like man…" [Second page of letter missing]

Letter to Maria Speer, 2000. Credit: Maria Speer.

Case 97. November 1974. Blackheath, NSW. Afternoon.

While bicycle riding near Blackheath aerodrome, a 12-year-old boy saw a large creature running through scrub about 50 metres away. More like a gorilla than anything else, and extremely muscular, it was covered in short brown hair with some patches of skin visible. Its fingers were long and held in a hooked position. Its large head was more densely hair-covered than the body. No ears were visible.

Witness interviews with Paul Cropper and Tony Healy, 1981 and 1982.

Case 98. Summer 1974. Between Casino and Whiporie, NSW. 1 am.

A close shave: the Allison brothers' yowie.

Having attended the Lismore Speedway, Michael Allison and his two brothers began driving home to Grafton. They took Summerland Way, a road they knew well. The night was perfectly clear.

As their vehicle negotiated a sweeping left hand bend, its powerful driving lights caught a huge ape-like animal about 200 feet ahead in the middle of the road, almost directly on the centre line. It turned its head to face them and, apparently dazzled, stood stock-still, which was just as well, because the speedway fans were travelling at around 100 miles per hour. They passed, Michael estimates, within one foot (30 cm) of the gob-smacked gorilla. Seconds later, several hundred metres down the road, the brothers turned to each other and said as one, "Did you see *that*?"

The animal was "enormous" – about three feet wide at the shoulders – and although "hunched over" was much higher than the roof of the car: "at least eight feet". It was covered in long, dark brown hair. The shoulders and face reminded Michael of an ape or gorilla, but he used the term "Neanderthal" several times.

Grafton Examiner, undated article, Apr 1990. Witness interview with Paul Cropper, Dec 22, 1996.

Case 99. May 1975. Below Cataract Falls, Blue Mountains, NSW. Midnight.

While camped in a cave beside a swimming hole, local resident Steve Croft and his future wife Doris heard something large pushing through the scrub. When Steve built up their campfire, he saw, on the far side of the pool, a six-foot-tall, man-shaped creature. It was so completely covered with long, straight, "brown to orange" hair that no facial features, hands or feet could be discerned. It turned and ran away.

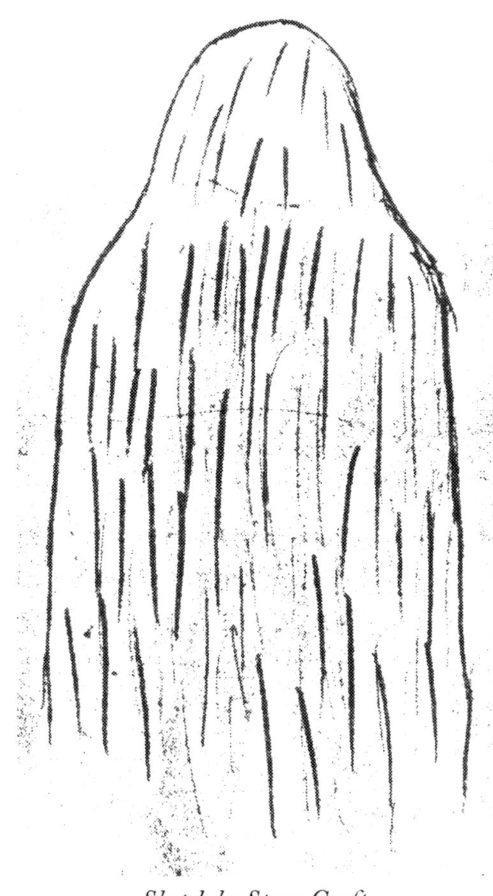

Sketch by Steve Croft.

Steve and Doris are the parents of Brad Croft, who saw a yowie near the O'Connor residence in 1999. [Case 232]

Witness interviews with Neil Frost and Ian Price, 1997; Neil Frost, "Encounters with Fatfoot" (MS), p. 16. Credit: Neil Frost.

Case 100. June 1975. Tailem Bend, SA. Night.

During a wave of UFO activity, two groups of people in separate vehicles allegedly saw a big, hairy, ape-like creature walking along the side of the road with what looked like a lantern in its hand.

Although, as UFO researcher Bill Chalker points out, the case isn't particularly well documented, it is interesting to note that local Aborigines have a strong belief in the existence of a huge, hair-covered, ape-like creature which they call *mooluwonk*. [See Case 47]

Frank Anderson, "The Yowie Mystery", in *Bigfoot – Tales of Unexplained Creatures and Psychic Connections*, Page Research, 1978.

Case 101. November 1975. Poatina Hills, east of Great Lake, TAS.

Two shooters sighted a black, seven-foot-tall, human-like creature that was running very fast. A nearby tree fell over shortly afterwards.

Keith Basterfield, "A Catalogue of UFO Entity and Humanoid Related Reports", 2001. Credit: Bill Chalker.

Case 102. 1976. Lake George, NSW. 1 or 2 am.

While working on "Currandooley", a large property on the eastern shore of Lake George, Billy Southwell lived in an isolated cabin. "It's not real heavily-timbered country, but it's not far from the top of the Great Dividing Range." One night he was woken by the sound of a cow bellowing. His dogs, out on the verandah, started to act up and he became aware of a "droning, real deep moaning sound, like an old aeroplane in the distance". After a while he realised the sound was a lot closer than he'd thought and that it was moving around just outside the house.

"The dogs became quite frantic, barking and carrying on, and there was a hell of a crash and all hell broke loose. I flew out of bed to tell them to shut up ... hit the light switch and opened the door. And the dogs weren't barking any more – they were screaming, and the first dog bowled through the door with its hackles up, smacked into my legs. I was just yelling, 'sit down!' then I saw my other dog was backed up, jammed in the corner of the verandah, screaming his head off, poor bugger. And this thing was standing on the end of the verandah [there was no rail]. It had thrown the chairs off and ripped down a heavy plank that was nailed to the wall.

Billy Southwell. (Tony Healy)

"It was about five foot eight to five foot ten [about 175 cm] and covered in gingery-browny hair. Quite even coverage, about two to three inches long. It was a strange shape; heavy set, but not so much in the shoulders – more in the hips area". The light, shining through the doorway, illuminated the lower part of the creature better than the top. "Thick neck, very heavy-set brow. Long arms – it looked a bit apey-like, the stance of it. It was not a human, not a deer or a 'roo. It stood there for about the count of four and then just spun around, took a couple of steps and jumped off the end of the verandah into the darkness.

"I've never in my life seen dogs that upset. It put me back a peg. I was pretty bloody shaken by it. Afterwards I used to sweat that my car would never break down out that way after dark."

Witness interviews with Tony Healy, 1986 and 1998. Credit: Bob Elgood.

Case 103. October 1976. Near Queanbeyan, NSW. Night.

Late one night, 20-year-old Mike Sillis and two mates were driving south along the Old Cooma Road. Having previously consumed a couple of beers, they stopped shortly before the Burra turnoff and got out of the car. His mates went to the right of the road and Mike to the left.

The night was a moonless, but there was enough light for Mike to see something huge which suddenly loomed up behind the roadside fence, just a few paces away. It was at least seven feet tall and, because it had broad shoulders and a round head, Mike is certain it was not merely a big horse standing front-on.

Suddenly filled with overwhelming terror, he ran, yelling, to the car. Strangely, although his mates had not actually seen the creature, they too, were overcome with terror. In retrospect, Mike admits, the scene must have been quite comical. As they drove off in a shower of gravel all three were crowded together on the driver's side of the front seat – as if to escape the clutches of the whatever-it-was.

Mike knew virtually nothing about the yowie legend and made no great claims about his experience. We found his story interesting, however, because we knew of another report from the same area at about the same time. On October 11, a fence-builder, Gary Costello, saw a six-foot-tall, dark grey creature near Googong Dam, just 3 kilometres to the east. It had, he said, a round head that merged into its shoulders and was "much bigger than any kangaroo".

Telegraph, Sydney (?), Oct 12, 1976; *Psychic Australian*, Nov 1976; witness interview with Tony Healy, Dec 1976.

Case 104. Mid-1976, South West Rocks, NSW. 2 am.

In 1976, Mamie Mason, of South West Rocks told *Macleay Argus* editor Patricia Riggs of several yowie incidents in the Kempsey district. Some had occurred decades earlier [e.g. Case 50] but one, involving a female cousin, was quite recent.

"She went out to the toilet … at 2 o'clock in the morning … and she said she saw this black thing standing with its arms down. It was covered in hair and the dogs were carrying on … yelping, barking … terrified, they were."

Macleay Argus, Oct 2, 1976. Credit: Patricia Riggs.

Case 105. November 1976. Woodenbong, NSW. About 1 am.

Woodenbong is a small town right at the foot of the McPherson Range in northern New South Wales. Remarkably, this incident occurred on Richmond Street, within the residential section of the village. On the night in question, 49-year-old Thelma Crewe had been unable to sleep, so at about 1 am she rose and went to her kitchen.

"I didn't turn on the kitchen light straight away because it was such a moonlit night, and stood at the open window looking at the view. Suddenly this creature walked onto our lawn from the next-door vacant lot. He stood there for two to three minutes just looking towards me. He was sort of flexing his arms in a circular movement in front of his face – first one, then the other. The creature then moved down the side of the house about ten feet [3 metres] towards the bedroom where my husband was sleeping. There was another creature of exactly the same height and appearance standing under our bedroom window." Both were about five feet tall and covered in tan-coloured hair.

"The hair on the arms was about six inches longer than the hair over the rest of the body. Their heads seemed to be sunk low into their shoulders, but I couldn't describe them properly ... couldn't see the facial features properly ... My first impression was that they looked like an Afghan hound because of the long hair on the arms, but ... they were too big and walked on two legs ... they had a shuffling kind of walk. I was much too close to mistake it ... It just isn't possible to compare them with any other animal."

Although she didn't feel frightened, Mrs. Crewe said she was "mesmerised" by the animals as they stared back at her through the window. When they moved into the street, she rushed to wake her husband, but by the time she returned they had disappeared. She checked the lawn early next morning but could find no tracks. One strange aspect of the incident was "the unusual quiet. We have a street full of dogs that usually bark at anything that moves, but there was no sound at all."

Ten months later a yowie was seen in another backyard less than 300 metres away. [Case 111]

The Northern Star (Lismore) Jul 5, 1977. Witness interviews with Bill Chalker and Paul Cropper, 1978 and 1980.

Under the direction of Thelma Crewe, Bill Chalker sketched the creatures' strange arm movements.

Case 106. 1977. Mt. Victoria, Blue Mountains, NSW. Day.

While trail bike riding, Ken Lindsay came within 100 yards of a strange creature standing on a log. It was four to five feet tall, very muscular, with a chest much broader than a man's. It had thick legs, long, thick arms, and was covered in orange or tan coloured hair. Its head seemed to merge into its shoulders without benefit of a neck. The young man and the (presumably) young yowie stared at each other for about three seconds, before the creature stepped unhurriedly down from the log. The teenager fled.

Witness interview with Paul Cropper, 1980.

Case 107. February 1977. Oxley Island, near Taree, NSW. 11 am.

Oxley Island is one of three large, flat, fertile landmasses in the estuary of the Manning River, separated from the mainland by narrow, twisting channels. In the 1970s it was occupied by only a handful of dairy farming families. Although largely cleared, it was criss-crossed by swampy, tree-lined creeks and scattered belts of scrub.

One morning, Mrs. Betty Gee happened to glance out the back door of her house towards her jetty. "I saw something standing there and I thought to myself, 'Who's mucking around on the wharf? It looks like a big ape!' I said to myself, 'that couldn't be!' Then I got out the binoculars." Because of the high riverbank, she could see only the upper part of the creature.

"When I first saw it, its hands were outstretched and it was turning side on ... its features were brown, dark brown ... through the binoculars all I could see was the back of a big, round-shaped head and shoulders with a lot of black fur." She couldn't make out any ears, and the head was set well down on the shoulders: "It was just a short [neck], you know, just like the back of a real ape-person standing there."

Although amazed, she wasn't particularly frightened. "I was going down to investigate when the telephone rang and when I returned the animal had disappeared." The "ape" had been visible from about the centre of its chest upwards, yet when Mrs. Gee's six-foot-tall son stood on the jetty, not even the top of his head could be seen.

About a week later, the family found that on a clear, still day, a large water tank had been pushed from its stand. "I thought, 'Oh gracious, who doesn't like me?' After that we found our fence knocked down. Our field had been freshly ploughed and these big footprints were coming right across it. The tracks were like a big, big footprint with toes, but I didn't count them because I didn't know what I was looking at. I just thought it was some sort of monster! I'm not the sort of person who lets their imagination run wild, but I believe in [yowies] definitely; they must be there, because I saw one."

Manning River Times (Taree), Mar 11, 1977; *Sun Herald* (Sydney), Mar 13, 1977; witness interview with Paul Cropper, Jan 1980.

Case 108. April 1977. Oxley Island, near Taree, NSW. Night.

Two months after Mrs. Gee's experience, a neighbour, Geoff Nelson, was spotlight shooting with his mate Alan Merrett. Everything went normally until they stopped their vehicle near Scotts Creek. There, to their amazement, a huge creature suddenly climbed up out of the creek bed, about 20 metres in front of them. It was on all fours and began to stand up just as Alan switched on the spotlights.

It was covered in long, shaggy, black hair and it remained in a crouched position for a moment, staring straight at the lights before dropping back down the bank. It then appeared to stand up and walk away. The face was almost entirely covered in hair. "The only skin that was visible," Geoff recalls, "was around the eyes, on the nose and maybe a bit of the mouth ... the skin was dark, near charcoal colour – but you could see a distinct difference between the hair and the skin because its skin was kind of shiny with the light on it". What most impressed Geoff was the intense brightness of the eyes. They were very widely spaced and not large, but were "vivid, red and glowing, like two flashlights shining back at you. I have never seen anything like them".

After they got over their initial shock, Alan suggested that, since they were armed, they try to follow the animal. Geoff, however, declined with the comment that the massive creature might decide to twist their puny .22 rifles around their necks.

When Geoff returned to the spot three days later, he was surprised to find a "strong, acrid, electrical smell, like burnt bakelite – like when you blow up an old radio – a sulphury stink" that seemed to permeate everything in the immediate vicinity.

Witness interviews with Paul Cropper and Tony Healy, Jun 1993.

Case 109. May 1977. Oxley Island, near Taree, New South Wales. Night.

About a month after seeing the yowie at Scotts Creek, Geoff Nelson almost collided with another one.

As he was driving down Cowans Lane, right beside Betty Gee's boundary fence, a huge creature suddenly charged across the road about 10 metres in front of him. It was over two metres tall and about the same size as the first creature, but was covered in pale grey or beige coloured hair. Having apparently vaulted the fence on Geoff's right, it crossed the road reserve in four or five huge bounds, took the opposite embankment and fence in its stride and seemed to continue at full speed – straight towards the Gees' homestead.

Geoff Nelson's sketch.

Meanwhile, Geoff had "hit the brakes, forgot

about the clutch", slewed to a halt and stalled the car in the middle of the road. His main impression of the creature was that unlike many of the yowie descriptions he has since heard, it did not have a slumped posture: "Its head was up and its chest was thrust out, just like an athlete heading for the tape."

In the mid-1990s Geoff and his family moved to a bush block to the south-west of Taree. Remarkably, he, his parents, his wife and their son have all encountered yowies at the new property, sometimes at very close range. Researcher Ashley Mills has also seen a yowie on that property [Case 234]. Geoff will detail all those incidents and many others from the Taree area in a forthcoming book.

Witness interviews with Paul Cropper and Tony Healy, 1993-2004.

Case 110. August 1977. Talbingo, NSW. 4:30pm.

While riding his motorcycle along a dirt road about four miles from Talbingo, John Crocker of Tumbarumba stopped for a short break. Glancing into the thick scrub beside the track, he was startled to see a huge animal standing motionless 30 or 40 yards away. Towering to about eight or nine feet, it was covered in brown hair and had arms which reached past its knees. John said it resembled depictions of the American bigfoot. It was, he said, like a gorilla, but taller, "like a monster".

After staring at the hairy horror for a short while, John became frightened and rode away. Although it was gone when he returned to the spot with his brother, 18-inch tracks and broken underbrush indicated that he hadn't imagined the incident.

Tumbarumba Times, Aug 31, 1977. Witness interview with Paul Cropper, Dec 12, 1993.

'... its chest was thrust out, just like an athlete heading for the tape.' Sketch by Geoff Nelson.

Case 111. 10 August 1977. Woodenbong, NSW. 2:30 am.

Ten months after Thelma Crewe's sighting [Case 105], another yowie was seen in the village of Woodenbong. This one appeared at the Maloney residence, about 270 metres from Mrs. Crewe's house. Because her husband wanted to avoid publicity, Jean Maloney kept quiet about the incident for a couple of days. Finally, however, she contacted the Lismore *Northern Star* because she felt the matter "was too darned interesting to keep to myself".

She had been awoken by the sound of her Australian terrier yelping. Intermingled with the yelping was a high-pitched, screaming sound. Jumping out of bed, she ran to the back porch and switched on the 200-watt yard light.

"I went down the stairs and ran into the backyard when I suddenly saw the creature directly in front of me. I was within six feet of the jolly thing and I think I stopped breathing for a moment because of the fright. It was sitting on its haunches and had my dog completely crushed up against its chest. The dog was almost completely covered by the creature's arms, which were wrapped around the dog, one above the other ... as though [it] was trying to crush the life out of her." It stood up, "looked straight up at me for a few moments then dropped the dog, which I thought was dead at this stage, because she fell to the ground and did not move."

As it stood, gaping at her, with its arms by its sides, Mrs. Maloney felt no fear: "It didn't seem vicious. The dog was very territorial – she might have tried to bite him." The animal was well over six feet tall. Its almost hairless face jutted forward and was ape-like, or "Neanderthal-like", with "very big, dark eyes", a heavy brow and no chin. Its head, which sat directly on its wide shoulders, seemed small in comparison to the rest of the body. Long ginger-coloured hair hung from its head, arms and legs. The very broad chest and abdomen were more sparsely covered, so that quite a lot of brown skin was visible. It had powerful-looking legs and slender arms that reached to its knees. It was male, and, unusually for a yowie, its genitals were quite apparent. The penis resembled that of an uncircumcised man, and was "quite large, maybe nine inches long". A human-like scrotum was also visible.

Mrs. Maloney noticed a strong, offensive odour: "The only thing I could compare it with would be a ferret." The creature then raised both arms and held them wide, with the palms of its "very square" but human-like, hands facing her, and "backed away ... towards the grape trellis, but it never took its eyes off me. It ... wrapped its right arm around the trellis post [and] stayed there for a few moments making these strange, deep, very loud, grunting noises." When a couple of other dogs – strays – entered the yard, the yowie bent forward and "ran off to the right, down the side of the house between the garage and disappeared towards the front street".

The terrier was bleeding from wounds to the chest and the back of the neck: "She could hardly walk, and if you moved her too much she would yelp. I also noticed that she was badly bruised on the chest." The dog's hair felt "very greasy, sort of waxy" and stank so badly that it had to be washed in antiseptic. It never recovered from its ordeal and died a short time later.

Before being disturbed by the dog, the creature had apparently been helping itself to a bucketful of fruit that had been left near the Maloney's garage. There were three distinct footprints alongside the house but two were soon destroyed by rain. *Northern Star* reporter Gary Buchanan examined and photographed the remaining print. It was 22 centimetres long by 11 wide [9 x 4 inches], with five toes of roughly equal size. Buchanan also spoke to neighbours who had heard both the dog's screams and the loud grunting noises.

Lismore Northern Star, Aug 15, 1977. Witness interviews with Paul Cropper and Tony Healy, Jan 2004 and Sept 2005.

Case 112. September 1977. Budd Island, Clyde River, NSW.

Oyster farmers found three huge footprints on Budd Island, just upstream from Batemans Bay. [See Ch. 4] Two years later Tanya Bowen saw a yowie less than a mile to the south. [Case 132]

The Examiner (Moruya), Sept 7, 1977.

Case 113. October 1977. Wellington, NSW.

An unnamed Wellington man claimed to have seen a yowie twice between Wellington and Yeoval. It was seven feet tall, wide and bulky.

Central Western Daily, Oct 12, 1977.

Case 114. October 1977. Tweed Heads, NSW. Afternoon.

Paul Cronk and Mark Gill twice encountered a huge ape-like creature in swampy bushland. [See Ch. 4] The site is less than three kilometres from Letitia Peninsula, where Aborigines have frequently encountered Hairy Men. [Case 185]

Witness interview with Paul Cropper, May 22, 2001; witness emails to Gary Opit. Credit: Dean Harrison and Gary Opit.

Case 115. October 22-23, 1977. Springbrook, QLD. Early afternoon.

At "Koonjewarre" campground about 20 students from Southport School experienced repeated sightings of a huge hair-covered, ape-like creature. One witness was future senator Bill O'Chee. [See Ch. 4]

Five months later, ranger Percy Window encountered a yowie only three kilometres from "Koonjewarre". [Case 123]

Gold Coast Bulletin, Nov 17, 1977. Witness interviews with Tony Healy, 1993.

Case 116. December 1977. West of Pambula, NSW. Dusk.

Kos Guines, of Frankston, Victoria, shot at "a huge black creature like a gorilla". Mr. Guines, originally from Greece, had never heard of the yowie legend. [See Ch. 4]

Melbourne *Sunday Press*, Jun 20, 1982. Witness interview with Tony Healy, Jun 1982.

Case 117. 1977 or 1978. Toonumbar Dam, 20 km west of Kyogle, NSW. Evening.

John MacLean of Innisfail said that when he was 14 or 15 years old, he and several friends were fishing when they were startled by large stones being thrown into the water. Looking up, they saw a huge figure standing 30 to 40 feet away. It seemed to be between 12 and 13 feet [over 3.6 m] tall.

Witness interview with Malcolm Smith, Dec 1996. Credit: Malcolm Smith.

Case 118. 1977 or 1978. Strathbogie, VIC. 10 am.

A farmer, his wife and a young workman watched a "gorilla" or "big ape" walk 150 yards across a swamp on their property. It was close to five feet tall and was covered in ragged, approximately two-inch-long, golden-brown fur.

Witness interviews with Roger Frankenburg, 1999, and Paul Cropper, Aug 11, 2002. Credit: Roger Frankenburg.

Case 119. 1977 to 1997. West of Gympie, QLD. Day and night.

Gympie historian Brett Green experienced three encounters with yowies between 1977 and 1997. On one occasion, he and some friends saw a small yowie steal a piece of meat off a bush barbeque. On another, he watched two yowies fighting. As he may write his own book about the yowie phenomenon, he asked us not to divulge any other details.

Witness interviews with Paul Cropper, Mar 14, 2000 and Tony Healy, May 8, 2000. Credit: Dean Harrison.

Case 120. 1978. Mt. Talawahl, 16 km south of Taree, NSW. Night.

Seventeen-year-old "Shane" observed an eight to nine-foot-tall, heavy-set, man-like figure standing eight metres away in the middle of an isolated bush track. Mt. Talawahl is less than 10 kilometres east of Krambach, where Julie Clark encountered a yowie in 1990. [Case 182]

Email to Paul Cropper, Mar 15, 2002.

Case 121. January 29, 1978. Springbrook, QLD. Night.

Scott "X" was shocked when a huge, foul-smelling ape-like creature attempted to enter his brother's house. He threw a chair at it, and it limped away. A friend, David Window, and others examined large, non-human footprints around the house. [See Ch. 4]

Gold Coast Bulletin, Feb 2, 1978. Tony Healy interview with David Window, Apr 2000.

Case 122. January 1978. Springbrook, QLD. Dusk.

A 23-year-old Sydney woman reported that she and her boyfriend encountered a "great big hairy beast", about 10 feet tall, with no neck, and giving off a "terrible smell" in Springbrook National Park. [See Ch. 4] This occurred near "Best of All" Lookout. [See below]

Gold Coast Bulletin, Jan 18, 1978.

Case 123. March 5, 1978. Springbrook, QLD. 2 pm.

Ranger Percy Window came face-to-face with a huge, hairy, gorilla-like creature near "Best of All" Lookout. [See Introduction]

Gold Coast Bulletin, Apr 7, 1978. Witness interview with Paul Cropper, 1991.

Case 124. July 1978, Narooma, NSW.

A Narooma man reported that he and his son repeatedly encountered large, bipedal, "bear-like" creatures to the west of Narooma. [See Ch. 4]

Moruya Examiner, Aug 3, 1978.

Percy Window. (David Window)

Case 125. August 1978. Nerang, Gold Coast, QLD. 2:30 pm.

Shaun Cooper saw an eight-foot-tall, hairy creature stripping bark off a tree. *Gold Coast Bulletin* staff later photographed several large three-toed footprints. [See Ch. 4]

Gold Coast Bulletin, Aug 25, 1978.

Case 126. August 23, 1978. Coomera Valley, Gold Coast hinterland, QLD. Midnight.

Retired security officer Leonard Rye, his wife Nan and several others were driving on Upper Coomera Valley Road when they noticed a huge creature at the edge of the road. The animal, which was covered in dark brown hair and between 10 and 14 feet tall, took two steps across the road, stepped over a four-foot fence and walked into a paddock. Mr. Rye said it resembled a "huge, hairy gorilla" or "huge ape".

Gold Coast Bulletin, Oct 18, 1978. Witness interview with Paul Cropper, Dec 20, 1993.

Case 127. October 1978. Towers Hill, Charters Towers, QLD. Night.

Nineteen-year-old Michael Mangan and a friend told police about repeated encounters with small, aggressive, "hairy men" on Towers Hill. [See Ch. 5]

Northern Miner, Feb 23, 1979; *Daily Bulletin* (Townsville), Mar 5, 1979.

Case 128. Summer 1978. Lamington National Park, QLD. Afternoon.

While hiking back to Binna Burra Lodge, 15-year-old Brendan Howard and a mate heard a noise like someone chopping wood. Shortly thereafter Brendan glimpsed a dark, possibly hair-covered creature that appeared to be over six feet tall. It was moving into the surrounding rainforest.

Witness interview with Malcolm Smith, Oct 22, 1994. Credit: Malcolm Smith.

Case 129. Summer 1978. Hollywell, Gold Coast, QLD.

Tyson Franklin, then 12 years old, and several friends watched a seven-foot-tall, dark tan, bipedal figure "like a gorilla" running, snorting, through the bush. Just prior to the sighting they'd noticed a strange, offensive odour. A week later the same group had an almost identical experience at a nearby location.

Gold Coast Bulletin, Jan 11, 1979. Witness interview with Paul Cropper, Dec 1993.

Case 130. 1978 or 1979. Canungra Land Warfare Centre, QLD. Late afternoon.

While fighting a bush fire at Mango Hill, Max Haimes and another soldier saw two very tall, dark, grunting figures running *towards* the fire. The men called out a warning, but then realised the creatures were not human. [See Ch. 4]

Paul Cropper interview with Pauline Haimes, Jan 17, 2001. Credit: Dean Harrison.

Case 131. August or September 1978 or 1979. North of Coonabarabran, NSW. 3 am.

On the Newell Highway, a Brisbane man, "Pat", and his mate "Bill" saw what they thought was a pig in the middle of the road. "We had powerful driving lights ... I flicked them off and on but it didn't move. I slowed down further and said, 'Crikey, look at the size of this thing!' I went down to second gear; had the lights right on it. It was on its hands and knees over a kangaroo carcass. Its head was bobbing up and down as if it was eating it.

"When we were about 60 metres away it stood up on its hind legs, and I went, 'Shit, have a look at this!' And Bill goes, 'What the f*** is it?' I'm six foot five. It would have been seven and a half, maybe eight feet. It was covered in brownish, slightly reddish hair. Broad shoulders ... [comparatively] small head ... no neck. We were stunned to the point we nearly drove into it.

"We had to go onto the wrong side of the road to get around and drove past pretty slowly, 20 miles per hour, if that. Bill could have reached out and touched it. As we went around, it turned and watched us and that's when we got the smell: like vomit. Its crotch would have been roof-high on the car; never saw any sexual organs, just a lot of hair. Hairy, solid arms and legs. I turned the car around real quick and this thing bent over, grabbed the carcass and ran off, dragged it into the bush.

"So I grabbed the torch out of the glove box and took off after it. I only went about 20 feet and then thought, 'What the f*** are you doing?' I went back and Bill had the windows up and the doors locked and I said 'Let me in', and he says 'No f***ing way. You're mad! Let's get out of here!'

"We went into the servo at Coonabarabran and the operator showed us hand-drawn sketches of yowies on the wall of the restaurant, and I said, 'That's what we saw!'"

Witness interview with Paul Cropper, May 27, 2001. Credit Dean Harrison.

Case 132. Circa 1979. Batemans Bay, NSW. About 3 pm.

About two years after large tracks were found on Budd Island [Case 112], Tanya Bowen saw a huge, hair-covered, ape-like creature at nearby McLeods Creek. [See Ch. 4]

Witness interview with Tony Healy, Apr 15, 2003.

Case 133. Summer 1979. Theresa Creek, 25 km north-west of Casino, NSW. Between 1 and 2 pm.

Leslie Davis said that while riding on her grandfather's farm she glimpsed a very large, hairy figure in rainforest. It moved off silently into thick undergrowth.

Witness interview with Malcolm Smith, Dec 1996. Credit: Malcolm Smith.

Case 134. April 1979. Wentworth Falls, Blue Mountains, NSW. Day and night.

While bushwalking near their home, Leo George and his wife Patricia found a kangaroo with all the flesh ripped from its hindquarters. A few minutes later they stumbled across a "massive" footprint in the sand. "It was very eerie," said Mr. George. "The print was four-toed, about 30 centimetres long and we started to get a strange feeling we were being watched. We walked further along a creek bed, then suddenly saw a shaggy grey mass disappearing through the trees."

"The beast was at least three metres tall," Mrs. George added, "and was too longhaired for a kangaroo. Its fur was about six centimetres long. It shambled away silently. I'm certain that what we saw was a yowie." A few days later, while spotlighting in the same area, Mr. George and his 17-year-old son illuminated a pair of red eyes "the size of tennis balls". "We were a bit afraid," he confessed, "but we left the spotlight on the car and walked toward the apparition." The eyes disappeared. "But once again we had the feeling we were being watched. Suddenly tremendous crashing noises came from a thicket behind us ... and we broke records getting away from the spot."

Sunday Mirror (Sydney), May 6, 1979.

Case 135. April 1979. The Pilliga Scrub, NSW.

Constable Wayne Warren of Coonabarabran police revealed that truck drivers were too frightened to sleep on the Newell Highway north of Coonabarabran. That stretch of road, which passes through the Pilliga Scrub, has been the site of many yowie incidents [e.g. Case 131]. The creatures had reportedly thumped on the sides of parked trucks and pulled tarpaulins from loads.

Northwest Magazine, Apr 1979.

Case 136. June 1979. Emerald, Dandenong Mountains, VIC.

A man identified only as "Vic" told workmate Douglas Bombadieri that while walking in bushland near a creek he'd heard strange noises and felt something was following him. He ran to his car pursued by something that sounded like "an elephant running in galoshes." As he tried to drive away he found that something had grabbed the rear of the vehicle (a Ford Fairlane). Looking through the rear window, he saw two black arms and a big chest. Putting the car into reverse, he knocked the creature down and sped away. Later he found two large handprints on the back of the car.

Mr. Bombadieri and Vic later found two five-toed tracks on a nearby building site. They were 21centimetres (eight inches) long but were much wider (16 cm across the toes) than normal human footprints. As both men were plasterers, they immediately made casts of the tracks. On another occasion Mr. Bombadieri found similar tracks near the creek. They were spaced four feet [122 cm] apart. A *Sunday Press* journalist dubbed the creature "The Black Hulk".

Sunday Press (Melbourne), Jul 8, 1979; *Melbourne Age*, Jul 9, 1979.

Case 137. June 24, 1979. Mount Victoria, NSW. 2:20 pm.

"Fact or fiction, the elusive 'Yowie' is on the prowl again. The 'great hairy man' of Aboriginal folklore was seen recently by Jenny Markam, of Bathurst St., Liverpool, while bushwalking. Jenny sighted the creature in dense scrub three miles from the Blue Mountains township of Mt. Victoria on Sunday, June 24, at 2:20 pm."

Liverpool Champion, Jul 11, 1979; *Daily Telegraph* (Sydney), Jul 16, 1979.

Case 138. July 1979. Macquarie Marshes, west of Coonamble, NSW. Afternoon.

When John Miller's father was a grader driver his family often travelled with him and camped in remote areas. One afternoon, between the Macquarie and Marthaguy rivers, somewhere north of Sandy Camp, John, then eight years old, wandered off alone. While walking through a dense patch of saltbush and trees he noticed a large animal about 30 metres away, stooped over, apparently eating berries. It stood erect and turned to face him, silent and still. Covered in "wild" reddish-brown hair, it stood six to seven feet tall. John can't remember seeing arms or legs, but the face, though "deep" and flat, with sunken eyes, was human-like.

He ran back to camp and told his mother (an Aborigine) that he'd seen a bear, but she told him it must have been "a Hairy Man".

Bathurst Advocate, Jul 11, 1979; witness interview with Paul Cropper, Feb 22, 1995.

Case 139. September 1979. Murderer's Hill, near Walhalla, VIC. 11 am.

While working on a bush road, surveyors John Macey and Sid Griffith saw, five metres away, an ape-like animal standing with it's back to them. It was about 1.2 metres [4 ft] tall, powerfully built with long arms, wide hips and buttocks. They noticed a patch of pale skin on the back of its neck, beneath the black hair that covered all of its body. It clambered over a log and disappeared into thick vegetation.

Witness interview with Gary Opit. "Understanding The Yowie Phenomenon" by Gary Opit, *Nexus* magazine, Aug - Sept 1999. Credit: Gary Opit.

Case 140. October 1979. 20 km from Kilkivan, QLD.

Mrs. Roy Locke, of Theodore, said she and her husband saw a one-metre-tall, broad-shouldered, ape-like animal standing by the roadside between Hervey Bay and Murgon. [See Ch. 5]

South Burnett Times, Oct 1, 1979; *Telegraph* (state unclear), Oct 2, 1979.

Case 141. December 1979. Kilcoy, QLD. Early afternoon.

While pig shooting at Sandy Creek, four kilometres from Kilcoy, Warren Christensen and Tony Solano, both 16, encountered a three-metre-tall yowie. They were having lunch when the huge creature, covered in dirty, chocolate-coloured hair suddenly appeared 20 metres away.

Warren grabbed his .22 rifle and fired from the hip. "I think I might have hit it", he said, "but it just took off." It left a faint sulphurous odour. The boys then followed the "thump, thump" of its giant strides along the creek until suddenly there was silence. "Then we heard the thumping noise behind us – it had doubled around. We jumped down into the creek bed so we could have a clear line of fire in case it attacked us."

Their biology teacher at Kedron State High School, Jenny Bolman, went to the spot with her husband John and made plaster casts of several 50 cm x 15 cm [19 x 6 inches] "distinctly three-toed" tracks.

Brisbane Courier Mail, Jan 4, 5 and 7, 1980; *People*, Feb 21, 1980.

Case 142. Late 1970s or early 1980s. Near Carlisle River, VIC. 2 pm.

As Wes Hodge, then in his early 20s, was sitting on a high ridge with five friends, a strange creature was noticed several hundred metres below. It remained stationary for nearly 10 minutes as they passed a pair of binoculars around. On viewing it, Wes's first thought was, "Bigfoot, from America!"

It was sitting on an embankment between a dirt road and a pine forest, with its knees drawn up to its chin and "its big elbows out" [like the Letitia Peninsula yowie, see Case 185]. They could see the side of its body and the back of its head. "A big hairy thing. Big back structure. Huge bloody legs ... way taller than a person, at least six and a half feet ... long reddish-brown fur. Shaggy, messy and matted. It had a long head, a big skull." It looked as massive as a grizzly bear. Wes compared it to both Chewbacca from *Star Wars* and the bigfoot from *Harry and the Hendersons*.

"It just looked so peaceful, like it was really enjoying the sun, the peace and quiet. It was looking down, then up at the pines." When a tractor appeared, coming down the road, it reacted instantly: "one leap and it was gone. It spun us right out – we never slept that night. I was thinking about it all the time. It's one thing I'll never forget in my whole life".

This occurred about 30 km north-west of where George Paras saw a smaller yowie in 1984. [Case 157]

Witness interview with Paul Cropper, November 2004. Credit: Dean Harrison.

Case 143. 1979 or 1980. Near Cowra, NSW. Approximately 11 pm.

Between Grenfell and Cowra on the Mid Western Highway, "Brownie" and his mate Gary sped past an enormous creature that was standing right on the edge of the road, just west of Broula.

Although the sighting was very brief ("we were going well in excess of the speed limit"), "Brownie" vividly recalls seeing "a huge, black, hairy, ape-shaped creature, approximately eight feet tall. It is still very clear in my mind". Although most of the surrounding land is devoted to wheat, there is extensive forest (in Canimbla National Park and the Broula Range) alongside that particular stretch of road.

Witness email to Tim the Yowie Man, Jul 2002. Credit: TYM.

Case 144. Circa 1980. Kowmung River, Kanangra Boyd National Park, NSW.

A family friend told researcher Pat Ryan that when he was about 19 years old, he was working with a party of cattlemen near the junction of the Kowmung and Coxes River. They camped a little upstream from a place known as "The Yards". Very early one morning, while lighting the campfire, he noticed a Hairy Man observing him from a few metres away. Its eyes were glowing red and its odour was revolting. It rocked backwards and forwards, made a thumping noise and took off into the scrub. Later, when he asked one of the older men why they never camped closer to "The Yards", he was told, "Well ... we just don't." The young man noticed several strange stick formations in the bush around "The Yards".

Pat Ryan added: "A number of bushwalkers have told me of unusual experiences in the area. One old bloke said he'd been tormented by 'Aboriginal devils' in that gorge, and would never camp there again." The place where Gary Jones and friends chased a hairy ape-man in 1989 is within a couple of kilometres of "The Yards". [Case 179]

Credit: Pat Ryan.

Case 145. January 1980. Coffs Harbour, NSW. Day.

"The hardy North Coast perennial, the legend of the Yowie, has broken into the spotlight again ... a middle aged itinerant worker claims to have spotted the elusive beast on a Boambee banana plantation. Although wishing to remain anonymous for fear of ridicule, his articulate description and obvious fear lend some weight to his claim. He said he saw a seeming half human, half-animal form bristling with masses of red hair ... about six months ago ... he thought he was seeing things but the next day he caught a fleeting glimpse of it again dashing through the plantation. That afternoon he collected his pay, vowing never to return there – and he hasn't."

Eastland Opinion (Coffs Harbour), Jul 9, 1980.

Case 146. 1981. Bodalla, NSW.

Several Aboriginal children saw a Hairy Man standing in the bush near a logging camp. The creature was dark and "easily seven feet tall".

Witness interview with Paul Cropper, Mar 15, 2001.

Case 147. Early 1980s. Springwood, Blue Mountains, NSW. Night.

A woman said that a huge, hairy creature frequently peered through high windows, broke tree branches and threw rocks at her property. This occurred only three kilometres west of the yowie-plagued Pendlebury property. [See Ch. 4]

Witness interviews with Neil Frost, 1995. Credit: Neil Frost.

Case 148. 20 May 1981, Dunoon, northern NSW. About 12.45pm

As three boys aged 14, 13 and 11 were exploring scrubby hills six kilometres west of Dunoon, a hairy, man-like creature crossed the track ahead of them. Then a similar creature appeared.

"The second animal seemed to stumble, stopped behind a tree ... and peeped around [it] towards us," said the oldest boy. "It kind of squatted ... and looked at us for about five seconds before running across the path behind the other animal, which had its back towards us and appeared to be waiting for its mate to catch up." Both then disappeared, but could be heard moving away. The boys' dog "went berserk", made a crying sound and briefly pursued the creatures.

The boys rejected suggestions that they had misinterpreted a sighting of ordinary bush animals, pointing out that the creatures were about two metres tall and walked on their hind legs. "I've never seen a wild pig walk down a hill on its hind legs," commented the eldest lad. "They weren't wallabies either. They moved quickly, but looked slightly awkward, and bent forward a little as they moved." Asked whether they resembled gorillas, he said, "Yes, but these weren't gorillas. Gorillas are black and bow legged. These had straight legs and were brown, more human like."

Lismore Northern Star, Jun 3, 1981.

Case 149. June 1981. Cooma, NSW. 1 pm.

In early 1981, at their property on the banks of Cooma Creek, two kilometres north of town, the Marion family found several of their horses slashed on the sides and back as if by a large animal. One mare almost died. Mr. Peter Marion later heard, coming from a scrubby hillside overlooking the horse paddock, a curious, deep throbbing "like someone shaking a big sheet of galvanised iron". Finally, on the same hillside, on a sunny afternoon, Adam Marion and Shane Goodwin, both 12 years old, ran into a hair-covered giant.

"It was just standing there beside a tree," said Adam. "We could see it real clearly. It was huge, very broad, solid." Its arms were longer than a man's and "the head was stuck straight into the shoulders". It was covered, but not densely, with dark brown to reddish hair. "You could make out the chest muscles."

'There's apes in them there hills'. (Lismore Northern Star)

From where the boys stood, 25 to 30 metres downhill, a tree branch obscured the creature's eyes (the branch was later found to be almost two metres high) but they could clearly see its mouth. "We just stood there", said Shane. "Then it bent to look at us ... and it stepped ... and we turned and ran straight down the hill without worrying about falling over."

Shane Goodwin and Adam Marion. The yowie's head was partially obscured by the branch. (Tony Healy)

Sketch by Adam Marion and Shane Goodwin.

Adam seemed to be in a state of shock. His mother said that after blurting out the story he went straight to bed and slept soundly for two hours, something he was never known to do in daylight.

Witness interview with Tony Healy, Aug 1981.

Case 150. June 1981. Porcupine Ridge, 40 km north east of Ballarat, VIC. 11:30 pm.

A young Melbourne man said that while spotlighting for rabbits, on foot, with three mates, he suddenly came within a few feet of a 6 foot 5 inch, man-like, hair-covered creature. It turned, jumped 15 feet into a creek bed and ran off. Another member of the party later saw the same animal, at a distance.

Witness interview with Tony Healy, 1981.

Case 151. May 1982. Mount Tamborine, Gold Coast Hinterland, QLD.

Three men were mustering cattle on the steep northern slopes of Mount Tamborine when their dogs took fright. They then spotted a "huge, grey creature" walking slowly through the bush.

Gold Coast Bulletin, May 19, 1982.

Case 152. November 1982. Bowen Mountain, Greater Sydney, NSW. Sunset.

When Barry Porter and Kelly Tromp (who are now married) were both 10 years old they lived on Maple Street, Mt. Bowen, which is on the extreme north-western edge of greater Sydney, two kilometres from Blue Mountains National Park. "There was nothing but bush to the horizon." One day, while playing handball in the street, they had the feeling "someone was watching us". Looking up, they saw a large creature standing about 70 metres away in the bush.

Barry, who is blessed with a near-photographic memory, recalls that it was about three metres [10 ft] tall, covered in dark hair, with broad shoulders and a narrower waist. Its long arms hung by its sides. "It looked like it had a hairy ball of a head squashed onto the shoulders. It was looking intently at us. We were just freaking, and then as quick as a flash it was gone."

Barry's mother-in-law told him later that throughout 1978 something had repeatedly thumped the walls of their newly built brick house, sending shudders through the entire building.

Report from Barry Porter to the GCBRO web site; emails from Barry Porter to Dean Harrison, Dec 21 and 22, 2004. Credit: GCBRO and Dean Harrison.

Case 153. 1982 to 1992, Vulcan State Forest, near Oberon, NSW. Day.

Between 1982 and 1992, when he was between 12 and 22 years of age, a Sydney man, sometimes accompanied by other witnesses, had six encounters with a strange animal in and around Vulcan State Forest. It resembled an upright gorilla, was seven feet tall and covered in "messy" two-inch-long black hair. No neck was visible; its arms were long and its chest solid. It could run uphill at amazing speed.

On one occasion, he and his sister found a huge five-toed footprint on the muddy verge of a dam. "We turned around and there he was, only about 30 metres away. As plain as day; hands by his sides. And then I was running over to the motorbike and we rode off. I wasn't even game to look back." On another occasion, he fired several shots from an air rifle [at extreme range] directly at the animal in a futile effort to make it move. It merely turned side-on. During nocturnal encounters its eyes were very reflective and green in colour. When the creature was nearby, the family dog "went off its head, crying and running away. It wanted us to get the hell out of there".

Strangely, during one incident, a well-maintained motorbike at first refused to start and then revved uncontrollably.

Witness interview with Paul Cropper, Dec 30, 1995.

Case 154. Summer 1983. Kalamunda, WA. 10:30 pm.

After having a couple of beers at his local pub, Tim Masser began walking home along a road that skirts Kalamunda National Park. As he walked he heard something in the scrub. Because he and a mate had once been followed through the same patch of forest by a large, unseen creature, Tim stopped to investigate.

"I was looking into the bush ... finally a four-wheel-drive approached and [by its lights] I could see a shadow beside a tree. As the car got closer I saw it was one of those half-man, half-ape things looking straight at me. It was about 50 metres away, about six foot tall, had bright red eyes. Large, well built, like a man. Well covered in quite thick hair; not really long, but shaggy. I only saw it briefly [but] the hair seemed a bit thinner on the face and chest". Because "it stared me out – not like an animal would", Tim had the impression the creature was more man-like than ape-like.

Searching the area later, he found "sort of beds, like mattresses made out of twigs and leaves; really springy – I tried them out". Although he knew of no other yowie sightings, he'd heard that some escaped cattle had been found in the forest "with their throats ripped out".

Witness interview with Paul Cropper, Jan 11, 1997.

Case 155. June 1983. Bonnyrigg, western Sydney, NSW. Dusk.

A Bonnyrigg resident and several neighbours had for some time been losing ducks and chickens to a nocturnal predator, which they dubbed the "phantom dog". One evening this resident heard a noise from a small creek behind his house, grabbed an iron bar, went outside and saw what he thought was a large black dog holding several of his ducks under the water. He struck it on the back, and was shocked when it rose up to a height of about seven feet, extended its arms and growled at him. His only thought before turning to flee was, "It's a bear!"

Sydney Sun, early Jun 1983. Witness interview with Paul Cropper, Jun 1983.

Case 156. 1983 or 1984. Mount Kembla, NSW. Between 10 pm and midnight.

A Sydney man, "R", then 19 or 20 years old, and his future wife were parked on the Mount Kembla Road when a huge animal approached the car.

"I was occupied doing something else," "R" recalled, "when I got this really bad feeling. It just struck me out of the blue ... and I looked over my shoulder." Through the side widow, he could see the outline of a huge head, "four times the size of a human head", about four metres away. "It was a roundish type head, almost like the top of a human head, with flat sides". He couldn't make out much of the body, which seemed to be behind a bush, but thought it must have been immense. Although it seemed to be crouching, it was still close to six feet high.

The most alarming feature, however, was the animal's eyes: set four to six inches apart, they were "a deep, dark red, but with a brightness to it ... it's hard to explain – like a bright glow. There was very little light in the area, only the interior light from the car, and light from down in the city, which was really a dull light. But these eyes just glowed – glowed by themselves!

"My immediate reaction was to flee. I had this feeling I was being hunted – an immense feeling of dread ... like someone had drained all the life or blood out of you ... an awful, awe-inspiring feeling of like ... all your senses come alive ... it is very hard to explain. I remember thinking, 'this thing's going to come over here and rip your head off!' I started up the car and revved it really hard, to hopefully scare this thing [because] I had to reverse towards it. I really gunned it out of there and never got dressed until about three kilometres down the road."

Witness interview with Paul Cropper, Apr 2001. Credit: Dean Harrison.

Case 157 February 1984. Near Apollo Bay, VIC. Night.

When he was 21 years old, George Paras, now Head Ranger of the Latrobe University Wildlife Reserve, encountered a strange animal in the Otway Ranges. Rain was pelting down and the track, off Old Coach Road, was narrow and slippery. George, his four-wheel-drive in first gear and lights on high beam, was creeping along very slowly. As he rounded a bend the lights fully illuminated, for about five seconds, the head of an animal standing behind a fern on the side of the track. He passed within just a few feet of it.

It was about five feet tall and its head was covered in long, coarse, brown hair, like that of an angora goat. George compared the "hairstyle" – swept back above the eyes and falling downwards below the eyes – to that of Chewbacca from the movie *Star Wars*. The face was "like that of an ape". Its eyes were huge – the size of a cow's – with dark irises and white surrounds. They followed the vehicle as it moved past. George frankly admits that "fear overtook me". He locked all the doors and kept driving.

This occurred about 30 kilometres south-east of Carlisle River, where a larger yowie was seen about four years earlier. [Case 142]

Witness interview with Paul Cropper, July 30, 2002.

Case 158. February 1984. Noojee, 30 km north of Warragul, VIC. Dusk.

As 25-year-old Jennifer Fiume and friends were rabbit shooting, their headlights illuminated a tall figure 50 metres away in a clearing. As they closed to 20 metres, they saw it was no ordinary animal.

"It was huge … the size of a doorway with a head on top. Covered in dark brown hair, fairly long. The head seemed rather small for the size of the body; just plonked on top of its shoulders. Very thickset legs, arms – like an upright gorilla, with probably a more human stance. It had its back to us, and its hands up in front of its face, evidently doing something to a pine tree. It turned sideways, looked at us as if it was really annoyed, dropped its hands to its sides, took two or three steps and disappeared into the trees. I was in front with the driver and was just dumbfounded. The driver got such a fright that he spun around and drove back to the farmhouse as fast as he could, went inside, slammed the door and wouldn't come out again. He wouldn't talk about it for a couple of years. The guys on the back [had been] shining their spotlights the other way and didn't see anything. They couldn't understand why we turned around so suddenly."

Witness interview with Paul Cropper, Aug 25, 2002. Credit: Dean Harrison.

Case 159. 1985. Port Hacking, NSW. 1 am.

Ten-year-old Adam Bennett and another boy encountered a three-foot-tall, growling, red-eyed, hair-covered "dwarf" in Royal National Park. [See Ch. 2]

A 12-foot-tall yowie was reported at virtually the same spot in 1856. [Case 5]

ARFRA Yearbook of Reported Events in 2001, p. 98. Witness interviews with Dorothy Williams, 2001 and Tony Healy, 2002. Credit: Dorothy Williams.

Case 160. 1985 to 2006. Blue Mountains, NSW.

For more than 20 years, Neil and Sandy Frost have experienced encounters with yowies on and around their property. [See Ch. 4]

Neil Frost, *Encounters with Fatfoot* MS, 1999. Witness interviews with Paul Cropper and Tony Healy, 1994 to 2006.

Case 161. July 1985. Woolamia, near Jervis Bay, NSW. 10 pm.

After setting up camp on their new bush block on Willow Ford Road, the Maron family drove to town for a shower and a meal. As Frank Maron, then 12 years old, remembers it, "We came back around 10. It was a long dirt road ... just widened to about 20, 25 metres; on the sides there were big stacks of logs. Dad was looking for the entrance to his property with the lights on high beam – and this thing started walking across the road. It was about 40 metres ahead at first, closing to 20. At first I thought it was a person in grey overalls, but when we had a closer look it was, like, '*What??*' and we're telling Dad, 'Stop! Stop!'

"[It was] covered all over with long, shaggy grey hair; about six foot tall. Lanky, but not thin. You know Chewbacca, the Wookie out of *Star Wars*? That's what it looked like, even the way it moved. It walked across taking quite large, but quick, steps, its arms swinging to and fro, almost as wide as a cross-country skier, looking straight ahead, not at us.

Dad slowed down and it disappeared behind logs on the other side. We pulled up and took off again to go into the driveway and ... it must have put its head out from behind the logs and its eyes reflected ... two eyes, not huge. I can't remember if they were reddish or just white. Then it tucked its head away again. We were too scared to go closer.

"That night – and previously and for years afterwards – there was this strange noise that used to freak us out, used to send shivers all over your body. Like a howling, a mourning type of call. One little sister wet the mattress that night because she was too scared to go outside to the toilet."

Witness interview with Paul Cropper, 2002. Credit: Dean Harrison.

Case 162. Winter 1985. West of Darriman, VIC. Day.

In 1985 the "X" family moved onto a farm that was bordered on three sides by Mullundung State Forest. One chilly afternoon, Mr. "X" told his bored 15-year-old son "Malcolm" to rug up and take a walk. A short while later the boy ran home in such a state of panic that his father thought he'd been bitten by a snake. Breathlessly, the boy poured out his story: he'd seen a "hairy lady" at Four Mile Creek.

The creature had been standing in the creek bed. It swung around, sniffed the air, coughed, and then looked directly at him. He immediately turned and ran. It was brown in colour, and had no neck, its head sitting directly on its shoulders. As it had sagging breasts it was clearly female.

Mr. "X" wanted to go straight back with him to the spot, but it was two hours before "Malcolm" could be persuaded to return. By then the creature had gone, but Mr. "X" noticed a curious odour "something like chicken broth". On several occasions over the next ten years or so he noticed the same smell around the property, generally in cold weather.

Although there have been no further sightings on his farm, he was told that a hunter took several shots at a small (two to three foot tall) yowie in 1986, in the pine forest at Longford. It may be worth noting that a stream just north of Darriman is called Monkey Creek. Its main tributary is Little Monkey Creek.

Email from Colin Coomber to the Yowiehunters.com website, Feb 4, 2000. Paul Cropper interview with Mr. "X", Nov 7, 2000. Credit: Colin Coomber and Dean Harrison.

Case 163. 1985 or 86. South-east QLD. Dusk.

John Mitchell saw a seven-foot-tall, dark, hairy figure on the road 200 metres ahead of his vehicle. It turned, noticed the car, and darted into roadside scrub. This happened between Tin Can Bay and the Gold Coast, but Mr. Mitchell can no longer remember the exact location.

Witness interview with Paul Cropper, Feb 26, 2001.

Case 164. March 21, 1986. Canungra, QLD. Dusk.

About three kilometres from Canungra Land Warfare Centre, soldier Lester Davison watched a huge bipedal creature bound across the road 30 metres ahead of his vehicle: Covered in shaggy, dark hair, it was two and a half to three metres tall. [See Ch. 4]

Witness interview with Paul Cropper, Sept 17, 2000. Credit: Dean Harrison.

Case 165. 1986. Bian Bian Plains, Barrington Tops, NSW.

American academic Burris Ormsby and an Australian tour guide discovered and photographed five or six large, human-shaped footprints on a remote wild horse trail. The prints were only size 12 or 13 in length, but were unusually wide.

Email from Burris Ormsby to Paul Cropper, Jan 20, 1996.

One of the Barrington Tops footprints. (Burris Ormsby)

Case 166. May or June 1986. Linden, Blue Mountains, NSW. 12:30 am.

While walking home, a 16-year-old Linden resident saw roadside trees moving, then watched a dark figure step onto the road about 50 metres away. Silhouetted by street lighting, it was about 50 per cent larger than a man and seemed completely covered in hair. It walked with a slight stoop. Springwood Police were called but found nothing.

Witness interview with Paul Cropper, Jan 27, 1994. Credit: Neil Frost.

Case 167. 1986 or 1987. Grace's Elbow, Abercrombie River, NSW. Night.

A National Parks and Wildlife Service ranger spoke to a party of fishermen who surprised a huge, hairy figure as it rummaged through their tent. It ran up a steep slope and disappeared. Later, rocks were rolled down the slope onto the tent.

Another fishing party was terrorised by a hairy giant at the Abercrombie in 1876. [Case 22]

Informant interview with Paul Cropper, Mar 14, 1994.

Case 168. 1986 or 1987. Eden, NSW. About 9:30 pm.

When George Fairweather was 16 or 17 years old, he, Michael Innes and Andrew Petrie had a frightening encounter on the southern outskirts of Eden. As they were walking along a narrow track accompanied by Michael's large, rather savage, bull terrier, they all suddenly seemed to sense something was amiss.

"It was almost like a bit of a psychic experience," said George, "because we all seemed to stop together – including the dog. Everybody knew something was not right ... we stopped dead – and *then* saw the thing." The "thing" was a "seven or eight foot tall loping animal" and it crossed the path only 10 or 15 feet ahead. Even 13 years later, just thinking about it gave George "a funny, wobbly feeling" in the legs.

"It took two or three good steps across the track ... towards a gully ... and was gone. It was just a silhouette, really; it was a full moon and a bit overcast, we couldn't see any hair or detail, but it was much taller than any man – even while stooped!" It seemed bent at the knees and had "a peculiar, non-human gait, difficult to describe." There was no vocalisation, or even the sound of footfalls. No unusual odour was apparent. We froze for a good couple of seconds, then turned and walked away about 20 metres, walking as if over broken glass – and then *ran!*"

Eight years later, Maria Speer saw another yowie just a little further south. [Case 200]

Witness interview with Tony Healy, Mar 30, 2000. Credit: Maria Speer.

Case 169. Summer 1986 or 1987, Lake Burrinjuck, NSW. 2 am.

The road between Wee Jasper and Yass crosses the Murrumbidgee arm of Lake Burrinjuck on Taemas Bridge, a sizable steel structure. For James Basham of Cootamundra, the site will always have special significance.

"When approaching the bridge I saw this set of orange-ish lights on the other side moving really quickly. I thought it was a car ... then it dipped down and disappeared. By that time I was on the bridge and I thought it was odd and I stopped to see what it was. I just was sitting there [close to the other end of the bridge] waiting for something to ... and from the left side of my car, probably less than two metres away, something ... I thought I was looking at a human face, but it wasn't, because I saw this huge set of eyes, hair and fangs. His mouth was closed but I could see two fangs from the top down, huge white carnivore fangs, like a tiger or something. And he had a really angry expression. An apeish face. It scared the shit out of me. It was all hairy and wet and I could see a sort of outline. It kept looming up quickly. I could see a big hairy hand grabbing one of the girders.

"This thing had been [hanging onto] the side of the bridge. He was probably assuming I would just drive past and he'd be able to hop out and [continue on his way]. He went up the pylons and down and started walking in front of my car ... he sort of hunched over a bit. He was so close that all my lights were shining into his legs ... I could see legs, one arm [and the body silhouette and eye shine]. Shaggy, like a wet goat. It was close to three metres, four metres [over 11 ft]. It wasn't a person; it was the size of a grizzly bear. He looked down at me and squinted his eyes – they went narrow. These eyes were looking at me – not at the car [and] I was looking up at him. The hairs on the back of me spine went up and, you know, I nearly died."

The creature eventually ran towards the end of the bridge. Then "he was gone ... off to the side. He was heading towards a hilly ravine. There's a few trees, but not many. I was thanking God he wasn't coming after me. I drove off".

It is worth noting that a yowie was reportedly killed about 10 kilometres to the north-east in 1847. [Case 2]

Witness interview with Paul Cropper, January 2005. Credit: Dean Harrison.

Case 170. September 23, 1986. Near Tuena, NSW. Between 2 and 3 am.

A Penrith man encountered an "ape-like creature" while camping. He noticed, with surprise, that the animal had a "shocked" look on its face.

Tuena is 23 kilometres south-east of Rocky Bridge Creek, scene of a dramatic 1876 yowie event. [Case 22]

Email submission to the Yowiehunters.com website, Oct 2000. Credit: Dean Harrison.

Case 171. 1987. 30 km south-east of Glen Innes, NSW. 5:30 pm.

While camped in a rainforest clearing on the banks of Henry River, Queenslanders Michael Beran, 24, and Lloyd Madison, 26, heard an extremely loud "grumbling roar". Then an "immense furry brown creature – half-man, half-ape and at least two and a half metres tall" emerged from the bush. "It stared as if it hated us, then reached up and snapped off an absolutely massive tree branch and hurled it to the ground … we grabbed our rifles and I ricocheted a bullet off a rock. It ran off … that night we sat around a blazing fire, too nervous to sleep." Next morning they found huge footprints.

The following night a cooked rabbit was taken – its alfoil covering neatly unwrapped. The men also discovered an abandoned farm behind their camp, where potatoes were growing wild. "We could see, from the craters everywhere, that someone (or something) had been regularly digging up those spuds."

Taemas Bridge. (Tony Healy)

John Pinkney, "Abominable Potatoman terrorises campers", *People*, Dec 28, 1987. Credit: John Pinkney.

Case 172. 1987. Near Gootchie, QLD. 10:30 am.

In the winter of 1987, Stan Pappin was working on a property situated between the old Maryborough Road and the Bruce Highway. "I was about 37 years old then, in the prime of me life. I was working with Billy Wilson. I'd just cut a tree down; I felled it into the bracken fern and walked up to the stump and then me dog, a bullmastiff, barked and ran back to the truck. And we looked up and this thing was coming down through the bracken fern. I must have nearly felled the pole right on him.

"He was on all fours, but his back was well and truly visible above the bracken fern, which was about a metre tall. It was reasonably black, with nice long hair, about three inches long. His head was down, but when he got within about a stone's throw, 20 or 30 metres, he stood up and walked on two legs to within about five metres of us. He was about 10 or 11 feet tall. I'm six foot, and he towered over me: I looked up at him. The only reason I wasn't running was I had the chainsaw in my hand, still idling. I didn't know whether to hit it or run. Virtually, I think, I froze. But it didn't look like it was going to have a go at us. It was more inquisitive than anything. It stood there, sniffing – you could see its nose moving.

"He was a *big* animal, robust; big and strong. He had muscles in his arms a lot bigger than mine ... he would have weighed a thousand pound or something – like a three quarter grown bullock. When it first stood up I thought it was a human, but it wasn't. It wasn't the right shape. But it walked on two legs like it was just natural – never swayed from side to side, just walked like a human.

"He looked like a big black bear, but his face was too short ... human-like. I reckon from his shoulder to the top of his head was at least 18 inches. His neck was nearly the same width as his head ... a short nose. Little ears, on the side like human ears. You could see the actual skin, like looking into a dog's ear. The eyes were quite shiny but I'm not sure of the colour – I was more interested in watching what he was going to do. His nose was pronounced; you could see his two nostrils no trouble, and he had a front lip, between his nostrils and his mouth, of about an inch. He had his mouth half open ... nice sharp teeth; maybe seven on each jaw, top and bottom. Each tooth was about an inch long. His top and bottom lip seemed more pronounced than his chin ... it's hard to describe. The skin was brown more than black.

"Its arms were a bit like a bloody gorilla ... hung down in front like a big grizzly bear, but he wasn't a bear. His front feet looked ... like hands. I could see these long finger nails or claws ... His shoulders were wide ... huge muscly shoulders, and you could see the muscles in his top shoulder part. There was short hair on its head. Its nose had no hair on it; nothing around his mouth ... he didn't have eyebrows. He had long hair from about his shoulders downwards. On his arms he had hair up to about three inches long. I noticed that on the inside of his arms, where they weren't touching the body, the hair was partly worn off. There was no smell about him.

"It *looked* like a male animal, by the look of its face. He never had a penis hanging out of his belly. Maybe he had one down below, like a human, but he was still standing in the bracken fern that was about a metre high. The bottom part of its torso was of a bear-type ... his body is a lot longer than his legs ... the top half was more like a gorilla, and his head was like a human – that about describes it. I didn't see its face from the side – it was looking straight at us, but its nose would have protruded a maximum of three inches, that's all.

"He stood there for at least three or four minutes before he turned around and walked off, and when he got far enough away he got back down on four legs and away he went ... he cantered off, and he can run reasonably good on four. He wasn't real quick on two. It's got no tail, we noticed that.

"And my mate had froze. He was as white as a sheet. You could have knocked him over, he was that frozen, that solid. He never said a word; his mouth was half open. I don't blame him – I might have been looking the same!

"No one would have believed me, only the old bloke who owned the place, Percy Coyne, he patted me on the shoulder and thanked me for seeing it, because he'd seen it only a week before, and he'd been too frightened to tell anyone about it because they'd think he was going around the bend. And he showed me a young steer that something had ate; pulled it down and ate half of it. This thing was big enough to pull down a half-grown animal, no trouble. It had eaten the rear half – bones and all. The leg bone was bitten clean through, so it must have terrific jaw strength.

"I'd heard people talk about the yowie before and didn't laugh it off. I thought they may be mistaken with something else, but this was probably what they call the yowie – it does exist."

Witness interview with Tony Healy, Nov 2000. Credit: Edwin Stratford.

Case 173. January 2, 1987. Oatlands, TAS. 2 am.

When Stella Donahue and Bill Johnstone, of Kew, Victoria, decided to camp beside tiny Lake Dulverton, on the outskirts of the village of Oatlands, they were expecting to enjoy a nice, peaceful night. They were wrong.

At 2 am Ms. Donahue was woken by a strange, screeching, bird-like noise. Looking out of the tent, she was shocked to see an enormous ape-like creature standing waist deep in the still, moonlit waters, only 18 metres away. It was about 2.5 metres tall, looked silvery in the moonlight, had a head "too small for its body" and was staring straight at her. "I screamed for Bill and he raced out with the torch. By this time the thing was [wading] towards us. We raced to the car and took off" [leaving all their belongings behind].

Several newspapers carried the strange story and rumours of previous encounters soon emerged.

Launceston Examiner, Jan 6, 1987; *People* Magazine, Mar 9, 1987.

Case 174. February 1987. "Yambah" Station, 50 km north of Alice Springs, NT.

When an Aboriginal family reported being frightened by a big, hairy entity at an isolated spot called Top Bore, the local press had a field day. In a lurid front-page article the *Centralian Advocate* said the family, Phyllis Kenny, Frank Burns and their three children, believed the creature was a yowie. Police later found a very tall, obese, longhaired, bearded, mentally disturbed man wandering naked at the spot, and the "mystery" was solved.

A disgusted local landowner, who was present when the man was taken into custody, told us that the story was a complete "beat-up": no one, at any time, thought the poor fellow was anything but human.

Centralian Advocate (Alice Springs) Feb 18, 1987; *The Sun* (Melbourne) Feb 19, 1987; *Sunday Press* (Melbourne) Feb 22, 1987. Witness interview with Tony Healy, Jan 2005.

Case 175. Late 1987. Near Woronora Dam, Heathcote National Park, NSW. 10 pm.

Seven young men reported seeing a two-metre-tall, hair-covered ape-man carrying the carcass of a kangaroo. [See Ch. 7]

Sydney Morning Herald Jul 27, 1987; The *Age* Jul 27, 1987; The *Sun* (Sydney) Jul 28, 1987; The *Telegraph* (Brisbane) Jul 28, 1987.

Case 176. 1987 or 1988. Near Stanthorpe, QLD.

While prospecting along Quart Pot Creek, Cecil Thompson and his sons Ronald and Ross came across two remarkable sets of footprints. Because a yowie had recently been reported in the area, and because he himself had seen one only two miles to the west in 1934 [Case 59] Cecil was pretty sure the tracks were those of the elusive Hairy Men.

"The prints weren't quite as long as I would have expected. Only about ten to twelve inches [25-30 cm] long and about eight inches [20 cm] wide, but they were very deep – and they seemed to have virtually no instep. There seemed to be four toes: the big toe, then a gap, then the other toes at a fork angle almost. They were three or four inches deep in dry sand and would have needed a weight of 250 to 300 kilos [550-660 lbs]. They were angled slightly out from the line of travel, about 10 or 15 degrees, like a person's. [The stride] was about 27 inches [69 cm]. The second set ran almost parallel and were plainly discernable [underwater] in a small running stream. There was no drag of sand or dirt in these tracks in the creek, as if cut with a knife."

The Stanthorpe Border Post, Nov 1, (probably1992). Paul Cropper interview with Cecil Thompson, 1997.

Case 177. April 6, 1988. Mount King Billy, VIC. 8:30 am.

Cryptozoologist Bernie Mace watched through binoculars as a dark, eight to 10-foot-tall creature walked in an "unhurried" manner down a slope several hundred metres away. Mt. King Billy was the site of an apparent yowie "hut invasion" some years earlier. [Case 264]

ARFRA Files report, Apr 7, 1988. Witness interview with Tony Healy, 2002.

Case 178. October 1988. Bridge Creek, 10 km north of Mansfield, VIC. 3 pm.

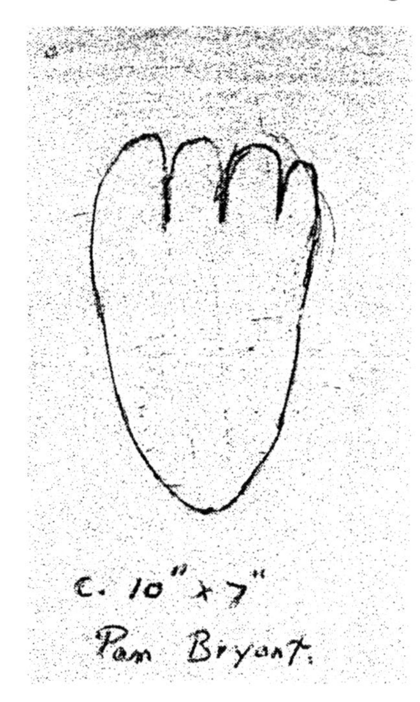

While 39-year-old Pam Bryant was leading trail rides in the area, a property owner showed her some strange footprints. Whatever had made them had come down a four-foot-high bank into a muddy gutter, crossed a dirt road and walked down into the bush on the other side.

"They were *huge*, flat, and only had four toes – big toes. I never contemplated what it could be. I'm a zoology graduate, and I'm ashamed of myself, because I did nothing. I just thought it was a bit of a joke [but] where they were I just can't see that anyone would think of setting up a fake situation. They were there for a very long time because the mud set hard."

A month later and about a kilometre away, Pam "was riding along at walking pace with a beginners' group … I was 50 metres ahead of the first kids, head down, watching for snakes. It was overcast, but excellent visibility. Then I just sort of had the feeling of something there and [looked to the right] and saw these head and shoulders above a bush about 25 metres down the hill. And I just thought, 'Oh! What's a gorilla doing here?' You know how you do a double take? It was like that – like '*What??*' And it just bobbed down behind the bush.

"I'm five foot nine, and it would have been about my height, maybe less. Shoulders as wide as an ordinary bloke's. The head was oval shaped. Long black hair all over everything except the mouth, nose, and the orbital protuberances – the eye sockets had tan on them. The eyes were set in quite deep – a heavy brow. I think the nose was flat, with the nostrils up and outward. A short neck." Her horse didn't react. "I thought about putting [it] down the bank but because I was the trail boss all the other horses would have followed me.

"When I got home – I still hadn't given a thought to the footprints – I rang Melbourne Uni and asked the Head of Zoology if a gorilla could have been let loose in that area. He said no, so I said 'Well, I must have seen a yeti then' – and he hung up. Remembering the footprints, I went to the library to look up gorilla tracks but found, of course, they were nothing like them. When I went to the zoo and looked at the gorillas I realised that what I saw was different – different nose and everything.

"Years later I saw a program about the Chinese wildman – and the Chinese villagers' sketch was the absolute spitting image of what I'd seen."

Witness interview with Paul Cropper, Jul 30, 2002. Credit: Roger Frankenburg.

Case 179. June 1989. Coxs River, NSW. About 4 pm.

While camping in the southern Blue Mountains, Gary Jones and two friends chased a tall, hairy, foul-smelling creature up a steep ridge. [See Ch. 7] Nine years earlier, a group of bushwalkers had encountered a similar creature about four kilometres to the north-east. [Case 277]

Witness interview with Paul Cropper 2003. Credit: Neil Frost.

Case 180. Early 1990s. Rivertree, 40 km south-west of Woodenbong, NSW.

Lismore Northern Star correspondent Rosemary Clarke reported that a truck-driver observed a yowie for five minutes when it came within 20 metres of his timber rig.

Lismore Northern Star, Jan 26, 2002.

Case 181. 1990. Cotter Dam, ACT. 11 pm.

At their campsite just above Cotter Dam, 17-year-old "Len" and two mates were startled by the sound of footsteps in the surrounding scrub. They yelled out – and something called back, making deep moaning sounds. The boys retreated to the park below the dam, where they spread their sleeping bags under a barbeque shelter.

At around 1 am they realised the creature had followed them; it was walking around, quite close by, shaking bushes. They retreated further – towards the nearest building, the old Cotter pub. As they walked they saw their stalker for the first time; it overtook them, running "faster than any human could possibly move" and disappeared into the shadows ahead.

It reappeared and moved into the centre of the road, directly under a streetlight about 50 metres in front of them. It was upright, roughly human-shaped, about the height of a man and seemingly covered in grey-brown hair. It took off again at amazing speed, going so fast that it seemed to leave a blur behind it, "like a cartoon character".

The boys were, by this stage, *very* frightened. Len found himself saying, over and over, to his mates, "they [yowies] are not meant to be here!" They reached the pub, which was locked and deserted, and huddled on the steps, clutching their penknives, unwilling to venture the 150 metres to the nearest forestry worker's house. They could still hear the whatever-it-was moaning and moving around. One boy said, "It's a werewolf!" They were so unnerved that, "L" admitted, they were actually *crying* with fear.

From the phone booth beside the door they called Len's mother, begging her to pick them up. As she arrived, the moaning stopped, but the feeling of dread continued. In fact, it proved contagious; as they drove away, Len's mother was also weeping with fear. All the way home, the boys kept glancing back through the rear window. Even back at home they were jumpy, looking nervously out the windows.

For some time afterwards Len suffered nightmares and, even three years later, felt very ill at ease in the bush.

Witness interview with Tony Healy, Oct 2, 1994.

Case 182. Winter 1990. Krambach, NSW. 2 pm.

As 13-year-old Julie Clark was riding through a steep mountain gully her horse became nervous and reared up. Glancing uphill, she was shocked to see an enormous, shaggy creature crouching only ten metres away. Covered in long, untidy, dark brown hair, it was roughly man-shaped, but immensely broad and muscular. The young rider froze, staring in disbelief, until the animal stood up, looming fully three metres [10 ft] tall, and moved towards her. Turning her horse, she galloped hell-for-leather downhill.

A few months later, on the same hillside, Julie and 11-year-old Jodie Betts fled from what appeared to be the same creature. On that occasion it "let out a terrible high-pitched scream".

Australasian Post, Feb 1991; "The Extraordinary", Channel 7 (Sydney), 1993. Witness interviews with Paul Cropper, Geoff Nelson and Tony Healy, Jun and Aug 1993. Credit: Geoff Nelson.

Julie Clark and Geoff Nelson. (Tony Healy)

Case 183. September 1990. Near Ewen Maddock Dam, 12 km west of Caloundra, QLD. Between 9 and 10 pm.

A couple camped on a bush track noticed that the many crickets outside their tent had suddenly stopped chirping. Stealthy bipedal footsteps then became audible, approaching the tent from the rear, then circling it and the campfire three times. As the male witness, "J" frankly admitted, he and his girlfriend were by this time thoroughly unnerved: "I had a shaking, hysterical female holding a camping knife and I was well on the way to being scared witless. I forced myself to gingerly crawl over and peer out the rear flap. What I saw [about five metres away] froze me where I was. I swear I could not move an inch. It was about three metres [10 feet] tall and very thickset in the shoulders. Also it was hairy. I could not move, I could not speak and I sure as hell thought that if this thing decided to go [attack] the tent it would have ripped it to shreds and us too. Its neck was as thick as its head and it had arms as thick as my legs. Its arms seemed a little long for its body size … it started to move around the tent again, with long, slow, hushed strides." Just then, mercifully, a friend approached the camp on a motorbike: "When the light of the bike [came] over the last rise, I heard a low, guttural growl and heard it bounding down the gully."

The incident had an odd effect on "J" and his girlfriend. Without mentioning it to their friend, they went back to bed immediately and lay awake all night clutching knives and a tomahawk – and never spoke of the matter again.

It was, says "J", "an extremely freaky part of my life. One that I was not able to talk about until recently. I've always been a bush person; I've hunted every type of feral animal … so what I'm trying to say is that nothing in this country puts the wind up me to the point that I become a ball of jelly unable to move due to a fear that I have never experienced since this encounter. Put it this way, I NEVER, NEVER EVER go into the deep bush by myself unarmed. I'm not talking a club or knife either! For what I saw you would want a firearm of some merit and size. These creatures are big, strong, mean and haven't survived all this time, with the Aboriginals before the white men came, up to the present, by being friendly, cute and stupid."

GCBRO Online Reports Submission, Nov 18, 2000. Witness interview with Paul Cropper, Nov 18, 2000. Credit: GCBRO and Dean Harrison.

Case 184. October 1990. Dandenong Mountains, VIC. Night.

A Clematis couple reported seeing a hairy, long-armed "orangutan" dart out from bushes beside a restaurant car park, then swing off into nearby trees. This occurred within two kilometres of the "Black Hulk" incident of 1979. [Case 136]

Hills Trader, Oct 31, 1990.

Case 185. 1991. Letitia Peninsula, north-east NSW. Dusk.

Sixteen-year-old Kyle Slabb and some mates walked to within one metre of a Hairy Man sitting beside Letitia Road. When it looked up at them, they panicked and ran.

Other members of Kyle's Goodjingburra clan have encountered yowies in the area. [See Ch. 4]

Witness interviews with Paul Cropper and Tony Healy, Mar 18 and Apr 30, 2001.

Case 186. Circa 1991. Barwon Falls, near Brewarrina, NSW. Night.

While camping, Bill Gibson, his brother Ron and a mate called Brad, all 10 or 11-year-old Aboriginal lads, saw something strange in a tree. It was "shaking the tree and it jumped out. It was a little Hairy Man; seemed about one and a half metres tall – very thickset. It ran behind the tree and the dog [an aggressive pig dog] chased it, and there was this big thump and the dog came back. The dog's shoulder was broken – it never really came good again."

Witness interviews with Paul Cropper and Tony Healy, Mar 15 and Apr 30, 2001.

Case 187. 1991 to 1998. South-east NSW. Day.

Over a period of about seven years a local businessman experienced a series of yowie encounters at a place in the south coast hinterland. [See Ch. 4]

Witness interviews with Tony Healy, 1998 and 2003.

Case 188. February or March 1991. 1 km north of Lake Buffalo, VIC. 1 am.

Colin Baker, of East Warburton, and two friends were spotlighting on the Yarrambulla Track when their lights illuminated a strange animal about 100 metres away.

"I had a one-million-candle power spotty and it was a clear, starry night. Visibility 100 percent. We had a brilliant view of it. It was like a cross between a great big hairy man and a gorilla, walking parallel with us. It didn't seem too worried at all. The small pines were very sparse and no undergrowth. It had black or brown hair from top to bottom; no bare skin. Egg-shaped head; looked like it didn't have a neck – just a head on very, very wide shoulders. Seven to eight feet; taller than the tallest man I've seen – very solid. Arms looked like they went right down to its knees. Walking like a soldier: arms swinging to the full extent. In the last few years I've read a lot about yowies. That's what I saw. Thinking about it still sends shivers up the back of my neck."

Witness interview with Paul Cropper, May 29, 2000. Credit: Dean Harrison.

Case 189. September 1992. Georges River, near Ingleburn, NSW. 2 pm.

While heavily camouflaged, lying in ambush during a war game, 16-year-old Cole Dwyer and a mate were amazed to see a huge, bipedal animal moving slowly through nearby scrub. "It had no tail or long snout, was at least eight feet tall and covered in dark brown-black fur, short fur like a dog's, very big body but ... no neck." They watched it for about four minutes before it noticed them. Then it immediately "ran uphill through dense bush incredibly fast. It seemed so scared of us, it seemed to be running for its life. We followed its trail. There were small trees snapped like twigs, two larger trees were uprooted.

"I think about [this event] all the time and would be happy to sit [for] a polygraph test, and I'm sure my friend would too."

Email to AYR website. Witness interview with Paul Cropper, Sept 2004. Credit: Dean Harrison.

Case 190. October 1992. Tewantin, QLD. 2 am.

A woman told Noosa police that she'd seen an eight-foot-tall, hairy creature in Forest Drive, near the Noosa-Tewantin golf course. Police searched the area but found no footprints. A police spokesman said there had been previous sightings, but this was the first for some time.

Noosa News, Oct 13, 1992.

Case 191. Circa 1993. Near Nowendoc, NSW. 11 pm.

Tamworth residents Buddy Knox and Ray Berry were driving slowly along a dirt road when, just beyond the reach of their lights, something crossed from left to right. As they got closer it retraced its steps.

"There was a big bushy tree on the left, and as the lights hit [the animal], it went behind the tree. We saw it from behind. It had no neck; big hairy back; big animal. It looked pretty solid. And we just *shit* ourselves. I'm six foot and 17 stone. It'd be bigger. It reminded me of a standing-up bear. More brown than black; streaks of black. Shaggy, like a goat. It didn't look at us. Crossing the road [the second time] it took three steps. One, two, three – *gone*. Huge steps. For its size it was light on its feet. Even thinking about it now makes you scared, you know."

Interestingly, even though both men are Aboriginal, it didn't immediately occur to them that the creature was a yowie. "We seen it and said, 'what the f*** was *that*?'"

Another yowie was seen near Nowendoc in 1979. [Case 276]

Witness interview with Paul Cropper, 2002. Credit: Dean Harrison.

Case 192. 1993. Between Port Macquarie and Walcha, NSW.

A surveyor told Port Macquarie wildlife consultant Kel McKie that he'd seen and chased a foul-smelling yowie on the Oxley Highway.

Informant interview with Paul Cropper, Feb 8, 1995. Credit: Kel McKie.

Case 193. 1993. Near Culburra, about 8 km south of Nowra, NSW. Day.

A Helensburgh woman and her husband saw a large, bipedal animal on a bush track. It appeared to be covered in beige-coloured hair, took "damn big steps" and seemed to be "moving in slow motion". This occurred within 10 kilometres of the sites of two other yowie incidents: one involving Henry Methven in 1901 and another involving multiple witnesses in 1985. [Cases 39 and 161]

Letter to Tim the Yowie Man, Jun 18, 2002. Credit: TYM.

Case 194. 1993. Blue Mountains, NSW. Night.

In bright moonlight, 500 metres from the yowie-plagued Frost residence [Case 160], Ian Price saw a yowie cross a fire trail 20 metres in front of him. [See Ch. 4]

Witness interview with Neil Frost 1993. Credit: Neil Frost.

Case 195. July 1993. Waste Point, Lake Jindabyne, NSW.

Two anglers reported finding a set of strange footprints one frosty morning on a remote shore of Lake Jindabyne. The bipedal prints were human-like, and about the size of a small man's, but instead of toes they exhibited the imprints of six claws. They wandered along the shoreline, entered and exited the water, then headed into the scrub of Kosciusko National Park.

About 14 years earlier, while fox shooting at the back of Grosses Plain (about 20 km to the south-west), three Jindabyne men found strange "size nine" tracks with indentations rather like claw markings. Whatever made the tracks appeared to have killed and eaten a rabbit. They dubbed the mysterious creature "The Buffer".

These odd tracks may or may not have been made by yowies, but the creatures have certainly been reported in the general area. [See Cases 202 and 215]

Undated Cooma newspaper clipping, c 1978-80; *The Chronicle* (ACT), Jul 19, 1993. Credit: Bryan Pratt.

Case 196. Late 1993. Blue Mountains, NSW. Night.

Nine-year-old Jeffrey "X" walked to within a few feet of a huge yowie. [See Ch. 4]

Witness interview with Neil Frost. Credit: Neil Frost.

Case 197. Circa 1994. Near Carnarvon Gorge, south of Springsure, QLD. Dawn.

Timber worker Paddy O'Connor said two very small Hairy Men approached his campfire, uttering "chirping" sounds and pointing at his billycan. [See Ch. 5]

Witness interview with John Pinkney; John Pinkney, *Great Australian Mysteries*, p. 32. Credit: John Pinkney.

Case 198. May 1, 1994. Mount Franklin, Brindabella Mountains, ACT. Dusk.

While engaged in environmental research, a university student who later became known as Tim the Yowie Man encountered a massive ape-like creature. [See Ch. 4]

Interview with Tony Healy, Apr 1995; *The Adventures of Tim the Yowie Man*, pp. 3-4.

Case 199. May 1994. Gangerang Range, Blue Mountains, NSW. 2 am.

While camped with his brother on the edge of Tiwilla Tops in Kanangra Boyd National Park, "M", a Bathurst schoolteacher, was woken by what sounded like human footsteps. The intruder stopped about six paces away and "M", propped up in his sleeping bag, could make out a five to five and a half foot tall figure, shaped like a very powerfully built man. After standing stock still for about five minutes, it moved to one side and stood still for a further ten minutes before skirting around the campsite and walking away.

In the morning the brothers tracked it 300 metres to Hundred Man Cave. When they returned two weeks later, they found "a superb sleep site, constructed of grasses, leaves and long saplings together under a slight overhang". In April 1995, they heard eight or nine screams in the same area.

This occurred within about 10 kilometres of where Gary Jones and party chased a yowie in 1989. Other sightings occurred nearby in 1891 and 1980. [Cases 144, 179 and 274]

Witness interview with Paul Cropper, 2000.

Case 200. 1994 or 1995. Eden, NSW. Day.

While fossicking in a quarry six kilometres south of Eden, Maria Speer, now of Morwell, Victoria, noticed a more than two-metre-tall creature watching her from nearby bush.

"It was brown, thickset and short-necked with powerful, solid shoulders. It was standing upright on two legs and when it saw me it crashed off into

Sketches by Maria Speer.

the bush. My impression was that I had seen a powerful man-like creature. America's Bigfoot would be an identical type."

Sunday Herald Sun (Melbourne), Feb 20, 2000; letter from Maria Speer to Paul Cropper, Feb 27, 2000.

Case 201. Summer 1995. 40 km south of Darwin, NT. 9 pm.

Shortly after Katrina Tucker's 1997 yowie sighting [Case 220], local businessman Richard Kingsley said he'd seen what might have been a similar creature about two years earlier. On the Stuart Highway, midway between the Crocodile Farm and the Elizabeth River Bridge, he saw "a black, shimmering, but not shiny black, figure move straight out from the bush and without hesitation cross the highway ... go directly into the bush and disappear from sight. It was about one-and-a-half-times the size of a man, say eight to nine feet tall, and it loped from one side of the road to the other without any hesitation". He couldn't make out whether or not it was hair-covered.

In about August 1996 he saw another "very, very tall", black, roughly man-shaped creature at the same place. The second creature, however, was extremely thin. "It wasn't a person," he said. "It gave me the fright of my life."

Litchfield Times, Aug 28, 1997; *Who* Magazine, Oct 27, 1997. Witness interview with Paul Cropper, 1997. Credit: Tim the Yowie Man.

Case 202. Summer, mid-1990s. Sawpit Creek, near Jindabyne, NSW.

An elderly couple told National Parks and Wildlife officers that they saw a black, hairy, seven to eight foot tall figure cross the road only 500 metres from the Sawpit Creek Visitors Centre.

Strange, bipedal, non-human tracks were reported in the same area in July 1993. [Case 195]

Informant interview with Paul Cropper, Oct 21, 2000.

Case 203. January 22, 1995. Ballengarra State Forest, Kempsey, NSW. 5.30pm

While walking along a dirt road, two local boys heard heavy footsteps, then saw a dark brown or black, hair-covered creature a few metres away in the scrub. It seemed to be "between a human and a gorilla", and was hunched over and facing away from them. When it straightened up, looming to eight or nine feet, and turned its head slightly as if sniffing, the boys fled. Two weeks later, Paul Cropper photographed and cast a trail of 16 bipedal imprints at the site. They had the general shape of large, flat, very wide human footprints. The toes were indistinct. Each track was 30 centimetres long, 18 wide [12 x 7 inches] and three to four centimetres deep. By stamping the heel of his boot Paul could make an impression no deeper than two centimetres. A Port Macquarie wildlife consultant who also examined the impressions estimated they were created by something weighing around half a ton.

The incident occurred within a couple of kilometres of the sites of several colonial and early modern era yowie encounters. [e.g. Cases 16 and 53] Later, a dog owned by local researcher Dave Reneke barked and snarled when close to the footprint casts, possibly reacting to odours in soil adhering to the plaster. At least one other dog has reacted to a cast in the same way. [Case 160]

Witness interview with Paul Cropper, Feb 4, 1995. Credit: Dave Reneke.

One of the imprints. (Paul Cropper)

Case 204. April 16, 1995. Millaa Millaa, about 40 km west of Innisfail, QLD. Day.

Ron Cairns of Edmonton and his wife saw what they took to be a grey-brown monkey 12 metres up a tree. The animal, which had long arms and a round face, watched them for about three minutes before walking back into the canopy.

Cairns Post, Apr 22, 1995.

Case 205. 1995-2004. Yellow Rock, Blue Mountains, NSW. Dawn.

For almost 10 years an elusive creature tore bark off trees, evaded an infrared surveillance system and roared like a lion on Lynn and Gordon Pendlebury's property. Mrs. Pendlebury eventually sighted a four to five foot tall, "overgrown monkey on two legs". [See Ch. 4]

Witness interviews with Neil Frost, Tony Healy and Mike Williams, 1999-2004. Credit: Neil Frost.

Case 206. Winter 1995. Near Bathurst, NSW. Noon.

On a remote, forested mountainside, nine-year-old "BZ" and three friends saw a dark figure climbing down a tree about 40 metres away: "kind of hugging the tree, like when people climb coconut trees [but] it was quite obvious it wasn't human. A great big hairy beast, like a gorilla. Its back foot hit the ground and it *stepped* back onto the ground".

It was more than six feet tall and, except for some of its face, was covered in dark brown hair resembling dog fur. It had a big head, large eyes, a thick neck and was as powerfully built as a wrestler. It stepped away from the tree, looked around, appeared to notice them, "slowly ducked down as if frightened and then stood up again, staring at us. We were no threat; we were small". The boys ran away. At one point "BZ" stopped briefly and looked back. The animal was still standing beside the tree.

Witness interview with Paul Cropper, Aug 20, 2002. Credit: Neil Frost.

Case 207. 1995 or 1996. Lake Wivenhoe, QLD. 8:30 pm.

A man wrote that a family friend saw a strange creature resembling a "large dog walking on two legs". The witness was outside his house when the hair-covered animal walked in front of him, jumped a barbed wire fence and ran into the darkness. He said its face was "strange or deformed".

Several other reports have come from the same area. [e.g. Cases 141, 238 and 248]

Email to Paul Cropper, Mar 12, 2002.

Case 208. Summer 1996. Clyde Mountain, NSW. Early afternoon.

Peter and Belinda Garfoot of Elmore Vale, Newcastle, saw a seven-foot-tall, hair-covered, bipedal animal cross the Kings Highway. [See Ch. 4]

Witness interview with Paul Cropper, 2002. Credit: Dean Harrison.

Case 209. Circa 1996. Tabulam, 50 km west of Casino, NSW.

Four Aborigines from Tabulam mission reportedly encountered several Hairy Men on the Tabulam Bridge.

Email submission to AYR, Jul 26, 1999. Credit: Dean Harrison.

Case 210. March 23, 1996. Upper Main Arm, north of Mullumbimby, NSW. Afternoon.

Ten-year-old Joshua Clark observed an upright creature running headlong down a steep, forested slope on a property rented by naturalist Gary Opit. When it stopped about 30 metres away he and his mother Lynn saw it was 1.5 metres tall, covered in thick black hair, with a round head and no tail. After staring at them for a few seconds, it dropped down, knuckle-walked briefly, then stood up again and ran away.

At 3:30 am on June 1, Gary Opit heard a long series of loud, bark-like vocalisations, interspersed with "a

Sketch by Joshua Clark.

disturbingly strange, soft, gurgling call". The sounds, he felt, had "a primate feel to them". Next day he found the impressions of three large toes, plus a five-square-metre area where many clumps of native grass had been uprooted "and then placed back exactly where it had grown".

In 2001 a larger yowie was seen on an adjoining property. [Case 272]

"Understanding The Yowie Phenomenon", Gary Opit, *Nexus* Magazine, August-September 1999. Witness interviews with Gary Opit and Tony Healy, March 23, 1996 and Oct 25, 2000. Credit: Gary Opit.

Case 211. April 1996. Near Aberfeldy, VIC. Night.

While some distance from his school campsite, 15-year-old Jordan Riley heard something pushing through the scrub. "This large ti tree thicket ... about three inch diameter trunks ... something was just pushing it out of the way – just like wading through water. Then this big figure just walked straight past ... casually, like it didn't even notice me. Just like it didn't even bother ... it would have been, like, 10 metres away, walking up the hill."

Although he didn't have a torch, Jordan was sure of what he'd seen: "Visibility was clear. I had it in view for about four or five seconds. I got the impression [that it was hairy] ... it was probably about six foot three tall."

Witness interview with Paul Cropper, Nov 2000.

Case 212. Late 1996. Near Townsville, QLD. 3 am.

During an escape and evasion exercise in the ranges near Townsville, "H", a 22-year-old soldier, had an unforgettable experience:

"We were hunting down participants, and had been laying up most of the night conducting periodic sweeps by twos ... wearing NINOX (monocular night vision gear). Sometimes to get attention in silent running, you click your tongue ... my partner 'N' had just done that ... I swung around to where he was pointing his weapon.

"About 15 metres, maybe a little more, stood the outline of a very large hairy man, standing still and looking, if not right at us, then in our general direction. 'N' slowly went down on one knee, as I did, with our weapons on the threat (loaded with blanks only, of course) and 'N' turned his head slightly and mouthed 'What the f***?' [It] was swaying slightly from side to side, there was no smell at all. It was about seven to seven-and-a-half feet tall. Short, tough-looking hair all growing downwards, powerful but not overly powerful looking, with proportionately larger legs than a human ... eyes somewhat closer together than I would have expected, but perhaps this was a trick of the NINOX, and it seemed to have its head tilted up, sniffing the air.

"Thank God we were upwind ... we observed it for about 25 seconds (remember, we are trained to be very observant in short periods). Then from off to our left there was the sound of strained voices and the rustling of undergrowth ... the yowie's whole body seemed to go tense and the air was full of that disgusting smell like rotting meat, but worse, more primal, somehow scary and sickening ... it gave a surprised sounding, but also menacing grunt, turned slightly towards the sound, then began to trot in a ground covering lope straight at us. 'N' must have panicked a bit then, as the next thing I saw was a bright flash as my NINOX blanked from the muzzle flash of his weapon. There was a very loud, growling scream from the yowie and in the towering silence ... we heard footsteps receding into the distance. We [investigated] to the left and caught two pilots in a small creek bed."

"H" added that he and his colleagues have also encountered the "rotten meat" smell during jungle training at Canungra Land Warfare Centre in the Gold Coast hinterland.

Email from "H" to AYR website. Credit: Dean Harrison.

Case 213. 1996 to 1998. Between Torrens Creek and Pentland, QLD. Night.

When he drove trucks from Mt. Isa to Townsville three times a week, Rossco Macrae saw yowies on three occasions. All the sightings occurred just east of Torrens Creek, where the Flinders Highway crosses the Burra Ranges in White Mountains National Park. They looked like "big apes". One "was standing beside the road trying to shake something off an electric light pole. [It] was about three metres tall and had hair all over it". Another "was carrying a dead 'roo".

One rainy night another driver, Wally Thorley, parked at the foot of the ranges, stretched out a tarpaulin and went to sleep on a folding bed beside his truck. "My dog was next to me and I was woken up about 3 am ... a huge hairy thing had jumped on me and tried to suffocate me. My dog [bit it and was] going berserk ... I managed to get free. It ran off up the highway and I tried to get a shot at it but it disappeared. I won't sleep there again."

White Mountains National Park contains a great quantity of ancient Aboriginal rock art, including, in a place called the Yahoo Gallery, several engravings of huge five-toed tracks.

"Wowie! It's a yowie!" by Alf Wilson, *Blues Country*, 1998; "Putting the yowie legend to the test" by Alf Wilson, *Sporting Shooter*, Apr 2005; Graham Walsh, *Australia's Greatest Rock Art*. Credit: Rebecca Lang.

Case 214. 1996 or 1997. Wentworth Falls, Blue Mountains, NSW. Night.

At the isolated Queen Victoria Memorial Hospital, a male nurse and his wife encountered a yowie on two occasions. [See Ch. 4] The hospital (now derelict) is located on Tableland Road, where Justin Garlick encountered three yowies in Aug 2002. [Case 250]

Witness interview with Neil Frost. Credit: Neil Frost.

Case 215. January 1997. West of Thredbo, NSW. About 4.30am.

After reading an article about the Hairy Man, John Lythollous of Albury broke three years of silence about his own encounter:

"I left Jindabyne at 4 am via the Alpine Way ... About five to 10 minutes past Thredbo, I was negotiating a sharp left hand turn with my lights shining on the upside of the mountain. Something came down the hill at pace towards the vehicle in an aggressive manner. In other words, it tried to attack the car. The 'something' appeared to be about six-foot-tall, very thin, small head, black hair all over, walking as we do but with very long arms. The only thing I have ever seen that slightly resembled what I saw, was a gibbon monkey at the Adelaide Zoo.

"I have spent a lot of time in the bush and have seen a large slice of Australia, but I have never seen anything like this before or since ... [it] scared the pants off me."

Email from John Lythollous to Tim the Yowie Man, Sept 12, 2000. Credit: TYM.

Case 216. April 1997. Brown Mountain, NSW. 11 pm.

Ranger Chris McKechnie of Pebbly Beach almost collided with a seven-foot-tall, ape-like creature on the Snowy Mountains Highway. [See Ch. 4]

This occurred about four kilometres from the site of Mr. Summerell's 1912 yowie encounter. [Case 42]

Witness interview with Tony Healy, Dec 6, 1998. Credit: Lisa Stack.

Case 217. Autumn or winter 1997. Camira, QLD. Day.

While playing with friends, 12-year-old Keane Wisniewski saw the back of a black, hairy animal like a "little gorilla" holding on to the trunk of a tree. When he returned later with his father, they found claw marks four or five metres up the same tree.

Camira, on the south-western edge of Brisbane, borders the 50 square kilometre Greenbank Military Camp.

Witness interview with Paul Cropper, Apr 30, 2001.

Case 218. July 1997. Ormeau, QLD. 11 pm.

Dean Harrison was stalked and then chased by a huge, roaring ape-like creature that seemed intent on ripping him to pieces. [See Ch. 4] In April 2003, only 2.5 kilometres away, a similar creature menaced Jason Cole. [Case 256]

Witness interviews with Paul Cropper and Tony Healy, 1999-2004.

Case 219. Mid- to Late 1997. West of Cowley Beach, near Innisfail, QLD. Afternoon.

While carrying two sets of equipment on a jungle training exercise, army reservist Richard Easton of Salisbury East, South Australia, lay down for a brief rest. Although a few rays of sunlight penetrated, it was dim under the jungle canopy. Lying on his stomach, with an M60 machine gun in front of him and a heavy radio on his back, he felt a touch on the back of his head. "It was very firm, like the tips of a few fingers. It pushed … a little pressure, but it was gentle." Almost simultaneously, he became aware of an extremely foul odour, "like a bin full of rotten meat".

He craned his head around "and saw its silhouette … then it moved around to about three or four metres in front of me, and I could see that it was short, and extremely, unbelievably, broad and muscular, about five feet tall and three feet wide … and reddish-brown hair all over – a couple of inches long. Its head looked like a half deflated [rugby] football. It was like an apex; it came down onto its shoulders. It didn't seem to have a neck. It walked funny: a bit of a waddle, from side to side, and dropping its head down as it was going. It must have been heavy, built like that [but] it was virtually silent; I didn't hear a thing; a few small sticks cracking – that was it."

Although most of the body was essentially just a silhouette, he could see the feet quite well. They were roughly human-like but extremely broad: "great big fat things. Huge. I can't say how many toes [but] some images stick in your head. I focussed on the big toe. It was, like, two inches wide and splayed out, way bigger than the other toes. The side of its foot, too, was really padded. The skin looked dark, but it could have had mud on it. I wasn't scared; I was very curious". Sweat ran into his eyes and he ducked his head in an effort to clear it. When he looked up again the creature was gone. He immediately examined the ground and undergrowth, but found no footprints or hair.

"The whole incident lasted a minute, max, and nobody believes me. I didn't think 'yowie' at first, not until much later. This is not bullshit, I know what touched me, and I have 20/20 vision. They *are* out there."

Witness interview with Paul Cropper, Jan 18, 2001. Credit: Dean Harrison and AYR.

Case 220. August 1997. Acacia Hills, NT. 3 am.

Starting in about 1992, Katrina Tucker and her husband heard strange cries emanating from a wild region to the east of their mango farm. They were high-pitched, like those of a howler monkey. Whenever they occurred, her Doberman and German shepherd guard dogs "started to cry". In mid-1997, cows began jumping fences; horses became spooked.

One night, Katrina was woken by the cries, which sounded closer and more distressed than ever before. Fearing an animal had become entangled in a fence, she jumped on a four-wheeler bike and rode out to investigate. Her dogs refused to accompany her, and she soon discovered

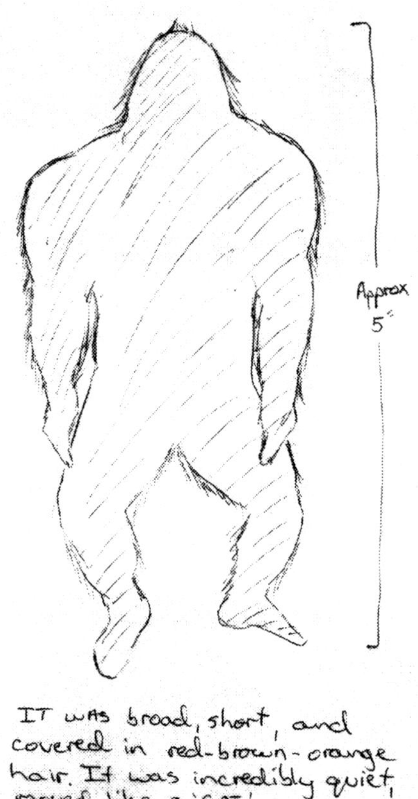

Sketch by Richard Easton.

why: speeding along the fence line, she almost collided with a huge, screaming ape-man.

It was hanging onto the fence, leaning back as if attempting to tear it down. As the bike slewed to a stop about eight feet away, it stood up, looming to nearly seven feet. Covered in matted, dull, dark reddish-brown hair, it had very long arms, sloping shoulders and no discernable neck. It seemed more animal than human. Katrina was hit with a smell that made her dry-retch. It resembled a cave full of bats, urine, or a badly kept hen house.

The animal turned and ran. Its arms did not "pump" as a human's would, nor did it bend its elbows. It ran by bending its knees, and swayed from side to side. Although it took large strides, it appeared almost to run in "slow motion". Katrina knew nothing of the yowie phenomenon and assumed the creature was some kind of great ape that had escaped from a menagerie.

Whatever it was, she hadn't imagined it; in the morning she found a line of fairly well defined footprints at the site. They were very strange indeed. 310 by 160 centimetres [12 x 6 ins], they displayed three long, sausage-shaped toes, a small "thumb" protruding from the instep and rather "squared-off" heels. They were inspected by local naturalist Ian Morris and cast by Karen Coombes of the Northern Territory Museum. Although neither doubted her honesty, Mr. Morris and Ms. Coombes both thought Katrina must have been hoaxed. The police

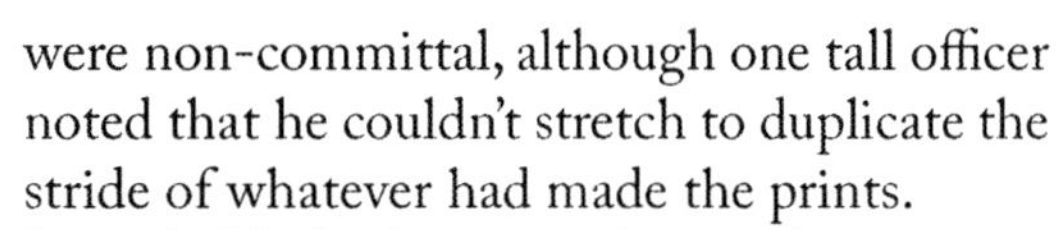

were non-committal, although one tall officer noted that he couldn't stretch to duplicate the stride of whatever had made the prints.

Improbable-looking as the tracks were, it is difficult to imagine how a hoaxer could produce screams calculated to unnerve dogs or generate a smell foul enough to make a person almost vomit. The hoaxer would also have to have been extremely tall and very well kitted-out with a fur suit and false feet. He must also have been very foolish and very lucky. Acacia Hills is a wild area and mangoes are valuable fruit. Katrina has been known to set her dogs onto intruders. She drove one away with gunfire.

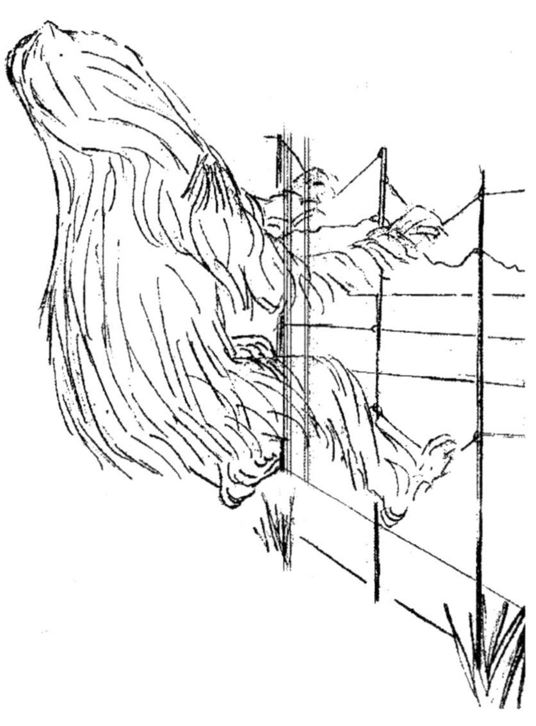

Sketches by Katrina Tucker.

It transpired that there had been at least three other sightings in the area. Another occurred soon afterwards. [Cases 201 and 229] The Tuckers continued to hear the yowie's cry from time to time and on one occasion it left them an unsolicited gift: very close to where Katrina saw the animal, they found a devastatingly foul, cigar-shaped dropping of gargantuan proportions.

Litchfield Times (NT), Aug 14, 1997. *Who Magazine*, Oct 27, 1997. Witness interview

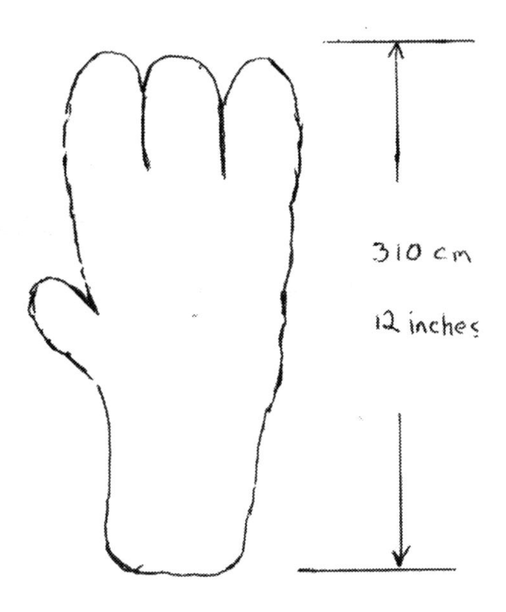

with Paul Cropper, Dec 8, 1997. Credit: Tim the Yowie Man.

Case 221. Mid-1997. Mount Lindsay Highway, northern NSW. Dawn.

Mark Pope saw a one-metre-tall, hair-covered, bipedal, chimpanzee-like creature not far from Woodenbong. [See Ch. 5]

Many reports of "normal-sized" yowies have come from the same area. Also worthy of note is the fact that Mark claims to have seen another, much larger, ape-like creature near Bexhill in 1973. [Case 91]

Witness interview with Paul Cropper, Jan 30, 2000. Credit: Dean Harrison.

Case 222. August 9, 1997. 20 km south of Alice Springs, NT. Sundown.

"On the late afternoon of Saturday, August 9 my wife and I stopped in our campervan to have a rest ... about 20 km south-west of Alice Springs ... there was still quite a lot of light in the western sky. We were walking not far from our van when we became aware of what seemed to be a very large person standing 30 or 40 metres away, the size made it hard to estimate [sic]. When I called out there was no answer. It was hard to say if it was looking at us or the sunset. Pam and I went quietly back to the van and got in and then found the large being had come closer. It had been on the other side of a fence ... it must have stepped over. It looked to be over three metres tall. Before we wound up the windows we were aware of a terrible odour. While I was fumbling to get the key into the ignition the very large person (?) crouched low then sprang sideways and ran very rapidly into thick trees and out of sight. We weren't able to see any features or colour but think it was quite dark and may have had short hair or fur. We left in a hurry."

Letter to the *Centralian Advocate* from Douglas Evans, of Dubbo, NSW, Aug 15, 1997.

Case 223. September 1997-2006. Blue Mountains, NSW.

Since moving into their house on the southern edge of a Blue Mountains township, Jerry and Sue O'Connor have experienced a wide range of yowie activity, similar to that experienced by Neil and Sandy Frost [Case 160] who live on the northern edge of the same village. [See Ch. 4]

Witness interviews with Dean Harrison, Neil Frost, Paul Cropper, Tony Healy, David Mc Bean and Mike Williams, 1997-2006. Credit: Dean Harrison.

Case 224. 1997 or 1998. Between Colac and Warrnambool, VIC. Midnight.

While driving towards Warrnambool, a Melbourne man spotted a "small hairy man" walking beside the road. About four feet tall and covered in brown or black hair, it glanced up with a "shocked look" in its eyes.

Witness interview with Paul Cropper, Mar 2001. Credit: Dean Harrison.

Case 225. 1998. South-east NSW. Night.

"Sue" encountered a foul-smelling yowie on the edge of her village. One of her neighbours saw similar creatures repeatedly over the course of several years. [See Ch. 4]

Witness interviews with Tony Healy, 1998.

Case 226. February 2, 1998. Gatton, QLD. 11:30 pm.

This incident involved a family that had recently migrated from the Middle East. They had absolutely no knowledge of the yowie phenomenon.

It happened as 54-year-old Mr. "N", his wife and their 21-year-old son were driving home to Gatton from Toowoomba. When they reached the outskirts of town, they drove along a sealed road that was bounded on the left by farmland and on the right by suburban housing. There was no street lighting, but visibility was good; the moon was out and the car's headlights were on high beam.

Suddenly a large animal appeared on the left, and, when the car was only 15 or 20 metres away, dashed across the road. It ran like a man, but much faster, leaning slightly forward, with its arms moving slightly but not pumping as a man's would. It was seven or eight feet tall, but rather slim. Its arms seemed proportionately longer than those of a man. The body was covered in sparse brown fur. They didn't notice any neck. There was nothing to indicate its sex, but they had the impression it was male. It ran lightly, apparently on its toes, and its feet seemed different to those of a man. The heels looked odd, but they couldn't say exactly why. It was all over in seconds, but, as the son put it, the image of the creature was burnt into their brains.

Interestingly, when last seen, the creature was running into the front yard of a normal suburban house. Because of the size and strangeness of the animal, and because they saw it only a couple of hundred metres from their own home, the family were so terrified that they didn't speak about the incident for days afterwards.

When researcher Malcolm Smith accompanied them to the site three weeks later, he saw that the field from which the creature emerged was planted with vegetables. The animal must have crossed a metre-high fence to reach the roadside. In the field close to where it was seen was a row of beehives. Malcolm also noted that a tree-lined creek on the far side of the field runs into nearby Gatton State Forest.

Sketch by Mr. 'N'.

Witness interviews with Malcolm Smith, Feb 21, 1998. Credit: Malcolm Smith.

Case 227. Winter 1998. Near Springbrook, QLD. Dusk.

During an Australian Yowie Research expedition, a team member, Steve Bott, walked to within 20 metres of a huge yowie crouched on a bush track. [See Ch. 4]

Witness interview with Dean Harrison, winter 1998. Credit: Dean Harrison.

Case 228. September 1998. Scotchy Pocket, QLD. 5:30 pm.

While working alone about 15 kilometres from Gunalda, 20-year-old Bradley Stratford twice observed a large, hairy, bipedal creature:

"The property is a cane farm but across the river there's a lot of thick scrub. It was getting on towards dark; the sun had gone down. I was working under a tractor when the hair stood up on the back of me neck. You know – the feeling like you've walked over someone's grave? And I turned around and this big, brown, bloody thing was running across the yard. And I thought, 'Oh, get into it, you idiot – you're dreaming. Imagining things'.

"So I turned around again and was hammerin' away again for about four or five minutes. I was thinking that I must have been working too long that day. And then me hair stood up again, and I looked, and it was running back the other way. It must have seen me, but it didn't seem to worry about me in any way."

Both times Brad saw the creature it was about 20 to 30 metres away. It seems to have emerged from a lightly treed paddock, crossed a fence and run 20 metres across a grassy area to the unoccupied farmhouse, which was partly obscured by shrubbery. There were no fruit trees or fowl yards, so Brad couldn't imagine what had attracted it. When it ran back, it disappeared into the shrubs from which it came, presumably crossing the fence again.

Bradley Stratford. (Tony Healy)

"It had reasonably straight hair ... nearly all covered, but I think it was longer on the arms than on his back. A tanny-brown colour, probably a bit darker than those rawhide boots, you know? I'd have to say it was about seven and a half or eight feet. It was pretty heavy built. I'm about 70 kilos ... it'd be two feet taller than me and at least twice my weight. It looked like a big, hairy person. It had a biggish head ... I didn't get a look at its face [but] hair was hanging down around the face.

"[It ran] probably as fast as a man, if not a bit quicker. It was pretty straight up and down like we run [not slumped]. I didn't see the shape of the feet but they looked pretty big – I could see them when he was running. They had long hair on them. [Its arms] were pumping. They were just a touch longer, I think [than a human's]. The yard is pretty heavily grassed so I didn't find any footprints.

"I worked there for another month or so but never saw it again. I told a few people around here about it, but they look at you silly ... Someone reckoned there was something up there a couple of months back around Scotchy Pocket or Gunalda area – I think it was supposed to be a yowie."

Witness interviews with Paul Cropper and Tony Healy, Nov 2000. Credit: Edwin Stratford and Dean Harrison.

Case 229. November 18, 1998. Arnhem Highway, NT. 10 pm.

YETI BEAST SPOTTED AT
ADELAIDE RIVER

"Carpenter Darryl Campbell 32, was driving along the Arnhem Highway on the Adelaide River floodplains at 10 pm last Wednesday when he sighted the yeti-like beast in the bush. He and a friend slowed down when their vehicle approached it. He said, 'It was like a bloody big gorilla or ape. It was crouched down on the ground and hobbled along holding grass and other junk in its hands'. He said it stood about the height of a man and he pulled his vehicle over as a group of European tourists had also stopped. He said, 'They saw it too and stopped to ask what it was. They were shaken up. I turned and drove back towards the animal but got scared and tore off myself". Mr. Campbell and the group of tourists reported the sighting to Transport and Works Department traffic controllers at the Adelaide River bridge."

This occurred about 20 kilometres north-west of Katrina Tucker's farm. [Case 220]

Northern Territory News, Nov 23, 1998.

Case 230. October 23, 1999. West of Caboolture, QLD. About 2.30 am.

After stopping his car on the D'Aguilar side of the Mount Mee Road, a Kallangur man heard a loud, echoing scream. About two minutes later, he saw a large, hairy figure about 50 metres away. Illuminated by moonlight, it seemed to be about 6 feet 5 inches tall. It ran into the bush. As he drove away, the man glimpsed what was apparently a similar creature kneeling at the side of the road.

Email submission to the yowiehunters.com website, Mar 14, 2000. Credit: Dean Harrison.

Case 231. November 1999. Inglewood, 95 km west of Warwick, QLD. 11 am.

Two brothers saw a strange creature while driving on Twin Lakes Road, about 30 kilometres south of Inglewood. As a truck was approaching from the opposite direction a large black figure emerged from the bush between the two vehicles. It began to cross the road, glanced at the truck, and retreated into the bush. It seemed to be at least seven feet tall.

Undated GCBRO website submission. Credit: GCBRO and Dean Harrison.

Case 232. Late November 1999. Blue Mountains, NSW. 8:45 pm.

Brad Croft saw a big, hairy, bipedal creature dash across a dirt road in front of his car. This occurred only 500 metres from the yowie-plagued O'Connor property. [Case 223]

Witness interview with Paul Cropper, Oct 2000. Credit: Neil Frost.

Paul Cropper interviews Brad Croft. (Tony Healy)

Case 233. 2000. Royal National Park, NSW. 1 pm.

On a narrow track through lantana, 20-year-old "Matt", of Helensburgh, and a mate encountered a huge creature:

"We must have woken it up. It got up and ... just crashed all the bush. It was probably only three or four metres away but we couldn't see it ... it paused, and then took off and ran right in front of us, within about five or six metres, and we seen clear vision of it: a massive animal running on two feet, with massive shoulders; about eight foot high. The scrub there is about five foot, and we seen about three foot of it. Its fur looked similar to a grizzly bear – same brown colour – but shorter and shinier. It would weigh about 300 kilos [660 lbs]. Its head was small in comparison to its shoulders. Its shoulders went halfway up to about its ears. A real stocky neck.

"We talked about it and then said, 'Righty-oh, let's see where this thing's gone'. You could see clear as day, a massive track through the scrub … and, like, a tunnel into the lantana, like an Eskimo's igloo, a big opening. And up there it was real dark. We were thinking about crawling up, but thought that might not be a good idea."

Above the noise of breaking vegetation, the men had heard a strange vocalisation: "It was quite a strange noise. It didn't sound like it would come from this sort of animal. It was like a horn, like a duck, but extremely loud – you could hear it echo right across the gully."

Matt, his mate, and a couple of friends returned to the area with a video camera on several occasions, playing cat-and-mouse with the creature in the undergrowth. Sometimes it crept about with uncanny quietness, sometimes, when spooked by thrown rocks, it ran headlong through the thickest lantana: "No ordinary animal could do that. It sounded massive, like a truck going through the bush with its engine off."

On two occasions they heard the same deep, hollow, thumping noises others have heard in similar situations: "About 10 metres away … these massive stomps. It vibrated the ground ... probably warning us, you know: 'Get out of here!'"

They never actually caught sight of it again but found plenty of physical traces: "It had pulled down a massive dead tree and pushed it over the entrance to its den so nothing could get in, and another big log across an approach path. A big tree was scratched to bits, like a panther had clawed it, and there were areas about two metres diameter where it lays in the daytime … rolls around like a pig. And a tree where it had ripped down a branch about six inches in diameter. There were tracks everywhere. Massive. But they were in leaves, which only indicates the size, not the [exact] shape. We never found any hair and didn't smell anything strange even when we were close to it."

On a couple of occasions they left food for the yowie, with surprising results: "We left a bit of steak, bread, lettuce, banana, snow peas and a tomato. And you know what was left? The lettuce and snow peas; it ate everything else. And it left the banana skin there. So we went and bought a live chicken … put a little knot around its neck so it wouldn't choke and tied it to the tree where [the yowie] lays. And the next day the chicken was there with its head missing and its guts slit open and an egg next to it. There were big prints everywhere. A fox would have run off with the chicken's body."

Since massive bushfires burnt huge swathes of Royal National Park in 2001, the men have seen no trace of the creature. Interestingly, some of the earliest colonial-era yowie sightings were reported within about 10 kilometres of the site of their adventures. [Cases 6, 9 and 10]

Witness interview with Paul Cropper, 2002. Credit: Dean Harrison.

Case 234. April 14, 2000. Near Taree, NSW. Day.

After Geoff Nelson reported several yowie events on his property [Case 109] a four-man AYR team visited the site.

As 23-year-old Ashley Mills, of Bribie Island, Queensland, was checking an area about 100 metres from the Nelson residence, a rock flew out of the scrub and landed at his feet. "Then another landed a bit further away. I walked down to where that came from and a kangaroo ran out. Then I heard this crunching sound – boom, boom, boom – but it was getting fainter. I ran around the lantana and saw this thing moving away through the trees and uphill. It was about 2.4 metres tall, had big shoulders and was covered in greyish-brown fur. He was going fast, and there was no way I could have kept up. I had a camera with me, but it was all over in about four seconds."

Ashley had the distinct impression that the yowie had been stalking the 'roo. He thought the rocks might have been lobbed to drive it from cover or to make it change direction. On the other hand, they may have been deliberately thrown in his direction to scare him off.

Witness interviews with Geoff Nelson and Dean Harrison, Apr 14, 2000, and Tony Healy, Feb 8, 2004. Credit: Geoff Nelson and Dean Harrison.

Case 235. April 22, 2000. Mount Yengo, west of Wollombi, NSW. Between noon and 1 pm.

A young Sydney man said that while he was bushwalking in Yengo National Park with four friends a "large hairy ape-man" crossed the track in front of them. It was covered in a "coarse reddish fur", had a strong smell and was at least two metres tall.

Email to the yowiehunters.com website, May 14, 2000. Credit: Dean Harrison.

Case 236. May 2000. Woodford, Blue Mountains, NSW. 4 am.

A 19-year-old woman and her father saw a huge, hair-covered ape-man on the Great Western Highway. They pursued it down a side street and observing it from as close as 20 metres. [See Ch. 4]

Witness interview with Paul Cropper, Sept 2002. Credit: Neil Frost.

Case 237. August 28, 2000. Brindabella Road, ACT. 6 pm.

Steve Piper filmed a dark, hairy, man-sized figure limping along a bush track to the west of Canberra. The authenticity of the film is open to question. [See Ch. 6]

Witness interviews with Tim the Yowie Man, Dan Perez, Neil Frost, Paul Cropper and Tony Healy, Aug 29 and Sept 24, 2000.

Case 238. September 25, 2000. Near Kilcoy, QLD. Day.

In thick scrub near an isolated campsite, Rhonda Kay and her sons heard rustling in the underbrush. Seven-year-old Tim saw a white, gorilla-like creature. It looked "ugly and angry" and walked with a slouch.

Soon afterwards, Dean Harrison led two expeditions to the area and on both occasions heard vocalisations from an apparently very large, aggressive-sounding creature.

The camp is within 14 kilometres of the sites of two other yowie encounters. [Cases 141 and 248]

Email submission to the yowiehunters.com website, Sept 26, 2000. Witness interviews with Dean Harrison and Paul Cropper, Sept 2000. Credit: Dean Harrison.

Case 239. November 2000. Near Woodenbong, NSW.

A woman told *Northern Star* correspondent Rosemary Clark that she had seen a bipedal animal resembling a "man-like ape". It was of medium build and was walking in the open beside a creek. It moved quickly but appeared "not stressed". The experience sent her into deep shock.

Letter from Rosemary Clark to the *Lismore Northern Star*, Jan 26, 2002; email from Rosemary Clark to Paul Cropper, Aug 12, 2002.

Case 240. February 2001. Canungra, QLD.

A local man discovered and photographed a single, massive, five-toed footprint on a bush track near Canungra Creek. It was over twice as long as his size 10 sandal.

Witness interview with Paul Cropper, Aug 2002.

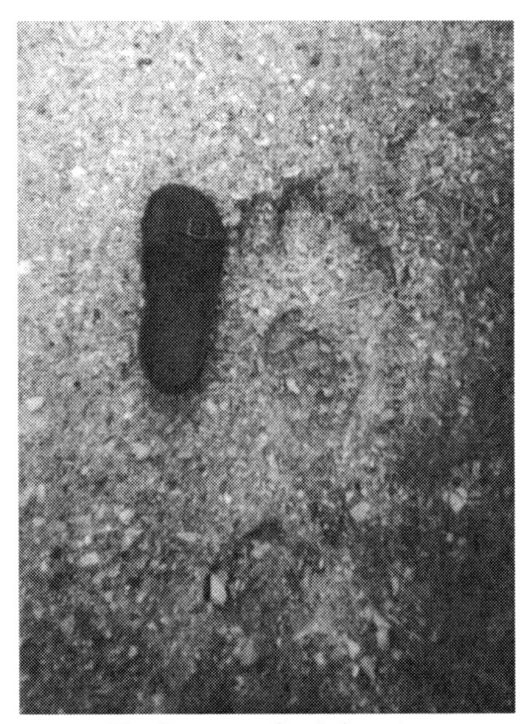

The Canungra Creek footprint.

Case 241. March 2, 2001. Near Hinze Dam, south-east QLD. 11 pm.

A "big hairy man about eight feet tall" ran across the road in front of Aaron Carmichael's car. The vehicle apparently struck the creature before spinning out of control. [See Ch. 4]

Witness interviews with Dean Harrison, Mar 3, 2001 and Paul Cropper, Mar 18, 2001. Credit: Dean Harrison.

Case 242. April 2001. Tamborine Mountain, QLD. 9 pm.

Twenty-seven-year-old David Holmdahl was chased by a huge, hair-covered, seven to eight-foot-tall figure. [See Ch. 4] This occurred only five kilometres from the boundary of Canungra Land Warfare Centre, scene of several other yowie incidents. [See Ch. 4]

Witness interview with Paul Cropper, Apr 30, 2001. Credit: Dean Harrison.

Case 243. April 2001. Woodford, NSW. 4.15am.

"D", an apprentice baker, lives on Glen Street, which, like many other streets in the Blue Mountains, backs directly onto wilderness. Early one morning, as he was leaving to go to work, he got the fright of his life.

"Up at the top of the driveway, around 60 metres away ... where the streetlight is ... there was this huge, dark figure. It was a moonlit night and it was in the middle of the driveway, in full view. It appeared to be looking at me, and we both turned and started to run at the same time, and I looked around and looked back, and it was gone – straight into the bushes.

"It was seven foot, seven foot three, maybe. Its whole body was huge. Big, beefy arms ... the head didn't look all that big, compared to its shoulders. I couldn't see any hair, just a black silhouette. But when it leant over it had, like, arms. It ran hunched over like an ape and [the arms] swung real dopey-like. When it turned to the side I saw it had a huge stomach ... a gut, that's for sure, and it actually hung like, say, a pregnant woman, but lower, and not as big. Like maybe a tucked-over-the-pants beer belly, you know? A bit similar to that."

"D" rejects any suggestion that he could have been hoaxed: "It would be impossible for a human to be that big ... there's nothing [other than a yowie] that it could have been."

"D" isn't the only witness to describe a large, sagging belly. [See Case 35]

Witness interview with Paul Cropper, Oct 2002. Credit: Neil Frost.

Case 244. August 2001. Mulgowie, 35 km south-west of Ipswich, QLD. Night.

A Mulgowie woman called Laidley police to report seeing an "ape-like creature with a bare bottom". The "orangutan like" animal, which was "larger than an Alsatian" ambled across Mulgowie Road on all fours in front of her car. *The Chronicle* reported that the creature had been sighted 18 years earlier in the hills behind Laidley.

The Chronicle, Aug 16, 2001; *Queensland Times*, Aug 16, 2001.

Case 245. November 2001. Woodenbong, northern NSW. Early morning.

Two women told *Northern Star* correspondent Rosemary Clark that while checking on a horse they saw something "between a man and an ape." Six feet tall and covered in black hair, it stood looking at them, then turned and moved off into dense forest.

While one woman went to fetch a camera, her younger companion followed the animal at a safe distance. She watched it move towards the rock face of a nearby escarpment and disappear into low mist and cloud. Oddly, the women mentioned that the creature had "tired eyes."

Letter from Rosemary Clark to the *Lismore Northern Star*, Jan 26, 2002. Email to Paul Cropper from Rosemary Clark, Aug 12, 2002.

Case 246. December 2001. Springwood, Blue Mountains, NSW. 6:30 am.

Mary Camden, a local schoolteacher, encountered a large, hair-covered, flat-faced, bipedal creature in Magdala Gully. [See Ch. 4]

Witness interview with Paul Cropper, Aug 2002. Credit: Neil Frost.

Case 247. March 29, 2002. Wollemi National Park, NSW. Day.

While on a remote, overgrown track near Wolgan River, a Sydney rock-climber heard a crashing sound and saw a large, black figure weaving through the trees. Nearly six feet tall, it ran in a "hunched" attitude and was very agile. When he called out, it hid behind a small tree. The man found large, fresh footprints across the muddy track.

Email to Dean Harrison, Apr 3, 2002. Credit: Dean Harrison.

Case 248. July 2002. Mount Brisbane, near Lake Wivenhoe, QLD. Mainly night.

"G" and his partner claimed their isolated property was repeatedly visited by several yowies, which they sighted on a couple of occasions. They heard what seemed to be yowie vocalisations for over 20 nights in a row.

Witness interview with Paul Cropper, Aug 10, 2002. Credit: Dean Harrison.

Case 249. July 22, 2002. Between Wilton and Mount Kiera, NSW. Day.

Between Cordeaux Dam and the Mt. Kiera turnoff, a Sydney man, "J", saw an eight-to-nine foot tall creature standing beside the road. It was covered in very dark brown or black hair, had a small broad neck, hunched shoulders, a head "the size of a watermelon" and a very pronounced single brow ridge. "J" estimated that it weighed at least 300 kilograms [660 lbs]. It ran across the road, stopped, turned, crossed back to where it had come from, then "went on all fours and dove into the bush".

"J" stopped his car, and "after about three minutes thought it safe to get out. I stood by the driver's door and froze. I heard trees snapping, and grunting, almost like a bark, but too deep … a horrible sound. I could smell an odour … bad, like a dead rat. Then I heard a distinct 'CHUD' sound repeating, like if you were to throw a house brick into wet ground. It originated about 20 metres in front and moved around, toward the rear of my car. Then I heard the bark sound again, like a bellowed deep grunt. I saw a figure move in the trees not far from me, tall and black. I freaked out. I didn't know if it was some sort of escaped gorilla or bear. I got into my car quick and was glad to be doing 100 kph again with whatever that thing was behind me."

Over the past 140 years, several other yowies have been reported in the same area. [e.g. Cases 8, 15 and 26]

Witness emails to Dave McBean and Paul Cropper July 27 and 30, 2002. Credit: Dave McBean.

Case 250. August 26, 2002. Wentworth Falls, Blue Mountains, NSW. 9:30 or 10 pm.

Three huge, dark, bipedal animals approached Justin Garlick's car on Tablelands Road. [See Ch. 4]

Witness interview with Paul Cropper, Sept 2002. Credit: Neil Frost.

Case 251. September 4, 2002. Hazelbrook, Blue Mountains, NSW. 11 am.

Two boys riding bicycles in Oaklands Road saw an animal they at first thought was a kangaroo. When one boy threw a stick, the creature stepped out from behind a tree and sprinted into dense bush. It was over two metres tall and dark in colour.

Witness email to Paul Cropper, Sept 21, 2002. Credit: Neil Frost.

Case 252. September 6, 2002. Beerwah, QLD. Day.

While photographing plants about 10 kilometres from Beerwah, "JL", a 23-year-old horticulturalist, heard a strange, high pitched yell or scream. Moving to a new location a few kilometres away, he was startled by a large black-furred creature moving from roadside scrub into a pine forest. Crouched down at first, it became upright as it ran. About six feet tall and slightly broader than a man, it moved at amazing speed. "JL", who had a good SLR camera in his hands, turned and took two quick photos, but failed to capture it on film.

Beerwah is seven kilometres south of Ewen Maddock Dam, where campers were terrorised by a yowie in September 1990. [Case 183]

Witness interview with Paul Cropper, Sept 6, 2002. Credit: Dean Harrison.

Case 253. October 2002. Glossodia, western Sydney, NSW. Night.

A 16-year-old girl reported several encounters with a yowie on her family's property, which adjoins a forested 40-acre reserve. On the first occasion, while in the backyard, she saw the huge creature, partially illuminated by a house light, emerge from the bushes only four metres away.

"It was sort of crouched, then stood up until it saw me and then hunched over and ran. I thought it was a person, until it stood up and I thought, 'OK, no one's that tall!' I'm five and a half feet, and it was double my size ... and big across the shoulders. I'd never seen anything like it. I just went completely blank and stood there frozen.

It was covered in brown or black hair that "moved in the wind when it ran ... more like a goat's, say, than a bear or dog. [The head] fitted the body, in proportion ... quite a large head.

There wasn't enough light to see the face ... a short neck. The ears were more like a dog's than a human's – more pointed than rounded; they weren't really big. It ran hunched over, but not on all fours. The front legs looked more like arms, long arms. It proceeded towards the bush up the hill ... going very, very quickly. I turned and ran."

Eight kilometres away, across sparsely settled hobby farming land, which is cut by several scrubby creeks, is the boundary of yowie-infested Wollemi National Park.

Witness interview with Paul Cropper, Oct 2002. Credit: Dean Harrison.

Case 254. Early 2003. Near Jimna, QLD. 7:30 pm.

While checking out a thickly forested "hot spot", yowie researchers Rob Millar and Dave Glen suddenly experienced, as Rob put it, "a weird feeling of being watched. A presence ... all the hairs on the back of my neck went up. It wasn't a hot night but we [began] sweating". As Dave turned to retreat, Rob's Mag-Lite illuminated a huge hairy figure crouching about 10 metres away.

"For a moment I thought it was the base of a shaggy-barked tree, but then it stood up. It had a powerful set of shoulders, arms longer than a human's, legs slightly apart. The whole thing was a reddish colour. I couldn't see any facial features because of all the hair hanging down, but it seemed to be looking at me." The hulking creature was over seven feet tall. With a torch in one hand and a two-way radio in the other, Rob had no chance of photographing it. "It was all over in seconds. I turned and ran."

As the men retreated to their vehicles, they heard the creature "crashing through the scrub on either side of the track – it followed us uphill for a good 100 metres". During the pursuit Rob felt an impact on his boot: "It struck really hard. I looked down and found, on top of leaf litter in the middle of the track, a stone about the size of a cigarette packet."

Stone me! Rob Millar with a souvenir. (Rob Millar)

Witness interviews with Mike Williams and Tony Healy, 2003 and 2005. Credit: Mike Williams.

Case 255. March 2003. Springwood, Blue Mountains, NSW. 9 pm.

Every night for 10 years, Moira Haley took her dogs for a walk down a steep track into the forest at the end of her street, Sassafras Gully Road. An experienced bushwalker, she knew the track so well she rarely bothered taking a torch. Her dogs had never displayed any fear of the people and animals they occasionally encountered, so one night, when one dog suddenly snarled, bristled and hid behind her legs, she was quite surprised.

As the other dog, which is almost deaf and blind, carried on, Mrs. Haley heard something move in the scrub. Then she saw an upright figure "*step* out onto the path next to the dog, which got very frightened, jumped, turned and ran back past me. The figure was dark, about five feet tall. It stepped out like a man but it was not a man, not an ordinary animal. Looking at it, I experienced goose bumps and cold chills. I turned and leapt up the steps – ran up them like a rabbit." Since then, her dogs are afraid of venturing down the path. She is, she admits, also very nervous about going there after dark.

One evening about six months later, two young neighbours glimpsed a nine-foot-tall yowie in almost the same spot. The site of Mary Camden's yowie encounter [Case 246] is less than 800 metres to the south-east.

Witness interviews with Neil Frost and Tony Healy, Aug 2003. Credit: Neil Frost.

Case 256. April 2003. Ormeau, QLD. About 3:30 pm.

While felling trees on a steep hillside, Jason Cole noticed an eight or nine foot tall, hair-covered "gorilla-man" glaring at him from the scrub. As he retreated uphill, it circled around, apparently attempting to cut him off. Raised in the USA, Jason had never heard of the yowie. [See Ch. 4]

This occurred within 2.5 kilometres of the place where Dean Harrison was chased by a similar creature in 1997. [Case 218]

Witness interview with Paul Cropper, Sept 2003. Credit: Dean Harrison.

Case 257. May 9, 2003. Blue Mountains, NSW. Dusk.

While walking in the forest, 11-year-old Drew Frost noticed a huge, hairy, ape-like figure just a few paces away. When it turned to look at him, he beat a hasty retreat. [See Ch. 4]

Witness interview with Neil Frost, May 9, 2003. Credit: Neil Frost.

Case 258. September 21, 2003, Blue Mountains, NSW. 9:50 pm.

When her dog began barking hysterically, Jenny "X" opened a second story window and heard what sounded like two or more large creatures crashing through the scrub towards her back fence. She is sure they weren't human because of "the loudness and frequency of tree and branch snapping ... quite substantial trees [were] being snapped over a period of about 5 minutes. Then I heard voice sounds ... from a few metres behind our fence. The voice quality was so foreign that I knew it wasn't human and my heart started pounding". [For details of the vocalisations see Ch. 6]

When the vocalisations ceased Jenny and her husband shone a powerful torch along the fence line and saw, about seven feet above ground level, a set of "bright, iridescent, greenish-blue" eyes. What may have been a different set appeared briefly about a metre to the right.

In the morning they found "a lot of trampled grass and many dead trees pushed over and some snapped. One ti tree had been snapped at a height of about four feet".

Jenny's house is within 1,500 metres of the yowie-plagued O'Connor property [Case 223] and backs onto the same forest.

Witness interviews with Neil Frost, Jerry O'Connor and Michael Williams, Sept 2003; written report by Jenny "X", Oct 17, 2003. Credit: Neil Frost.

Case 259. September 27, 2003. Blue Mountains, NSW. 10 pm to sunrise.

While camped about one kilometre north of the swinging bridge in Megalong Valley, 17-year-old "K" and two friends heard something in the surrounding scrub. Switching on a torch, they saw a pair of very large red eyes and the silhouette of a very solidly built, apparently bipedal creature, which, given the uneven terrain, could have been between 1.5 and 2 metres tall.

It ran off at high speed, but with a "funny" irregular gait. Then, while keeping mainly out of sight, it circled the campsite, shaking trees and making extremely deep growling noises. It withdrew into the bush, but returned three or four more times before sunrise.

Witness interview with Neil Frost, Oct 16, 2003. Credit: Neil Frost.

Case 260. Dec 20, 2003. Near Neville, NSW. 5:20 am.

Darren "X", Cameron "Z" and another man encountered a yowie while pig hunting in a pine plantation about 19 kilometres southwest of Neville, near Wyangala Dam.

"We were up early and got to a large wallow where we had sighted pigs the previous day and sat waiting. About five minutes later our oldest dog Bonnie was going crazy, wanting to get off the leash ... so we let them [all] off to look around ... about five minutes later, Cameron heard a groaning yell ... the dogs had a medium size figure bailed up in a large blackberry bush. As we approached it tore out at us. The dogs were holding onto its shaggy fur. It ran right to us screaming in a deafening roar. Cameron yelled 'Jesus, what is it?'

"Well, what we were face to face with was a yowie/big foot/yettie whatever they're called. It shook our dogs off like they were not even there, and ran into the forest ... we called the dogs off and regathered our thoughts. A few hunters from nearby towns have witnessed similar sights."

This occurred within three kilometres of the site of a dramatic 1876 yowie incident. [Case 22]

Email from Darren "H" to Dean Harrison, Dec 27, 2003. Credit: Dean Harrison.

Case 261. May 23, 2004. Blue Mountains, NSW. 9:10 am.

While attending to chores, 13-year-old Avril Frost saw a black, hairy, seven-foot-tall figure standing, almost unobscured by underbrush, about 20 metres away. [See Ch. 4]

Witness interviews with Neil Frost, Mike Williams and Paul Cropper, May 23, 2003.

Case 262. August 15, 2004. Blue Mountains, NSW. 5 pm.

Margo Braithwaite, a next-door neighbour of Neil and Sandy Frost [Case 160], saw the head and shoulders of a very tall, dark figure standing in rain-soaked scrub on the edge of her backyard. Because it was late afternoon on a dull day, she could see only a silhouette. It stood immobile for around 10 minutes until she went indoors. Minutes later it was gone. Later investigation indicated it was between seven and eight feet tall.

It was standing in exactly the same spot as the creature seen by Avril Frost three months earlier and within a few feet of the spot where, in 1993, Neil Frost's automatic camera photographed what could be an ape-like face. [See Ch. 4]

Interviews with Neil Frost, Paul Cropper and Tony Healy, Aug 15 and 20, 2004.

Margo Braithwaite. (Tony Healy)

Case 263. Sept 25, 2004. Bungonia Gorge, Shoalhaven River, NSW. 2 am.

As 42-year-old "Grebo" and his mate camped near the gorge, they heard a "screeching and howling sound" about 50 metres from their tents. When they went to check it out, they noticed a "filthy stench", then heard something running through the scrub. "Grebo" switched on his torch and "20 metres away … seen this hairy, ugly monkey thing that was about four foot high. It had what looked like a dead possum in its hairy hands. It screamed like a cockatoo that had been hit with a cricket bat.

"My mate was just standing there and I was frozen to the spot, packing it. [It] reared up and made a move towards us, before scrambling up a tree with the possum in its mouth. We took off and locked ourselves in me ute [pickup truck]. We never slept that night and in the morning found a patch of some stinky stuff near the tree."

Witness email to AYR, Sept 27, 2004. Credit: Dean Harrison.

STOP PRESS: LATE INCLUSIONS

The following cases came to light after the footnoting system and the statistical analysis were completed.

Case 264. About 1965. Mt. King Billy, near Mt. Buller, VIC. Night.

When cattleman John Lovick was about 19 or 20, he accompanied his father Jack and four or five other people on a fishing expedition into the high country. They bivouacked at King Billy Hut, a one-room shack in trackless country close to Mt. King Billy. As there were only two double bunks, most of the party had to sleep on the floor. After "lights out" the hut was in darkness apart from moonlight shining through the door, which was wide open. Although John had scored one of the top bunks, he didn't have a particularly restful night.

At around midnight he was woken by "this bloody heavy thing pressing on my chest. It had its hands around my throat, trying to strangle me. It had a strong animal smell like you'd run across in a zoo. Its hair felt coarse and I could feel its hot breath in my face. I yelled and screamed and tried to throw it off, and we fell onto the floor". The creature broke loose and although John didn't witness it himself, "a couple of blokes saw its silhouette as it walked [upright] out the door. It was so broad it filled the doorway, but they said it wasn't very tall, five foot at most.

"The dogs were making a bloody ruckus and the old man went out and tried to sool them onto [the creature] but they wouldn't go out more than five or 10 metres: they kept coming back with their bristles up." In the morning they could find no tracks on the hard, dry ground, but there were discernable bruises on John's neck – like those that might have been left by strong fingers. John speculated that the creature had been used to sleeping in the hut, which was unattended for months on end, and became angry when it found the place full of snoring interlopers. But he is still wondering why it should have stepped around the men on the floor and attacked him in the bunk farthest from the door.

Yowies have been reported in the area on at least three other occasions. [Cases 79, 177 and 178]

Witness interviews with Bernie Mace and Pam Bryant, c. 1990, and with Tony Healy May 23, 2005. Credit: Bernie Mace.

Case 265. July 23, 1985. Palmerston National Park, QLD. Day.

In 2004, after telling his own story of having found a possible yowie nest near Russell River [see Ch. 6], Major Les Hiddens (ABC-TV's "Bush Tucker Man") presented a report written by Corporal J. Webster of the Training Squadron, Special Air Service Regiment:

"On the 23rd July 1985, while working in the Downey [Dowrey?] Creek area of the Palmerston National Park, I observed a creature, not animal or man. A number of us had arrived about an hour earlier by Land Rover ...We had camped 15-20 metres off the track among the trees

... I had gone back to the vehicle to gather some stores. Upon arriving there, I saw the creature about 50 metres away, on the side of the track. It was rubbing itself against a tree, as it saw me it stood up on its two legs, stared at me for a few seconds then walked off into the scrub. 29 Jul 85 J Webster."

Sketch by Cpl. J. Webster.

Major Hiddens commented: "It was indeed a brave move by Corporal Webster to put pen to paper and make this statement. He was also motivated enough to provide us with a sketch of the creature." Beside his sketch Cpl Webster wrote: "Height: 5 ft min; Weight: 80-90 kg; Covered in fur or hair; Walks upright". He apparently referred to the yowie as "Melon Man" – a term used by other servicemen and civilians in the Innisfail to Cairns district.

The "yowie nest" discovered by Major Hiddens and party was about 25 kilometres north of the site of Cpl. Webster's encounter.

Written report by Cpl. J Webster, Jul 29, 1985, cited in Les Hiddens, "Who Goes There?", *The Townsville Bulletin*, Dec 8, 2004. Credit: Mike Williams.

Case 266. July or August 1987. North of Rockhampton. Day.

One night during a church youth camp, the supervisors saw two four-and-a-half-foot-tall figures creeping around near the cabins. As all the children were found to be in their allocated places, it was assumed the figures were kangaroos.

The following day, however, as 12-year-old Trent Todd and a mate were exploring the surrounding rainforest, they came across a couple of places where dead wood had been "assembled so it looked like small huts". A little further on, Trent's mate "froze and pointed ... about 30 metres up the trail, about two metres up a tree, was this rather big monkey-looking thing with reddish-orange hair looking right at us. It was easy the same size as us ... it jumped ... and you could distinctly see arms and legs [that it] jumped down upright upon". The boys fled.

Although the supervisors told the excited lads they must have seen a tree kangaroo, a planned night hike was cancelled. The following morning the camp was abruptly terminated, a day earlier than scheduled.

Email from Trent Todd to Dean Harrison, June 19, 2005. Credit: Dean Harrison.

LATE INCLUSIONS

Case 267. 1994 or 1995. Near Gunnedah, NSW. 10 or 11 pm.

As 17-year-old "Angus" and three mates were returning to Gunnedah after a day's fishing, a strange animal "ran straight across the road, probably 15 or 20 metres in front of us. It ran in a real awkward-looking, uncoordinated gallop, hunched right over, almost on all fours – and as it ran its head was turned to the side, looking at us.

"We were all pretty silent for a while and then one of the blokes, Mark, said, 'That must have been a bloody yowie!' We all grew up in the bush, shooting, fishing and all that, but we'd never seen anything like it." The man-sized, solidly built animal had "really shaggy, dark grey hair all over, maybe three or four inches long". It was visible for only a couple of seconds, but "Angus" thought its head seemed wider at the top than at the bottom: a vaguely triangular shape.

This occurred only about two kilometres east of Gunnedah on the Oxley Highway. The creature "ran from the right, from the Namoi River flats, heading towards more of a hilly area". The yowie-haunted Pilliga Scrub [see Cases 131 and 135] begins about 50 kilometres west of Gunnedah.

Witness interview with Paul Cropper, May 2005. Credit: Dean Harrison.

Case 268. 1997. Warrimoo, Blue Mountains, NSW.

While two 14-year-old boys were hunting snakes, they saw a huge, hairy, red-eyed creature in a cave near the Warrimoo Bush Fire Brigade station. Thirty years earlier a man's headless body was discovered in the same area. [See Ch. 6]

Witness interview with Neil Frost, 1997. Credit Neil Frost.

Case 269. Circa 1999. Mission Beach, near Tully, QLD.

One morning, after having his sleep disturbed by high-pitched cries "like a woman screaming", farmer Billy Jepson found some of his goats dead – crushed, "as if they were made of foam rubber". The following night, on hearing the same calls, he crept outside and saw, in his torchlight, a huge, black, hairy animal walking on all fours towards the remaining goats. His savage pig dogs, at his urging, approached it, but they soon turned and ran. The creature then rose onto its hind legs and walked into the forest.

In August of the same year, in a jungle swamp 15 kilometres to the south, near Tully Heads, Les Holland stalked, and then fled from, a large unseen creature that shook the vegetation, thumped the ground, and emitted high-pitched whines and loud, aggressive grunts. Aboriginal friends told him they'd encountered yowies in the same locality in about 1979.

Some years earlier, Les had found three large, strange footprints in a banana plantation. They were similar to those of a human except that they lacked an instep and had only four toes that were "squared off" rather than receding at an angle from largest to smallest.

Gary Opit interview with Les Holland, August 11, 1999. Credit: Gary Opit.

Case 270. June 1999. South-west of Ipswich, QLD. 11:15 pm.

While stargazing just off Mt. Walker West Road, Ray Doherty of Geebung and three friends were suddenly assailed by "a foul stench that floated in from the south-west. It was just awful – at first we thought it was a dead cow". As the stench intensified, nearby cattle began to bellow in a panicky way. "Then came this high pitched, ghostly shriek, like a cross between a scream and a yell."

Switching on a powerful spotlight, the men saw cattle milling around before running away. Just then a pair of brightly reflecting eyes [Ray can't recall the exact colour] came into view. Through a primitive night-vision device, one of the party saw that they belonged to a large creature lying belly-down on a large earthen mound 40 to 80 metres away.

After a minute the eyes rose up, "much, much higher than ground level". No one could make out anything other than a tall, dark shape, but Ray is sure it was about seven feet tall. As the creature began to move around the shadowy paddock, "it appeared juvenile in its fascination with our lights. [When we] tried to replicate the sound we heard previously, it stopped and looked, but never responded". Eventually the spotlight's batteries failed and the creature disappeared behind some trees.

Two years later, another yowie was reported 15 kilometres to the west, near Mulgowie. [Case 244]

Witness email to Dean Harrison, June 27, 2005; witness interview with Tony Healy, July 3, 2005. Credit: Dean Harrison.

Case 271. February 2001. Near Scotts Head, NSW.

Commencing in February, a couple living on the slopes of Mount Yarrahapinni claimed regular nightly visits over several months by a very large, very elusive, bipedal creature that walked along the verandah of their house. Although the creature was only glimpsed in less than ideal conditions, the episode is of interest because of certain similarities it bears to yowie "yard invasions" in the Blue Mountains. Another interesting aspect is that a horse was killed on the property in mysterious circumstances and apparently dragged into nearby scrub.

Mt. Yarrahapinni is only four kilometres from Eungai, where yowies have supposedly been seen, and within 45 kilometres of at least nine yowie reports from the Kempsey area.

Witness interview with Paul Cropper, 2001. Credit: Dean Harrison.

Case 272. September 2001. Upper Main Arm, NSW. 6:30 am.

Jill and Dan "M" own a property adjoining the one on which Lynn and Josh Clark saw a yowie in 1996. [Case 210] One overcast morning, Jill walked onto her back verandah and noticed a wallaby grazing on her back lawn. As she watched, it stood up and stared fixedly into the surrounding scrub. Jill then heard three loud thumps and saw a very large creature about 20 metres away. It had apparently been crouching down in the underbrush, had stood up, and was now walking away.

About six feet tall, heavily built and completely covered in black hair, it was visible from the top of its head down to its thighs. The head sat directly upon broad shoulders, the torso was long and straight and the arms hung straight down.

Strangely, for ten to fifteen minutes after the yowie departed, the wallaby stood staring after it, as if mesmerised. Two days later, Dan and Gary Opit retraced the creature's steps and found freshly broken branches, at shoulder height, along the path of its retreat.

Witness interview with Tony Healy, August 2002. Credit: Gary Opit.

Case 273. April 7, 2005. Great Western Highway, Blue Mountains, NSW. 12:20am.

While driving between Valley Heights and Warrimoo, 28-year-old Craig Holmes came to a section of road with no settlement on either side. Approaching a patch of fog, he slowed down and flicked his lights to low beam. "That's when I saw somebody step out from the left, directly in front of me. I thought, 'You idiot!' I jumped on the brakes; they locked up.

"[The figure] was only 10, 15 metres ahead. My car slid, and by the time I'd covered that distance it had stepped into the middle of the next lane only about four metres [from the driver's side window]. I just felt it was all in slow motion [as he slid past]. I saw its back – I was looking up at it. And that's when I've gone, 'Hang on – that's not a person!' Because it was huge, absolutely *huge*. I'm six foot; this was seven and a half easy. Big boy; fairly solid; long legs. The body was very straight. It was dark, dirty brown from top to bottom ... a raggedy look, like it had that hair coating. I have the image just burnt into my head of the back, the legs, the right arm. The arm was long and straight [swung back] like when you're walking fast. I didn't see the head. I went past it, hit the gutter, jumped out – and there was nothing: it had gone."

To left of the road there was a scrubby gully; to the right more scrub, then the railway tracks, then dense forest. Still coming to terms with what he'd seen, Craig called out a few times and heard, from the right, "rustling, like someone was fidgeting in the bushes. I started to get that weird, stomach turning feeling. Something inside was telling me, 'Just get in the car and go!' I thought, 'I've had enough of this!' – and took off."

Witness interviews with John Appleton, Paul Cropper and Mike Williams, April 8, 9 and 10, 2005. Credit: John Appleton and Neil Frost

Case 274. 1891. Blue Mountains, NSW.

Two men from the Oberon area reported a strange encounter in the southern Blue Mountains. After discovering "the imprint of a huge foot", they "were startled by the noise of timber breaking, and a low, growling, grating sound", and then saw before them, on the edge of a cliff, "an animal of the baboon species" which, when advanced upon, "swung himself over the cliff by a huge vine, and descended in that manner until he disappeared in the gorge beneath".

This occurred in an area now covered by Kanangra Boyd National Park, where several yowie sightings have been reported in the modern era. [e.g. Cases 144, 179 and 199]

The Bathurst Times, Jul 9, 1891, cited in "What is a Yowie?" Blue Mountains City Library Local Studies Section web site, http://www.bmcc.nsw.gov.au. Credit: Rebecca Lang.

Case 275. 1932. West of Mittagong. Night.

Chris Bagnall told his grandson, Mick Stubbs, about something he experienced while camped in mountainous country near the Wollondilly River. One night, while washing dishes in the stream, he looked up and saw "a tall, hairy man" illuminated by moonlight, watching him from the top of a large rock on the opposite bank. He rushed back to camp to grab his rifle, but on his return the creature had vanished.

Information supplied by Mick Stubbs to Gary Opit, 2005. Credit: Gary Opit.

Case 276. 1979. Near Nowendoc, 60 km north-west of Gloucester, NSW. Day.

While pig hunting in remote country near the Nowendoc River, Mick Stubbs encountered a "big hairy man". It moved away into the scrub, but Mick soon heard its footsteps again as it circled around behind him. He felt sure it was observing him from cover. That night he was severely frightened by horrible screams emanating from the scrub. Another yowie was sighted near Nowendoc in 1993. [Case 191]

Information supplied by Mick Stubbs to Gary Opit, 2005. Credit: Gary Opit.

Case 277. 1980. Kanangra Boyd National Park, NSW. Early morning.

After setting up camp in the Wild Dog Range, in the southern Blue Mountains, a church group "retired to their sleeping bags and slept soundly until, in the early hours of the morning, they were disturbed by something moving around their camp. When the priest and several of the boys investigated, they were 'confronted by a creature about 2.5 metres [8 ft 2 ins] tall'. It casually made off into the bush, dropping one of their cooking pots as it departed".

This occurred about four kilometres north-west of the spot where Gary Jones and friends pursued a yowie in 1989. [Case 179]

"What is a Yowie?" Blue Mountains City Library Local Studies Section website, http://www.bmcc.nsw.gov.au. Credit: Rebecca Lang.

Case 278. January 1983. Walgett, NSW. Night.

After a party, "SC" and her partner cleaned up his parents' backyard, packing all the rubbish in an old 44-gallon drum. They retired early but couldn't sleep because of the humidity. "So we laid there talking in the dark. There were no curtains on the windows ... whilst we were laying there, there was a noise, it sounded like a dog was mucking about with the rubbish, so I looked out the window to shoo the dog away, but it looked like a small boy with a lot of hair all over him and he was searching in the drum and throwing rubbish all over the backyard ... there was a full moon and you could see everything in detail ... I laid there trying to comprehend what I was actually seeing. The small hairy boy was rolling around in the grass and scratching his back exactly like a dog would but it was evident that this was no dog. We got up quietly (I should mention that my partner witnessed these events) to go wake other members of the household but he must of saw us and disappeared. I am indigenous Australian ... my people have quite a few tales to tell of sightings over the years."

This occurred in a residential area but very close to a Namoi River levee bank, beyond which was bushland.

Witness email to Dean Harrison via Barry Porter of GCBRO, August 2, 2005. Credit: Barry Porter and Dean Harrison.

Case 279. 1995-2004. Glen Innes region, NSW. Day and night.

In 2006 we were fortunate enough to make contact with Paul Compton of Glen Innes, a researcher who has single-handedly gathered more tangible evidence of these creatures than any other yowie hunter we know of.

Since 1995, Paul has experienced three yowie encounters. Because he is maintaining a camera trap at one site and because he has already had to contend with irresponsible shooters at another, he has asked us not to reveal precise locations.

His first sighting occurred beside a country road at about 5:30 one midsummer morning. While checking his vehicle and trailer, he suddenly felt that he was not alone and turned to discover that he was being watched by a massively built animal standing some 25 metres away. Showing no sign of alarm, it turned and walked unhurriedly through some low wattle scrub, giving Paul plenty of time to observe it.

Paul Compton in one of the yowie beds. (Paul Compton)

It was covered with reddish-brown hair, "about the colour of the old Telecom overalls", that was three to four inches long on the body and eight to 12 inches long under the forearms. "He walked with a slight stoop with his arms motionless by his side. I estimate this animal to be approx 220-300kg [485-661lbs] and 2.1 metres [7 ft] in height."

Not surprisingly, after that experience, Paul became very interested in the yowie mystery and, after making enquiries throughout the district, was directed to a couple of locations where the creatures had been seen repeatedly. While searching those areas, he has discovered, photographed and preserved some very valuable evidence: four beds resembling primate nests, two large four-toed tracks and a huge (18-inch-long) dropping. [See Ch. 6]

In June 2003, after searching one of the locations, Paul and his brother-in-law returned to their camp at about 4:30 pm, set up a couple of chairs and had a beer. As they sat talking, Paul noticed, about 200 metres away, a large creature apparently observing them through the fork of a tree. Although he saw only its silhouette, he could make out "the waist, chest, arms, big shoulders and a head". As he stood up to get his camera and binoculars, the form quickly moved out of sight.

One night in January 2004, while camped beside a river in another yowie hot spot, Paul and his father-in-law heard a great commotion on the opposite bank, followed by a loud splash. Switching on a powerful torch, they saw that an eastern grey kangaroo had made a flying leap well out into the river. On the bank above it was another 'roo, "going in circles, panicking. I could see the rocks it was dislodging. It was scared and cornered. On looking further ... we could see a five-to-six-foot-tall yowie standing by some granite boulders ... about 30 metres away but we could not get a photo." In the torchlight the creature's wide-set eyes displayed a reddish-orange shine. After a few minutes it turned and moved away.

Email from Paul Compton to Tony Healy, Dec 8, 2005; interviews with Tony Healy, Dec 3 and 9, 2005.

Case 280. 1983 or 1984. Near Mingela, QLD. Afternoon.

When Aboriginal man Russell Smith was about 10 years old, his family lived on a property on the banks of Kookaburra Creek, about 35 kilometres east of Charters Towers. While there, they often experienced an eerie feeling that "... something was watching you".

One hot afternoon, as he and two older white girls were taking a horse across the creek, "... we came upon this thing by accident – only about five, ten metres from us. We saw it from the side. It was looking up at our property, kneeling on the riverbank, one knee up and the other down; back hunched over. It was hairy: longish, really thick hair; darkish brown to black. We could see arms; the hair seemed a bit lighter colour on the arms. It had a big head. No neck – that's one thing I remember. Like a body builder, you know? It didn't stand up but it would have been much bigger than a person – seven foot, no dramas at all. It was behind little bits of scrub, so I missed seeing its face.

"It was a fright for him and for us. It growled and we just got the hell out of there. The whole thing lasted only a couple of seconds but I'll never forget it. The girls said not to worry about it – 'It's one of those things that hangs around sometimes'."

Russell believes the yowie was close to the property because "The creek was dry, but we had dams and vegetables growing." Later, on two or three occasions, in the dead of night, heavy footsteps were heard very close to the house, and once something barged in and ate food out of pans in the kitchen.

Witness interview with Paul Cropper, Oct 22, 2005. Credit: Dean Harrison.

Case 281. 1997. Near Wee Jasper, NSW. 2 am.

After experiencing car trouble, Brett Young, Brian McLean and an Aboriginal friend were walking, under a clear, starry sky, along a dirt road near Buccleuch Forest. After a while they realised they were being followed by something about 20 metres back, in the roadside scrub. "It stopped when we stopped", Brett recalls, "and began again when we continued ... heavy footfalls ... undeniably bipedal ... loud, heavy breathing – like a person almost out of breath.

"After about 30 minutes we became almost bold, turned and moved towards the sound. I climbed the embankment and shone a very weak torch around. At the limit of the light, about 20 metres away, a very bulky, massively shouldered creature was crouching, hunched over behind a tree, peeping at me. Chin tucked into chest; big, round, high-domed head. It was a brown-red colour, almost chestnut. There was only very faint eye-shine, maybe reddish, in the weak light. I could hear it breathing – deep, barrel-chested sounds. I challenged it – yelled out, 'I can see you!' but my mates were saying, 'Let's go! Let's go!' so we continued on our way.

"We then heard a second, but seemingly smaller one advancing from an intercepting direction. Both then followed us. It became almost ludicrous: we sang and talked out loud to them and then after about another half hour we were suddenly alone. They had gone ..."

Yowies have been reported in the Wee Jasper area since the 1840s. [e.g. Cases 2, 28 and 169]

Strangely, Brett and Brian – both very knowledgeable bushmen and naturalists – had a very similar experience in northern NSW in 1995 or '96. On that occasion, near Mt. Warning, a single bipedal creature followed them, remaining out of sight, breaking large branches and producing hair-raising vocalisations: "A loud cough, followed by a long drawn out, spine-chilling half scream, half yell that tailed off, ending in several guttural grunts."

Brett thinks it beyond the laws of chance that the same two men could experience two such similar events in localities separated by many hundreds of kilometres. He's inclined to believe, therefore, that there is something paranormal about the hairy ape-men.

Email from Brett Young to AYR website, Apr 20, 2004. Interviews with Paul Cropper and Tony Healy, 2004 and 2006. Credit: Dean Harrison.

Case 282. 5 Feb 2006. Cessnock, NSW. 11.20pm.

Aaron "S" lives on the eastern edge of Cessnock, across the street from a forest reserve that is contiguous with a large area of sparsely populated, forested country.

One night, as he was closing his front gate, he was startled to see a huge figure, partly illuminated by streetlights, about 30 metres away on the grassy verge next to the tree line. Fully eight feet tall, and seemingly three feet wide, it was covered, head to foot, in long, light brown, matted hair. Aaron couldn't discern any arms, but he could make out long legs as, after a few seconds, it walked away into the scrub. Although it was only walking, its strides were so long that it covered the ground very quickly.

The following day, as he and his wife searched for tracks (finding only ill-defined imprints), they became aware of a very foul odour and heard what sounded like a large creature moving through the scrub. Aaron later heard loud, frightening vocalisations in the same area.

Witness email to AYR website; interview with Paul Cropper, Feb 6, 2006. Credit: Dean Harrison.

Appendix B
Yowie-related Place Names

Although we're confident that many of the following place names relate directly to the Hairy Man, in some cases the connection, though likely, hasn't been fully established.

Baboon Gully. Between Ballandean and Eukey, QLD.

Gulaga (aka Mt Dromedary). Near Bermagui, NSW.

Hairy Man Rock. Near Boonoo Boonoo Falls, north of Tenterfield, NSW.

Hairy Man's Bend. On the Stanthorpe to Mount Tully Road, QLD.

Jerrawarrah Flora Reserve. West of Grafton, NSW.

Jingera Hill. South of Captains Flat, NSW.

The Jingeras. Part of the Great Dividing Range, south of Captains Flat, NSW.

The Jingera (aka Egan Peaks). Near Pambula, NSW.

Monkey Creek. Near Darriman, VIC.

Monkey Hill. Near Hill End, NSW.

Monkey Mountain. Between Ulladulla and Batemans Bay, NSW.

Mullumbimby, NSW.

Nimbin, NSW.

Orang-utan Gully. Near Blackheath, NSW.

Yahoo Brush. About 6 km northwest of Wingham, NSW.

Yahoo Creek. A tributary of Nicholson River, north of Bairnsdale, VIC.

Yahoo Island. In Wallis Lake, near Forster, NSW.

Yahoo Peak. In the parish of Warraberry, near Molong, NSW.

Yahoo Range. In the parish of Benya, near Molong, NSW.

Yahoo Valley. A branch of Araluen valley, south-east NSW.

Yowrie. Locality. Sometimes spelt "Yourie". South-east NSW.

Yowrie River. Sometimes spelt "Yourie". South-east NSW.

Yowie Bay. Port Hacking, NSW.

Yowie Gap. Near Mullumbimby, NSW.

The Yowie Track. A geological fault resembling a road, on the side of a bluff beside the Macquarie River, north-west of Bathurst, NSW.

Bibliography

Alley, Robert, *Raincoast Sasquatch,* Hancock House, Surrey, British Columbia, 2003.

"A Squatter", *Reminiscences of a Sojourn in South Australia*, Kent and Richards, London, 1849.

Attenbrow, V., *Sydney's Aboriginal Past*, University of New South Wales Press, 2002

Beck, Fred, *I Fought the Apemen of Mt St Helens*, the Author, Washington, 1967.

Bicknell, Arthur, *Travel and Adventure in Northern Queensland*, Longmans, Green and Co., London and New York, 1895.

Bord, Janet and Colin, *The Bigfoot Casebook*, Stackpole Books, Harrisburg, Pennsylvania, 1982.

Chittick, Lee and Fox, Terry, *Travelling with Percy, A South Coast Journey*, Aboriginal Studies Press for the Australian Institute of Aboriginal and Torres Strait Islander Studies, Canberra, 1997.

Coleman, Loren, *Mysterious America: The Revised Edition*, Paraview Press, New York, 2001.

Coleman, Loren, *Bigfoot!: The True Story of Apes in America*, Paraview Pocket Books, 2003.

Collin, Captain William, *Life and Adventures (of an Essexman)*, H.J. Diddams & Co., Brisbane, 1914.

Cropper, Paul, "The Yowie 1840 – 1985, A Catalogue of Cases", MS, Sydney, 1985.

Cunningham, Captain Peter, *Two Years in New South Wales; A Series of Letters, comprising Sketches of the Actual State of Society in that Colony; of its Peculiar Advantages to Emigrants; of its Topography, Natural History, &c. &c*, Henry Colburn, New Burlington Street, London, 1827.

Derrincourt, William, *Old Convict Days*, T. Fisher Unwin, 1899, reprinted by Penguin, Ringwood, VIC, 1975.

Egloff, Brian J., *Wreck Bay: An Aboriginal Fishing Community*, 1981.

Ellis, Netta, *Braidwood, Dear Braidwood*, N.N. & N.M. Ellis, Braidwood, NSW, 1989.

Favenc, Ernest, *The History of Australian Exploration from 1788 to 1888*, Turner & Henderson, Sydney, 1888.

Flannery, Tim, *The Future Eaters*, Reed New Holland, Sydney, 1994.

Frost, Neil, "Encounters with Fatfoot", MS, 1999.

Gale, John, *An Alpine Excursion*, serialised in the *Queanbeyan Observer*, Feb 13 to Mar 17, 1903.

Gale, John, *Canberra, History Of And legends Relating To The Federal Capital Territory Of The Commonwealth Of Australia*, A.M. Fallick & Sons, Queanbeyan, NSW, 1927.

Gillespie, Lyall, *Canberra 1820-1913*, AGPS Press, Canberra, 1991.

Gilroy, Rex, *Mysterious Australia*, Nexus Publishing, Mapleton QLD, 1995.

Gilroy, Rex, *Giants From The Dreamtime, The Yowie in Myth and Reality*, URU Publications, Katoomba, NSW, 2001.

Gilroy, Rex, *Australian UFOs Through The Window Of Time*, URU Publications, Katoomba, 2004.

Green, John, *On the Track of the Sasquatch*, Cheam Publishing, Agassiz, BC, 1968.

Green, John, *Sasquatch, The Apes Among Us*, Cheam Publishing, Saanichton, BC and Hancock House, Seattle, 1978.

Gregory, Denis and Manciagli, Alf, *There's Some Bloody Funny People on the Road to Broken Hill*, Snow Gum Books, Orange, NSW, 1993.

Gresser, P., *Manuscripts Relating Principally to the Aborigines of the Bathurst District*, Bathurst, 1964, pp.167-71, MS 21/2, Australian Institute of Aboriginal and Torres Strait Islander Studies Library, Canberra.

Guttilla, Peter, *The Bigfoot Files*, Timeless Voyager Press, Santa Barbara, California, 2003.

Halpin, M. and Ames, M. (eds.), *Manlike Monsters On Trial*, University of British Columbia Press, Vancouver, 1980.

Harris, Alexander, *An Emigrant Mechanic, Settlers and Convicts, or Recollections of Sixteen Years Labour in the Australian Backwoods*, London, 1847, reprinted by Melbourne University Press, 1969.

Hassell, Ethel, *My Dusky Friends*, C.W. & W.A. Hassell, Fremantle, WA, 1975.

Healy, Tony, "Monster Safari", MS, Canberra, 1983.

Healy, Tony and Cropper, Paul, *Out of the Shadows, Mystery Animals of Australia*, Ironbark/ Pan Macmillan, Sydney, 1994.

Heron, Ron, *The Dreamtime to the Present, Aboriginal Perspectives*, (MS), College of Indigenous Australian Peoples, Southern Cross University, Lismore, 1991.

Johns, Glenda, *Nature's Weather Watch*, the Author/Queensland Complete Printing Service, 1998.

Jones, Max, *A Man Called Possum: The mystery man who became a legend*, S.M. Jones, Renmark, South Australia, 1984.

Joyner, Graham, *The Hairy Man of South Eastern Australia*, Union Offset, Canberra, 1977.

Joyner, Graham, "More Historical Evidence for the Yahoo, Hairy Man, Wild Man or Australian 'Gorilla'", MS, Canberra, 1980.

Joyner, Graham, "The meaning of *Yahoo* and *Dulugal*: European and Aboriginal perspectives of the so-called 'Australian gorilla'", *Canberra Historical Journal*, Mar 1994.

Karna Sakya, *Dolpo, The World Beyond The Himalayas*, Sharda Prakashan Griha, Kathmandu, 1978.

Kroeber, Theodora, *Ishi in Two Worlds, A Biography of the Last Wild Indian in North America*, University of California Press, 1961.

Lapseritis, Kewaunee, *The Psychic Sasquatch, and Their UFO Connection*, Blue Water Publishing, North Carolina, 1998.

Lawson, Henry, T*riangles of Life and other Stories*, The Standard Publishing Company, Melbourne 1913.

Lawson, Henry, "The Spooks of Long Gully", in *The Stories of Henry Lawson*, Cecil Mann (ed.), Angus and Robertson, Sydney, 1964.

McAdoo, Martin, *If Only I'd Listened To Grandpa: Recollections Of The Old Days In The Australian Bush*, Landsdowne Press, Sydney, 1980.

Murphy, Christopher, *Meet the Sasquatch*, Hancock House, Surrey, B.C. and Blaine, WA, 2004.

Naseby, C., *The Aborigines of Australia; stories about the Kamilaroi tribe*, as told to John Fraser, Maitland Mercury Office, 1882.

Nicholls, C., (ed.), *The Pangkarlangu and the Lost Child, A Dreaming narrative belonging to Molly Tasman Napurrurla*, Working Title Press, Kingswood, SA, 2002.

O'Reilly, Bernard, *Green Mountains and Cullenbenbong*, 1940, reprinted by W. R. Smith & Paterson, Brisbane, 1962.

Organ M., (ed.). *Illawarra & South Coast Aborigines, 1770-1850,* University of Wollongong, 1990.

Patterson, Roger, *Do Abominable Snowmen Of America Really Exist?* Northwest Research Association, Yakima, Washington, 1966.

Persinger, Michael and Lafreniere, Gyslaine, *Space-Time Transients and Unusual Events,* Nelson-Hall, Chicago, 1977.

Pilkington, Doris (Nugi Garimara), *Rabbit Proof Fence,* University of Queensland Press, 1996.

Pinkney, John, *Great Australian Mysteries,* The Five Mile Press, Victoria, 2003.

Povah, Frank, *You Kids Count Your Shadows - Hairymen and other Aboriginal folklore in* New *South Wales,* Wollar, NSW, 1990.

Robinson, Roland, *Black-Feller White-Feller,* Angus and Robertson, Sydney, 1958.

Smith, J., *Aboriginal Legends of the Blue Mountains,* the Author, Wentworth Falls, 1992.

Smith, Malcolm, *Bunyips & Bigfoots, In Search of Australia's Mystery Animals,* Millennium Books, Alexandria, NSW, 1996.

Swift, Jonathan, *Gulliver's Travels,* first published 1726, republished by Penguin Books, Harmondsworth, Middlesex, 1986.

Telfer, William, *The Early History of the Northern Districts of New South Wales* (also known as *The Wallabadah Manuscript*), c.1900, University of New England Archives A147/V213.

Theiberger, N. and McGregor, W. (eds.), *Macquarie Aboriginal Words,* Macquarie University, NSW, 1994.

Thomas, Guboo Ted and Mumbulla, Percy, *Umbarra, Introduction to the oral traditions of the Yuin people of the Wallaga Lake region,* Wallaga Lake Cultural Centre, NSW, 1999.

Tim the Yowie Man, *The Adventures of Tim the Yowie Man, Cryptonaturalist,* Random House Australia, 2001.

Trezise, P. and Roughsey, D., *Turramulli the Giant Quinkin,* Collins, Sydney, 1982.

Walsh, Grahame, *Australia's Greatest Rock Art,* E.J. Brill-Robert Brown, Bathurst, 1988.

Woodford, James, *The Wollemi Pine,* Text Publishing, Melbourne, 2000.

Index

P

Q

R

S

T

V

W

Y

Printed in the United States
62227LVS00002B/25-42